COMPUTER-AIDED DESIGN,
ENGINEERING, AND MANUFACTURING
SYSTEMS TECHNIQUES AND APPLICATIONS

VOLUME
III

OPERATIONAL METHODS IN COMPUTER-AIDED DESIGN

COMPUTER-AIDED DESIGN,
ENGINEERING, AND MANUFACTURING
SYSTEMS TECHNIQUES AND APPLICATIONS

VOLUME
III

OPERATIONAL METHODS IN COMPUTER-AIDED DESIGN

EDITOR
CORNELIUS LEONDES

CRC Press
Boca Raton   London   New York   Washington, D.C.

Library of Congress Cataloging-in-Publication Data

Catalog record is available from the Library of Congress.

This book contains information obtained from authentic and highly regarded sources. Reprinted material is quoted with permission, and sources are indicated. A wide variety of references are listed. Reasonable efforts have been made to publish reliable data and information, but the author and the publisher cannot assume responsibility for the validity of all materials or for the consequences of their use.

© 2001 by CRC Press LLC

No claim to original U.S. Government works
International Standard Book Number 0-8493-0995-6
Printed in the United States of America 1 2 3 4 5 6 7 8 9 0
Printed on acid-free paper

Preface

A strong trend today is toward the fullest feasible integration of all elements of manufacturing including maintenance, reliability, supportability, the competitive environment, and other areas. This trend toward total integration is called concurrent engineering. Because of the central role information processing technology plays in this, the computer has also been identified and treated as a central and most essential issue. These are the issues that are at the core of the contents of this volume.

This set of volumes consists of seven distinctly titled and well-integrated volumes on the broadly significant subject of computer-aided design, engineering, and manufacturing: systems techniques and applications. It is appropriate to mention that each of the seven volumes can be utilized individually. In any event, the great breadth of the field certainly suggests the requirement for seven distinctly titled and well-integrated volumes for an adequately comprehensive treatment.

The seven volume titles are

1. Systems Techniques and Computational Methods
2. Computer-Integrated Manufacturing
3. Operational Methods in Computer-Aided Design
4. Optimization Methods for Manufacturing
5. The Design of Manufacturing Systems
6. Manufacturing Systems Processes
7. Artificial Intelligence and Robotics in Manufacturing

The contributions to this volume clearly reveal the effectiveness and great significance of the techniques available and, with further development, the essential role that they will play in the future. I hope that practitioners, research workers, students, computer scientists, and others on the international scene will find this set of volumes to be a unique and significant reference source for years to come.

Cornelius T. Leondes
Editor

Editor

Cornelius T. Leondes, B.S., M.S., Ph.D., is Emeritus Professor at the School of Engineering and Applied Science, University of California, Los Angeles. Dr. Leondes has served as a member or consultant on numerous national technical and scientific advisory boards and as a consultant for a number of Fortune 500 companies and international corporations. He has published more than 200 technical journal articles and edited or co-authored more than 120 books. Dr. Leondes has been a Guggenheim Fellow, Fulbright Research Scholar, and IEEE Fellow, as well as a recipient of the IEEE Baker Prize Award and the IEEE Barry Carlton Award.

Contributors

Balaji Bharadwaj
Visteon Corporation
Allen Park, Michigan

D.Y. Chao
National Cheng Chi University
Taipei, Taiwan

B.T. Cheok
National University of Singapore
Singapore

Siva R.K. Jasthi
Structural Dynamics
  Research Corporation
Arden Hills, Minnesota

Hsu Chang Liu
Solid Works
Concord, Massachusetts

Han-Tong Loh
National University of Singapore
Singapore

Christopher McDermott
Rensselaer Polytechnic Institute
Troy, New York

A.Y.C. Nee
National University of Singapore
Singapore

Ken-Soon Neo
National University of Singapore
Singapore

Bart O. Nnaji
University of Pittsburgh
Pittsburgh, Pennsylvania

Heui Jae Pahk
Seoul National University
Seoul, Korea

P. Nageswara Rao
MARA Institute of Technology
Selangor, Malaysia

Utpal Roy
Syracuse University
Syracuse, New York

Bijan Shirinzadeh
Monash University
Clayton, Victoria, Australia

Yoke-San Wong
National University of Singapore
Singapore

Contents

1

Intelligent Computer-Aided Design (ICAD) Techniques for Design Automation

Balaji Bharadwaj
Visteon Corporation

Utpal Roy*
Syracuse University

Abstract

Conventional Computer-Aided Design (CAD) systems have dramatically reduced the workload of the human designer and reduced the duration of the product design cycle. However, the magnitude of effort and volume of information required to use these tools limit their application in various product design activities. Intelligent Computer-Aided Design (ICAD) techniques have been sought to assist in the development of Integrated Product Design (IPD) systems, thus providing a more "complete" design tool to assist the designer in all phases of the product design.

In this chapter, the role of ICAD techniques in various types of product design and development activities will be examined. The viability and utility of applying ICAD techniques for design analysis, process planning, and tolerance assignment tasks within the framework of an IPD system environment are described and demonstrated through the implementation of various prototype design systems. The objective behind

---

*Author to whom all correspondence should be addressed.

the implementation of the IPD system is an integrated approach, wherein all the primary components, such as the CAD tools, numerical analysis tools, data bases, knowledge bases, and expert systems, work cooperatively in order to perform various tasks associated with that design task. The techniques described in this chapter are relatively easy to implement, and are well suited to industrial needs.

Keywords: Integrated Product Design, Intelligent Computer-Aided Design Techniques, Automated Design System, Automated Process Planning

1.1   Introduction

Advances in product design methods coupled with the growing trend toward shorter product development cycles have resulted in an increased emphasis on the following issues: (a) reduction in product design lead time and costs, (b) improvement in product quality, (c) design automation, and (d) integration between product design and manufacturing tasks. The above objectives can only be realized through the synergic development of the following key product design-enabling methodologies (Figure 1.1): (a) development of product design methodologies, (b) techniques for synthesis of product specifications, (c) advanced information support infrastructure for management and distribution of product design information, (d) decision support mechanisms for various tasks associated with product design, and (e) development of frameworks for product design. The above five enabling methodologies represent a multitude of tasks and activities with far reaching implications on current product design practices. Their implementation within the framework of an *Integrated Product Design (IPD) system* is the subject for various current research initiatives (discussed further in Reference 1).

A careful examination of these five enabling methodologies reveals the following characteristic features that should be considered during the development of an IPD system:

- Suitable mechanisms for incorporation of different product design perspectives for the generation of product design information in a timely and useful manner
- Provision for suitable interpretations of the product design information for the different product design tasks
- Mechanisms for the simulation of the performance of the product from different product design perspectives
- Suitable algorithms and procedures for performing design in the presence of incomplete and uncertain information about product design attributes
- Provision for sharing of information among different product design tasks (e.g., process, assembly, and inspection planning)

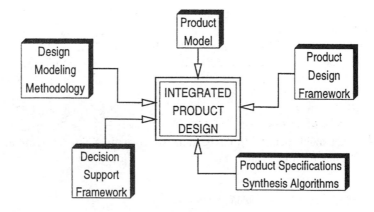

FIGURE 1.1   Enabling product design methodologies for IPD.

- Provision for support functions and methodologies for design critiquing, recording of design rationale, and planning and execution of design changes
- Facilitation of group interaction between different groups of designers and software systems used for product development activities
- Facilitation of synchronization between different product design tasks
- Mechanisms for resolving conflicts among competing multiple product design perspectives
- Provision for infrastructural support for the integration of the different product design modules
- Development of a suitably structured product model for every stage of the product design process
- Algorithms for mapping functional requirements of the component into appropriate types and associated values of product specifications (e.g., tolerances, surface finishes, geometry specifications, etc.).

Which of the above tasks an IPD system ought to do and how it should go about doing them have been the subject of continued debate.[2-4] As a first step toward realization of IPD systems, researchers have focused their attention on the use of *Intelligent Computer-Aided Design (ICAD) techniques* for the development of IPD systems. The domain of ICAD techniques among other things includes the organization, integration, information management and transfer, and control of knowledge bases, expert systems, and data bases within the framework of the IPD system.

A conventional CAD system with drafting, solid modeling, and design analysis capabilities often forms the basis on which an IPD system is built. Knowledge-based expert systems and data bases are the other key elements of an IPD system. In the past, the role of ICAD techniques was limited to the task of enabling interoperability between different CAD systems and design analysis tasks.[5-8] However, the issue of an effective integration of all product design tasks has still not been solved even with the introduction of such knowledge engineering systems. The issues that need to be addressed are mainly from the standpoint of a generic framework for design data representation, effective integration and coordination of design tasks, information processing, retrieval of similar designs, and design modification.

The objective of this chapter is to report on the application of ICAD techniques for the implementation of IPD systems for various tasks associated with product design, such as design analysis, process planning, and tolerance assignment. Several case studies are presented to demonstrate the salient features of the application of ICAD techniques for product design activities.

1.2   Elements of IPD Systems: ICAD Techniques

Expert System/Knowledge Base

As outlined in the previous section, one of the most useful functions of an IPD system is to embody the knowledge available for a given design task. Currently, this task is performed by a technical specialist who uses various high-level intelligent functions such as intuition, creativity, association, induction, recognition, and deduction, as well as low-level functions such as computation and information retrieval. Among the above list of functions, only a few low-level functions involving computation and search have been analyzed and algorithms have been devised for automatically performing these tasks. To facilitate higher levels of knowledge processing in an IPD system, the design artifact needs to be defined completely along with its full functional description. However, there are two obstacles in embedding such knowledge into an IPD system. First, the expert knowledge is characteristic of an individual or an organization based on the past experiences and design practices and is often difficult to encapsulate in appropriate design rules and guidelines; furthermore, this knowledge is extremely context sensitive and cannot be generalized. Additionally, when several such sources of knowledge are available within an organization, there is no algorithmic procedure or guideline to resolve the conflicting suggestions. In spite of these difficulties, expert system shells, together with knowledge bases, can be used for various types of inferencing tasks such as exploration of the space of all design solutions, generation of design alternatives, etc.

In our IPD system implementations, we have provided various mechanisms through which the designer can specify a complete description of design requirements for the object to be designed. Various design primitives, which are generalizations of past designs, are provided in order to assist the designer in the description of the design artifact. Each design primitive is represented as a frame consisting of a set of attributes that denotes various design parameters (geometry, physical, and material properties), design specifications, and requirements. These design primitives can be built using the various geometric and information primitives provided by the Concept Modeller system[26] and can be made available for subsequent designs.

Design Intent

Capturing the design intent during the design process is an uphill task, although it provides an important insight into how the design evolves. This is because design intent is not quantifiable, and it refers to the actions and their rationale used in arriving at the final design. However, it is possible to capture the design intent of the designer by recording the reasons and the conditions that prompt a change in the original design. This can be facilitated in an IPD system, since a deviation from the suggested design can be recorded along with the accompanying explanations. This information is useful in developing new alternatives, should a similar situation arise in the future. This concept of knowledge induction and reuse is similar to the case based reasoning (CBR) approach,[9] although IPD differs from CBR because CBR requires the indexing of a large number of designs in the design knowledge base, and the stored designs can only be used for a product design very similar to the present design. To record the design intent in the first place, one needs to carefully develop a faithful representation of all necessary design attributes and functions that are present in the design object. Moreover, object representation in a design synthesis stage is a dynamic process and involves the following steps: (a) build a tentative model, (b) evaluate it and compare the results with the design requirements, and (c) modify the model and then go back to step (b) to see if the model fails to meet the requirements, or to terminate the process otherwise. The representation scheme must be such that this dynamic procedure, based on the defined model, is performed conveniently. In our IPD system implementations, the above issue is supported through the development of: (a) a geometric information representation scheme to represent the shape and topology of the object; (b) a descriptive information representation scheme to represent design attributes, object behaviors, and functions of the design object; and (c) an information management representation scheme to store information pertinent to the management of product data within the IPD system and the coordination of various design subtasks. The object-oriented programming environment of the Concept Modeller CAD system, along with its knowledge processing capability, offers the necessary information representation infrastructure to support each one of the above representation schemes.

Feature- and Function-Based Modeling

A large number of design artifacts can be regarded as a combination of a base part and various types of features. Features have been traditionally used by designers to correlate the design intent of the design object with other manufacturing functions and tasks. However, this concept has only recently been implemented and advocated for use within CAD systems. Feature-based design systems allow the design object to be defined in terms of a set of attributes or features and not merely as a desired geometric shape. The feature-based design process provides a useful method for assessing the design intent. Since features are primarily classified based on their functionality, the choice of one feature over another clearly identifies the functional intent as desired by the designer. Moreover, parametric modeling of features would allow for the design object to be defined by a set of parameters or attributes that could be associated with either dimensional or functional information. The parametric-based feature modeling approach would allow for the incorporation of implicit and explicit relations between different design and geometric parameters. In our IPD system implementation, we have developed a unique representation scheme for associating design function information with the different constitutive geometric entities of the features of a component. This explicit association of design function information with features assists in various tasks associated with the assignment of product specifications such as tolerances and surface finishes. IPD systems could make use of these two approaches to quantify the design in terms of its features.

Information Management

The applicability of the IPD system for product design relies on how well the design information is managed, controlled, and utilized during the product design process. In one of our IPD system implementations, the blackboard architecture[10] concept has been utilized, wherein the state of the design is posted on the system blackboard data base. Whenever the information required to execute a particular design module is present in the blackboard data base, that module is automatically activated. These modules could be either analysis modules, which provide performance data on the design, or configuration assessment routines, or checks on whether the design meets requirements and constraints such as manufacturing, cost, environmental impact, etc. Within such a framework, the control of the design process could be coordinated either by the IPD system or by the human designer. The IPD system drives the design process, automatically performing various design checks and informing the designer of the design status at appropriate intervals of time. Based on the generated results, the IPD system can modify the state of the design to reach the desired level of performance.

1.3   Application Domain: Design Analysis

Design analysis often involves the application of the finite element method for analyzing the effect of design loads (structural, electromagnetic, thermal, fluid, etc.) on complex structural and mechanical systems. A key issue in design analysis is the automation of various tasks associated with the finite element analysis (FEA) procedure. This in turn involves the integration of CAD tools where geometric designs of mechanical structures are modeled, along with different numerical analysis tools that perform FEA. The main objective for this unification is to shorten the product development time, thus reducing the development costs and achieving high product quality with minimum product rejection.

In this section, we demonstrate the utility of ICAD techniques in automating various design analysis tasks and outline the dimensions of an Integrated CAD/finite element method (FEM) system under the overall framework of an IPD system.[11] The Integrated CAD/FEM system takes the support of symbolic computing techniques to draw conclusions about a finite element model and its analysis results, explains its reasoning, and interacts with the designers in terms of the qualitative descriptions of the system behavior. The main objective of this research work is to exploit the benefits of recent advances in knowledge engineering and to apply some of its powerful tools in the integration of conventional CAD and FEA systems. The potential of applying ICAD techniques in enhancing the automated analysis capabilities of current FEA packages used in a knowledge-assisted design environment is demonstrated in the prototype implementation of a system for structural design of aircraft wingbox structures.

Role of ICAD Techniques in Design Analysis

FEM offers a versatile procedure for performing design analysis that enables designers to tackle a wide variety of engineering analysis problems. The application of FEM is largely influenced by the designer's experience in carrying out the following sequence of steps effectively:

1. FEM modeling—constructing an appropriate FEM model from the given object model per the requirements of FEA and the given problem definition;

2. Specification of analysis-specific information—providing all details that are required for performing an effective and efficient analysis of the given design problem;

3. FEM model discretization—discretizing the entire FEM model into a properly connected mesh of suitably sized and shaped elements;

4. Specifying design conditions—specifying loads and boundary conditions at appropriate nodes and elements of the FEM model;

5. FEM Analysis—conducting the FEM analysis using specialized codes (in-house software programs) or commercial codes such as NASTRAN, ANSYS, etc.;

6. Interpretation of FEA results—assessing the validity of the results for the specified design requirements;

7. Repeating the steps until acceptable results are obtained.[12] In the recent past, most of the research efforts[5-7] were directed toward automation of Step 3. In order to achieve complete automation of design analysis tasks, one would need to address issues pertaining to the individual requirements of all of the above steps.

Since the FEA procedure would have to be integrated and driven by a CAD system, the most obvious requirement relates to modeling the object in the CAD system and transferring geometric as well functional requirements information from the CAD system to the finite element preprocessor. "The more complete the geometric information passed to the preprocessor, the more automated the finite element analysis process can be made."[5] Current CAD systems that are developed based on conventional information processing technology rely mostly on procedural representation of 3-D objects (solid models) and do not have the capability to undertake tasks such as automatic FEA. In order to break this barrier for integration, steps should be undertaken for developing a "complete" representation of the design object and the corresponding object model which should include: (a) geometric information to represent the shape and form of the design structure; (b) design-specific functional information to represent attributes, properties, behaviors, and functions of the design artifact; and (c) information management attributes and directives in order to ensure the smooth integration between the CAD and FEM systems. Moreover, the object model cannot be a static entity (e.g., data base data structure), and it needs to adopt flexible object and attribute representation schemes to aid in the FEA process. The system should attempt to utilize as much design-specific knowledge as possible in order to obtain a successful CAD-FEM interface.

Based on the above mentioned list of requirements, the object-oriented approach for object model definition has been adopted for modeling the FEM analysis attribute information.[12] Frames have been selected as the means of knowledge representation in the object-oriented programming (OOP) environment. In OOP, self-contained pieces of code called "classes" (at the information concept level) and "objects" (at the implementation level) represent the informational attributes of an artifact associated with an application or function. For instance, the Design-Information class shown in Table 1.1 represents a class template that contains information pertinent to the specified loading and boundary conditions on the design component, as well as the stated design requirements for FEA. Since the modeling of the original geometric model to an appropriate FEM model is necessary for reducing the computation time, a considerable amount of knowledge needs to be represented and consulted with in order to reason about: (a) simplified geometric representation of the object—submodeling the object model based on the inherent symmetry of the component geometry and design loads and restraints, and (b) removing subcritical geometric features (e.g., small holes, fillets, etc.) from the FEM model. Once the task of model building is complete, the geometrical, topological, and material data needs to be transferred from the CAD package to a standard FEA package such as ANSYS, etc., and package-specific knowledge is required for creating the appropriate data set for running the object model in the specific analysis package. The analysis of FEA results obtained is then assessed. Similar to FE modeling, the analysis of FEA results is also a knowledge intensive task, and often the designer's judgment and past experience play a critical role in successful execution of these tasks. Furthermore, based on the FEA results, the redesign of the design object is another task where the designer's experience and the redesign procedures adopted for

TABLE 1.1   Template of a Typical Design-Information Class

(	
define-part Design-Information-Class	/* *Design-Information-Class definition* * /
• inherit-from Information-class-mixin	/* *Inherits the properties of this class* * /
• Geometric-defn-object	/* *Pointer to the geometric definition (B-Rep model) of the object* * /
• Load-spec-Lx, Load-spec-Ly, Load-spec-Lz	/* *List of specified load locations (point & pressure, fluxes, etc.) along the global x, y, & z axis direction* * /
• Load-magnitude-Lx, Load-magnitude-Ly, Load-magnitude-Lz	/* *List of specified load magnitudes (point & pressure, fluxes, etc.) along the global x, y & z axis direction* * /
• Restraint-location-Lx, Restraint-location-Ly, Restraint-location-Lz,	/* *List of specified restraint locations along the global x, y & z axis direction* * /
• Restraint-magnitude-Lx, Restraint-magnitude-Ly, Restraint-magnitude-Lz	/* *List of specified magnitudes of restraints along the global x, y & z axis direction* * /
• Design-requirements	/* *Specification of design criteria and design variables, which are important from the standpoint of the design analysis to evaluate the design* * /
• Design-requirement-bounds	/* *Acceptable range of parameter values under which the design could be termed satisfactory* * /
• FEM-Mesh-Model	/* *Pointer to the Finite Element Mesh of the object* * /
• FEM-Model	/* *Pointer to the Finite Element Model of the object* * /
)	

earlier designs play an important role. Consideration of issues that influence the representation, organization, integration, and control of knowledge bases (which contain knowledge pertaining to FE modeling, FEA result interpretation, and design object modification) is an important task during the formulation of the system architecture for an Integrated CAD/FEM system. In this research work, we propose to use four knowledge bases—object modeling, finite element modeling, FEA result analysis, and evaluation knowledge base—for automating various tasks within the FEA procedure. The automation of the FEA process involves tasks that require very high-level intelligence (for decision making from abstract knowledge bases) together with lower-level tasks that can be processed by the successful deployment of various knowledge engineering techniques. The features offered by OOP environments would be very useful in achieving some of these objectives.

System Architecture and Case Study

Framework of an Integrated CAD/FEM system

The developed framework of the Integrated CAD/FEM system is shown in Figure 1.2. The system essentially consists of an OOP-based CAD system (Concept Modeller), FEA packages (I-DEAS and ANSYS), an expert system shell (CLIPS), and necessary intelligent interface modules along with their knowledge base (KB) units (which contain FEA domain-specific knowledge). The aim of the Integrated CAD/FEM system is to provide an automated design environment for the design of engineering components.

As shown in Figure 1.2, the designer specifies the initial design requirements through interaction with a conceptual design unit (CDU) built in the Concept Modeller system. The CDU is used to translate the specified initial design requirements into an appropriate object model of a component design with the help of a knowledge base, consisting of a library of design instances (i.e., previous design models) and links with domain-specific "experts" for simultaneous engineering. The domain-specific experts assist in establishing the preliminary design from a set of design alternatives created earlier from the knowledge

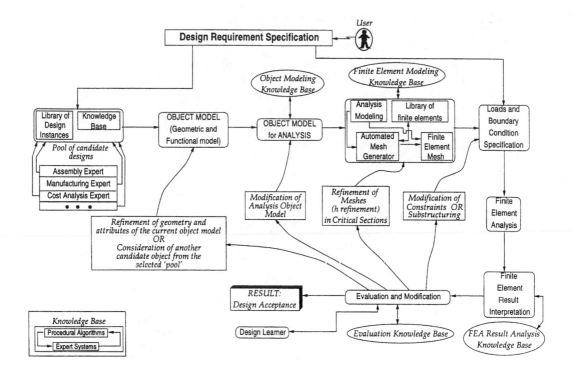

FIGURE 1.2   Framework of the Integrated CAD/FEM system.

base or the design instance library. These experts assist in pruning out the nonfeasible design alternatives during the preliminary design phase.

The CDU helps capture the functional intent of the component design in the form of design rules, spatial and functional relationships between assemblies and parts, and part features. Since the object model and the underlying geometric entities are represented as individual objects (in OPP's sense), subsequent changes in spatial relationships, assembly-part relationships, etc., can be easily undertaken without undesirably affecting the other parts in the assembly. At first, the Integrated CAD/FEM system builds an FEM model from the object model developed by the CDU for FEA in conjunction with an object modeling knowledge base based on the object's shape, size, and other design conditions. The FEM knowledge base is used for making decisions regarding mesh generation for the object model and selection of appropriate element types and element sizes, depending on the particular FEA package to which the finite element (FE) model is to be transferred and the specified design analysis conditions. At present, an automated mesh generator (2-D and 3-D), based on the Delaunay triangulation scheme, is used for generating the FEM meshes. The automated mesh generator also provides the capability for checking the quality of the FE meshes. The user also has an option to take advantage of the automated mesh-generation capability of I-DEAS and ANSYS packages.

After the FE meshes are completely generated for the FE model, the loading and boundary conditions are automatically specified on the FE model based on the initial design specifications. The whole FEM model would then be transferred as a single piece of information (input file) to the FEA package without any user interaction. The analysis of the FEM model in the FEA package would be controlled by means of a master control program running on the UNIX operating system. Currently, the master control program has been implemented using the CLOS-C foreign function call facility. Additionally, the control program also provides a restricted capability for window programming. The control program opens a window (X-window system) on the terminal screen where it executes the

FEA package, thus enabling the user to have a view of the analysis run. The master control program is capable of executing FEA runs in about 6 seconds; an expert designer would take about 15–20 minutes to do so. This provides an indication of the time that could be saved by the effective automation of FEA tasks.

Next, the results obtained from the FEA run are interpreted using an FEA result analysis knowledge base. The results obtained from any FEA run consist of a large amount of analysis-specific information, which has to be sorted intelligently before it can be passed over to a human designer or to a knowledge base. An intelligent interface performs the task of the human designer in discarding noncritical information and providing an input in an appropriate format to the downstream knowledge base unit for further processing. In conjunction with the stated design requirements specified by the user, an evaluation and modification module (as shown in Figure 1.2), aided by the evaluation knowledge base, diagnoses the filtered FEA output data, and makes a decision regarding the necessity for modification of the FEM model (i.e., geometry, FEM model attributes, design conditions, etc.). A design learner module keeps tabs on the values of all variables undergoing modification in repetitive design iterations, looks for specific patterns in their values, tries to reason these on the basis of earlier patterns, and provides input to the evaluation knowledge base for controlling the product model modification process. The design learner provides an automatic reasoning mechanism for controlling and evaluating the various design iterations. At present, the system analyzes the FEA results and provides the necessary suggestions to the users to carry out the desired design modifications. The user is then expected to interact with the initial object model in an appropriate fashion and reinitiate the design iteration process. The involvement of different symbolic tools (i.e., Concept Modeller, CLIPS) and numerical tools (i.e., I-DEAS, ANSYS and different utility programs) in the design activities of the proposed Integrated CAD/FEM system in realizing a product model is summarized in Figure 1.3.

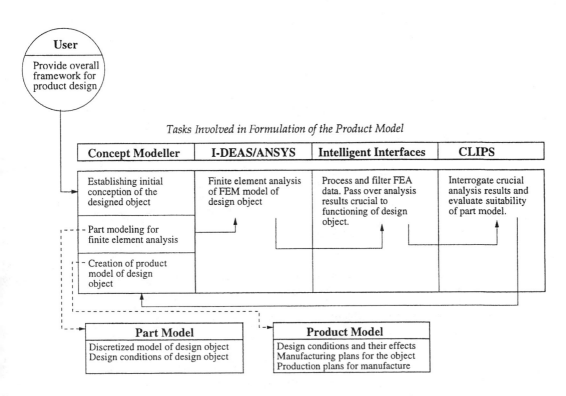

FIGURE 1.3   Different steps associated with the creation of a product model in the Integrated CAD/FEM system.

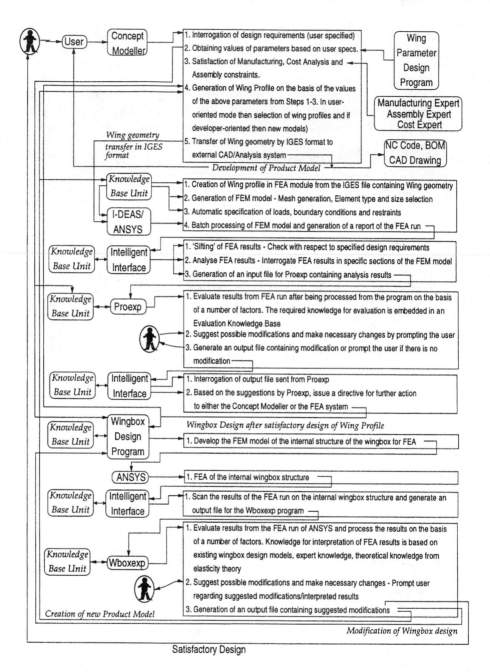

FIGURE 1.4   System architecture of the Integrated CAD/FEM system for the structural design of an aircraft wingbox structure, along with the different stages of design analysis.

Case Study: Aircraft Wingbox Design

A case study involving the structural design of an aircraft wingbox in the proposed Integrated CAD/FEM system is described in this section. Figure 1.4 describes the information flow, along with the different phases of design analysis in the Integrated CAD/FEM system during the design of the aircraft wingbox structure. The Integrated CAD/FEM system assumes the role of an intelligent structural design system and the end product of the design process is a result of cooperative design efforts between different symbolic and numerical analysis tools.

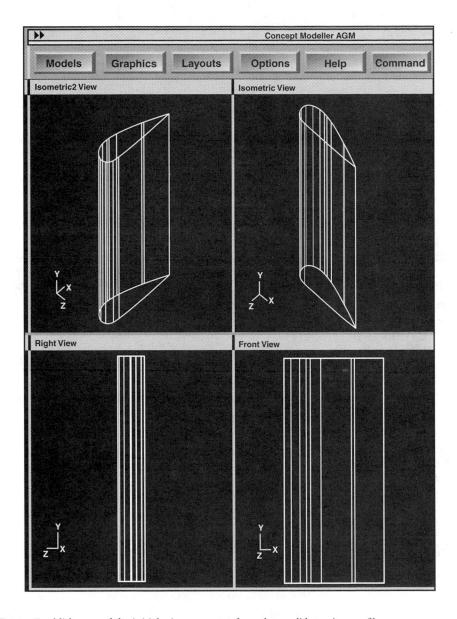

FIGURE 1.5   Establishment of the initial wing geometry from the candidate wing profiles.

The initial design requirements as specified by the user in the CDU are used as the design specifications for the aircraft wingbox. After processing the design requirements, an external procedural program (the wing parameter design program written in FORTRAN 77) is invoked by the CDU for establishing the wing's initial geometry. The system determines various important geometric parameters which define the wing (based on Roskam's aircraft wing design set and standard airfoil shapes) on the basis of the design mission profile (DMP) statement of the aircraft. Various geometric parameters that define the wing: wing area, wing span, wing chord, root chord, sweep, thickness ratio, etc., are also determined based on expert rules provided in the aerodynamic expert's knowledge base. The wing parameter values are checked by different experts (e.g., manufacturing expert, assembly and cost expert) for determining a suitable candidate wing profile (Figure 1.5) from the library of profiles available in the CDU knowledge base. Though the wing profile and the aircraft wing structure involve a complex geometry, a simpler FEM model of the geometry in the form of a beam is considered for FEA. The geometric information pertaining to the aircraft wing profile is then transferred from the Concept Modeller system to the I-DEAS package.

The meshing pattern for the FEM model in I-DEAS is automatically generated by using a set of predefined commands. In the working version of the prototype, we have only considered uniform cross-section wings, and the entire wing is modeled as a beam. The appropriate design loads and restraints are also automatically specified at the appropriate locations on the FEM model, which is then sent over to ANSYS for Analysis. The FEA results are passed on to an intelligent interface unit that extracts the necessary information that is further used by Proexp, an expert module (built in CLIPS expert system shell), for verification of initial design analysis results. Depending on the analysis results, Proexp would suggest the appropriate design modifications (sample suggestion is shown in Figure 1.6). The suggested modifications are passed over to an intelligent interface unit that automatically records the changes in the FEM model (either in I-DEAS or ANSYS) or conveys the results of a satisfactory design to the user. After a satisfactory design of the initial wing configuration has been achieved, the wingbox (the major internal load-bearing structural component of the wing) is designed. The major structural components of the wingbox are spars, ribs, and associated components such as rib and spar webs. The structural configuration forms a truss structure and is covered by the upper and lower covers, which form the external surfaces of the wing. To start the wingbox design, a separate design program is invoked and a finite element model of the wingbox is automatically generated. The wingbox design program automatically determines the element types (e.g., 3-D shell elements, SHELL63[13] for upper and lower covers and for spar and rib webs, the element BEAM4 for stiffeners, and spar and rib caps, etc.). As the entire wing model is intended to be loaded with a pressure distribution along its lower surface to simulate actual flight conditions, appropriate design conditions are also specified in the wingbox finite element model. The finite element model is then transferred to ANSYS where the analysis is performed, and the ANSYS output results are interrogated by Wboxexp, another expert module (CLIPS based) for verification of the stated design conditions. Any changes suggested by the Wboxexp are sent to the wingbox design program, and this loop continues until a satisfactory design is achieved. Upon achieving a satisfactory design the system prompts the user, and control is once again passed to the Concept Modeller system to develop the final product model (Figure 1.7) on the basis of the design parameters previously obtained. The final product model information can also be used for other activities such as numerical code (NC) code generation, bill of material (BOM) report and process planning, and scheduling schemes.

1.4   Application Domain: Process Planning

"Process planning" is defined as that task associated with the manufacturing phase of the product realization process that establishes the manufacturing processes. This includes process parameters, which are to be used for transforming a product from its initial configuration (i.e., shape and form of the raw stock) to a predetermined configuration (i.e., shape and form of the finished workpiece) based on the functional intent of the design as represented in an engineering drawing or solid model.[14] The process plan is also alternatively referred to as an "operation sheet" or a "route sheet" and provides instructions for sequencing of the manufacturing processes, selection of process parameters, machine tools, cutting tools, etc.

Automatic (i.e., generative) process planning involves (a) an explicit representation scheme of the workpiece model, different machining operations, and their effects on the workpiece, (b) reasoning about the effects of sequences of machining operations and the interaction of operations that may take place concurrently, and (c) the development of an algorithmic approach for an optimal process plan that can be adopted with reasonable efficiency for manufacturing the component. The generative process planning approach for generating process plans is largely dependent on the understanding of the manufacturing processes and their application in formulating the process knowledge bases of the planning system. In this research work,[15,16] we have demonstrated the application of ICAD techniques for automatic generation of process plans by using the blackboard architecture concept[10] of problem

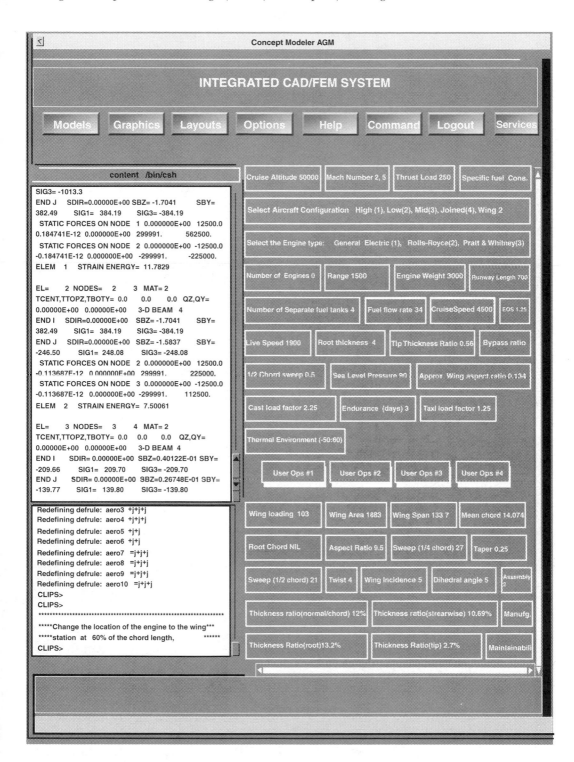

FIGURE 1.6   Generation of suggestions for modification of initial wing configuration.

solving, which utilizes the expertise of different domain experts in the field of process planning. The domain experts use the geometrical and technological information of the raw stock and the workpiece, which is available from the solid-modeler-based CAD data model, in order to develop the process plan. The system allows for the process plan to be developed in a hierarchical fashion with control executed

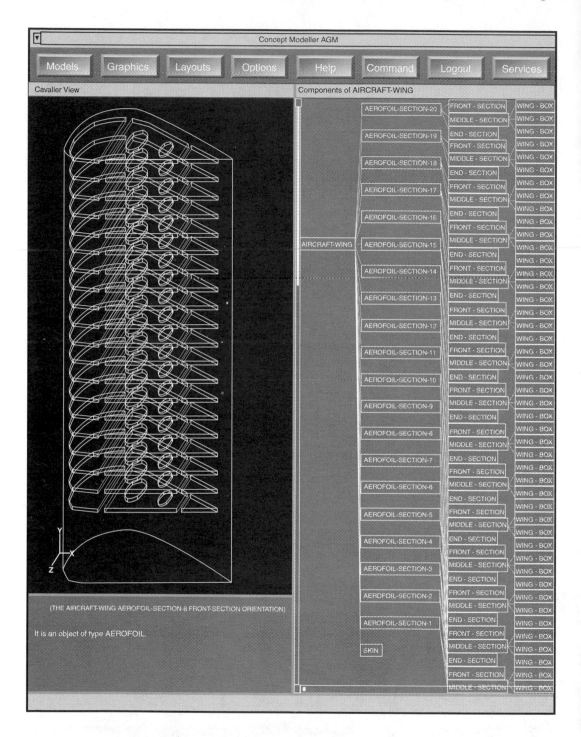

FIGURE 1.7   The complete product model of the aircraft wing along with its wingbox.

through the control expert (blackboard architecture terminology) resident in the process planning system. The control expert incorporates the appropriate heuristics for reasoning about resources and constraints and allows for concurrent exploration of alternate plans. The blackboard-architecture-based process planner (BBPP) was implemented using the CLIPS expert system shell in a prototype expert system.

ICAD Techniques in Process Planning

Process Planning Data Bases and Knowledge Bases

The ability of the process planning procedure to generate process plans is largely dependent on the availability and accessibility of the required information during decision making. Thus, it is essential to have an effective representation scheme of the process knowledge along with the knowledge about the capabilities and specifications of available machine tools, cutting tools, and other accessories. The data base component of the process knowledge in a generative computer-aided process planning (CAPP) system should consist of the following:

1. **Machine tool/machining center** data base: This data base contains the specifications and technical information pertaining to the capabilities of the different machine tools. Selection of a feasible set of machines from this data base is dependent upon the choice of a particular feature to be machined. The information includes details such as machine identification number, machine name, type of machinable features, horsepower rating, efficiency, size, speeds and feeds of the machine spindle, orientations of the spindle with respect to its table, repeatability, surface roughness and the accuracy that the machine can produce for different machining processes.

2. **Cutting tool** data base: The data base for cutting tools is divided into two parts: (a) cutting tools, and (b) cutting tool holders. Cutting tools are related to a set of manufacturing features that can be generated by cutting tools and their respective machining processes, whereas tool holders are only associated with the geometry of the cutting tools and the machine tools in which they are used. For the proper selection of cutting tools, information concerning their geometrical and technological parameters is analyzed. The data base for cutting tools contains information regarding tool identification number, tool name, cutting tool holders, tool material specification, tool geometry specification, cutting condition requirements, tool wear characteristics, accuracy, and type of manufacturing features that it can generate in a particular machine tool.

3. **Fixture** data base: Clamping devices (fixtures) are characterized in terms of their "resting" and "clamping" points, which are objects of the fixture's geometry. Each resting/clamping point is described by its function (position or fixing), principle of its action, and the nature of its contact with the workpiece. The fixture data base contains information regarding the fixture identification number, fixture type, their generic configurations along with their geometric specifications, and nature of clamping forces at the points of contact with the workpiece. The nature of contacts of the resting/clamping points determines the constraint relationships on the tool approach direction of the machine tool with respect to the fixture for a particular fixture-workpiece setup.

4. **Machinability** data base: Proper determination of the machining conditions and the required machining parameters depends on the selection of cutting tool geometry and its material specification, along with the type of workpiece material to be machined. This data base contains information pertaining to optimum machining conditions for different workpiece materials when specific tool materials with certain tool geometries are used within certain ranges of specified speeds, feeds, and depths of cut.

5. **Material** data base: This data base provides information regarding the characteristics of a particular type of material and its performance under a particular machining process.

6. **Cost** data base: The cost data base provides input regarding the standard costs (tool setup, tool change, etc.) incurred at any stage of the manufacturing process.

The process planning system also requires appropriate knowledge bases for performing the following two tasks:

1. **Integration and sequencing of machining operations:** A component of this knowledge base deals with the issue of determining groups of manufacturing features that can be machined by a single machining operation (in other words, can be "simultaneously" machined). These simultaneous operations can be achieved in three different ways: (a) by using multiple tools in a single spindle

machining facility (e.g., gang milling), (b) by using multiple spindle machine tools, or (c) by using more than one machine during the same operation time (e.g., automatic transfer line). The knowledge base consists of a set of constraining rules that identifies all those features that can be machined in one operation. The constraining rules depend on the feature accessibility criteria, machinability conditions, and fixturing considerations for machining different features. For simultaneous machining operations, rules are defined mainly to guarantee an interference-free operation. The main criterion to allow multiple features to be machined simultaneously is that all of the involved machining operations must be mutually exclusive, i.e., the condition of generation of one feature should not depend on the existence of another feature of the same group. Furthermore, knowledge regarding the constraints imposed by fixturing devices and clamping positions for simultaneous machining operations is also provided in this knowledge base. The second component of the knowledge base deals with sequencing of machining operations. Most manufacturing features require more than one manufacturing operation for obtaining the desired functional properties of the finished workpiece. Some precedence relations always exist among the machining operations, which are also stored in the knowledge base. These precedence relationships are established on the basis of the following criteria: (a) precedence requirements due to a unidirectional chain of process operations, (b) precedence relationships between machined surfaces due to surface quality and datum considerations, and (c) precedence relationships among part features depending on accessibility and clamping considerations.

2. **Generation of workpiece orientations for machining:** Stable workpiece orientations are determined on the basis of the workpiece configuration and the set of machining operations that are to be performed on it. The orientation must comply with the standard fixturing rules and should allow interference-free, concurrent operations. This knowledge base consists of two components; one part corresponds to the determination of orientation and location of the workpiece with respect to the machine spindle, and the other determines the required resting and clamping points on the workpiece. This knowledge base should also be capable of dealing with the issues of fixture selection from a standard fixture library.

All of the above data bases and knowledge bases have been incorporated in the BBPP system.

Knowledge Representation and Reasoning

Frames and rule-based knowledge representation schemes are considered to be the most appropriate means of storing the different data types pertaining to the knowledge and data bases (described earlier). Factual knowledge is best represented by frames while problem solving knowledge is best described through declarative statements in the form of production rules. A forward chaining inference strategy using the blackboard architecture concept for problem solving has been adopted for developing the process plan. The process plans are evolved through the interactions of different domain experts (in the process planning domain) who use the blackboard data base for developing different segments of the process plan. The blackboard approach to problem solving reflects the cooperative problem solving approach normally followed by process planners (shop floor personnel). The process plan starts with the raw stock and involves different machining processes to transform the raw stock into the finished part. The process planning system requires an explicit representation of the geometrical and technological information of the part in order to make different process planning decisions. Thus, a product data model (PDM) of the part is required at each stage during the formulation of the process plan. It uses a feature-based representation scheme for representing the part geometry.[17] Along with the geometrical and topological information of the part, it also provides the surface finish and variational information (tolerance information) for the process planning procedure. The PDM of the part has been specifically designed for this purpose and is composed of two segments. The first segment consists of specifications of the raw stock used for manufacturing the part while the second segment consists of the specifications of the finished workpiece model of the part. The finished workpiece model consists of a collection of specifications, which is used for characterizing the different form features of the part (Table 1.2).

TABLE 1.2   Part Product Data Model

Raw-Stock Template

Type	Raw-stock type (bars slabs, etc.)
Material	Raw-stock material
Raw-size	Raw-stock dimensions
Cut-size	Finished workpiece dimensions
Finished-piece-hardness	Hardness of finished workpiece
Raw-stock-hardness	Hardness of raw stock

A Feature Template in the Finished Part

Feature-id	Index number of feature
Feature-surface	Index number of surfaces of the feature
Feature-type	Type of feature
Dimension	Dimensions of the feature
Location	Location of the feature
Attachment-faces	Index number of surfaces that are adjacent to the surfaces of the feature
Tolerance-info	Tolerance information associated with the feature
Finish	Desired surface finish quality for the surfaces of the feature

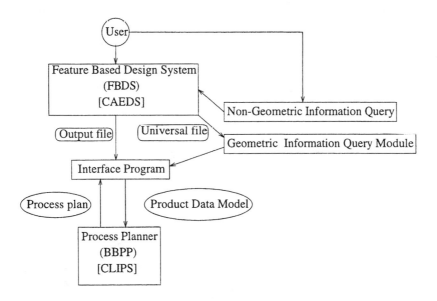

FIGURE 1.8   System architecture of the BBPP system with the FBDS system.

System Architecture and Case Study

System Architecture

The process plan for the part is developed on the basis of the geometrical and technological information of the raw stock and the workpiece that is available from a solid-modeler-based CAD data model. This necessitates the integration of the process planner with a feature-based design system (FBDS) where the necessary part geometry can be created in terms of its manufacturing features, and the appropriate technological information can be specified on the component geometry. With this objective in mind, the blackboard-based process planning system (BBPP) has been integrated with a FBDS such as the CAEDS package (Figure 1.8). The process planner derives all of its information (i.e., required for process planning) from the FBDS. The user also specifies various technological information attributes such as surface finish and hardness specifications of the part in addition to the task of creating the part geometry. The technological information attributes are written to an output file while the geometric information of the part is written to a universal file formatted CAD output file.[18] The CAD output file is interpreted by an interface program and all relevant geometric and technological information associated with the features is translated to the product data model of the part. The process planner interprets the PDM and develops the process plans for the part. The process plan is then written in a suitable format for use in manufacturing. Subsequently, the process plan information is used for generating the NC tool path for machining the features on an NC machine tool, utilizing the services of a dedicated NC tool path planning module.

Blackboard-Based Process Planner

The blackboard (BB) architecture-based process planning framework has been implemented in a prototype expert system BBPP, using the CLIPS expert system shell. (The current capabilities of the BBPP system are summarized in Table 1.3.)

The domain expertise of the various participating experts (knowledge sources in BB terminology) that have been incorporated in the BBPP system is shown in Table 1.4. The control expert (control component in BB terminology) consists of a set of rules that performs the following functions (Figure 1.9): (a) correlating and interpreting data provided by different experts, (b) directing the solution generation process between two subsequent process planning phases, (c) activating the participating experts on the basis of the current state of the BB data base, and (d) directing output from the BB in each one of the process planning phases to the appropriate sections of the route sheet. The rules for the control expert have been developed on the basis of some heuristic control algorithms. In the BBPP system, the process planning task is partitioned into five separate phases that correspond to certain activities being performed during process planning.

The BB data base is a data structure in which slots are provided for the specification, retrieval, and modification of different types of data pertaining to the different segments of the process plan (Table 1.5). The different participating experts refer to the BB data base for the data required to perform their tasks

TABLE 1.3 Current Capabilities of BBPP System

Process Plan Feature	Capability
Manufacturing features	Slots, Pockets, Protrusions, and Holes
Machining process	Drilling, Boring, Shaping, Planning, Slab Milling, End Milling, and Sawing
Finishing operations	Reaming and Grinding
Heat treatment operations	Quenching, Carbonitriding, Cyaniding, and Carburizing

TABLE 1.4 Domain Expertise of Different Experts in the BBPP System

Name of Expert	Fields of Expertise in the Process Plan Domain
Machining	Machining strategies
Machine-tools	Machine-tool capabilities
Cutting-tools	Cutting-tool information
Material	Material behavior
Cost	Machining and manufacturing cost
Surface finish	Finishing operations
Heat treatment	Heat treatment operations
Machinability	Machining parameters
Setup	Machining setups
Sequence planning	Setup sequencing

TABLE 1.5   Template of the Blackboard Data Structure

Blackboard Data Structure	
Feature	Description
ID	Feature identification number
Manufacturing-process	Set of all possible manufacturing processes that can be used for generating the feature
Manufacturing-setup	List of setups that are required for machining the feature using different operations
Machining-parameters	Machinability parameters for machining operations
Finishing-op-parameters	Machinability parameters for finishing operations
Hardness-operation	Specifications of heat-treatment operations for imparting the required hardness in the part

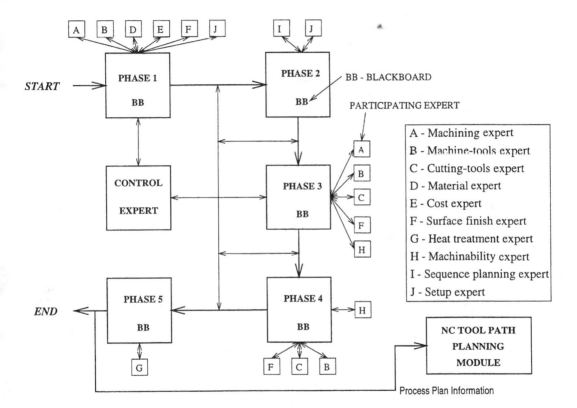

FIGURE 1.9   Process plan control flow.

and to modify the BB data base after the completion of their tasks. The BB data base in each one of these phases is closely monitored by the control expert.

At the end of each one of the five phases, the data present in the BB data base are written to an output file. These data constitute a segment of the process plan. The working of the BBPP system consists of the following steps:

1. The control expert initiates the process planner through the specifications of the PDM and the initialization of the phase 1 BB (Figure 1.9).
2. Next, the machining expert determines the set of all possible machining processes that can be used for generating the shape and form of each one of the features in the part. This selection is based on prior machining knowledge or can be obtained through geometric reasoning algorithms.

Depending on the type of feature to be machined, the machine-tool, material, cost and machining-setup experts are also activated. The activation of these experts is controlled through the control expert.

3. For each one of the machining processes, the machine-tool expert determines the number of setups and the setup times required for machining the feature on a machine tool capable of generating that feature; the material expert estimates the tool life of a standard cutting tool for machining the feature; the cost expert determines the production cost for machining the feature.

4. Next, the surface-finish expert estimates the amount of surface roughness zone that can be achieved by the machining process for a desired surface quality. This expert also establishes the machining cost required by the finishing process in order to achieve the desired surface roughness zone after the feature has been machined using a roughing operation.

5. The control expert then determines the machining index, which is a weighted sum of the inputs provided by the machine-tool, material, cost and surface-finish experts. Subsequent to this, the control expert selects the machining process with the best machining index. The setup expert determines the setup face to be used for machining the feature.

6. After establishing the machining processes for each one of the features, the control expert passes control from the phase 1 BB to the phase 2 BB. The setup expert sequences the setups so that the maximum number of features can be machined in one setup. The sequence-planning expert then studies the setup sequences and determines the sequence in which the different part features would be machined. The phase 3 BB is next activated by the control expert.

7. For each one of the machining processes to be used for machining the features, the surface-finish expert determines the amount of material to be removed in the individual machining processes. The different experts from the machining, machine-tools, cutting-tools, and machinability domains determine the machinability parameters using standard data handbooks and machining process models. On the basis of the data posted on the phase 3 BB data base, the control expert determines if the features require any additional finishing operations. In the even that the feature does not require any additional finishing operations, control is passed over directly to the phase 5 BB. On the other hand, if the feature requires additional finishing operations, then the phase 4 BB is initiated.

8. The surface-finish expert determines the finishing operations to be used for generating the desired surface roughness zone of the form feature. This selection is based on the capability of the finish machining processes. Next, the machinability parameters for the finishing operations are determined using standard data handbooks and machining process models in conjunction with the machine-tools, cutting-tools, and machinability experts. These data items are posted on the phase 4 BB, and the control expert then activates the phase 5 BB.

9. Depending on the desired surface hardness of the part, the heat-treatment operations expert estimates the depth of hardening and the heat-treatment operations that can achieve it.

10. After the successful generation of the process plan, the NC tool path planning module is activated in order to generate the NC tool path to be adopted for machining the features on an NC machine tool. The NC tool path planning module requires the following input information: (a) feature geometry information as obtained from the FBDS system, and (b) feature processing information as obtained from the BBPP. The four stages involved in the generation of the NC tool path (Figure 1.10) are

   • Selection of appropriate machining approach (either contour or edge-based), depending on the type of feature to be machined

   • Generation of the NC path depending on the type of feature to be machined, geometric configuration of the feature base surface, and cutting-tool specifications for both contour and edge-based machining approaches

   • Generation of machining volumes that represent the volume of material to be removed from the raw stock for generating the specific form feature, depending on the NC path, machining criteria, and cutting-tool specifications

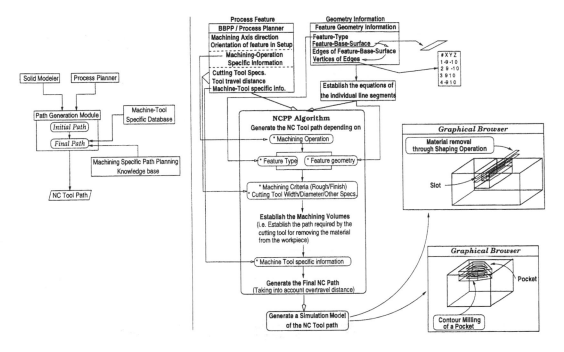

FIGURE 1.10   Framework of the NC tool path planning module.

- The final NC path is generated by taking into account machine-tool-specific information such as tool travel distance. The machining volumes generated in the earlier step are completely dependent on the geometry of the features but independent of the machine tool. However, the final NC path would have to account for machine-tool-specific machining instructions in order to generate an NC path that is feasible for machining on a certain machine tool.

Case Study: Process Plan for Sensor Bed

In this section, a case study is presented to demonstrate the working principles of the BBPP system. A sample part as shown in Figure 1.11 is created in the FBDS system using ten volumetric features (one square slot, two square holes, and seven cylindrical holes). The test part is a sensor bed for a wear testing machine. The different form features in the sensor bed require different levels of form and size accuracy. In addition to specifying the geometry of the part, the user also specifies the surface finish values for various features of the part. The PDM for the raw stock and the part is shown in Table 1.6.

The BBPP system first determines that the raw stock needs to be machined to the dimensions of the finished workpiece using the slab milling machining operation. Subsequently, the BBPP system determines the process plan for machining the different features in the part. A high-level view of the process plan for machining the different features in the finished workpiece is shown in Table 1.7. The NC tool path for machining the slot feature (Feature 1) of the test part is shown in Figure 1.12.

1.5   Application Domain: Allocation of Tolerance Specifications

The tolerance specifications assigned on a component govern the variability in the functional performance of the component, the accuracy and effort required by the various manufacturing processes to fabricate the component, the deviation allowed in component assembly, acceptable ranges of size, form, and locational variations of the components, and the ease of interchangeability of the components. Thus, tolerances influence a number of factors such as design function, manufacturing conditions, component assembly, serviceability, etc.[16]

TABLE 1.6   Feature Characteristics as Obtained from the PDM of the Raw Stock and the Finished Workpiece

Product Data Model—Raw Stock

Number of features—10, Raw stock dimensions—210 $\times$ 90 $\times$ 12
Finished workpiece dimensions—200 $\times$ 85 $\times$ 10, Raw stock hardness—200 BHN

Product Data Model—Finished Workpiece

Specifications	Feature 1	Feature 2,3	Feature 4,5,6,7,8,9,10
Type	Square slot	Square hole	Cylindrical hole
Surface-id	S11, S12, S13	S21, S22, S23, S24, S31, S32, S33, S34	S41, S51, S61, S71, S81, S91, S101
Dimensions	5, 5, 200	20, 10, 100	5, 10
Location	(0, 0, 0)	(40, 0, 0) (−40, 0, 0)	(5, −98, 0), (−5, −98, 0), (5, 98, 0) (5, 98, 0), (3, −58, 0), (3, 38, 0) (−3, 83, 0)
Finish	4.0	2.0	0.005
Face-attachment	2	3	2
Tolerance-info	Std. Tol.	Std. Tol.	Size: $+/-0.09$, $\perp$: 0.02

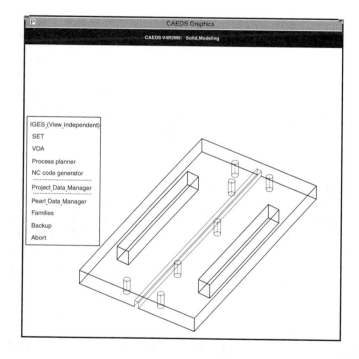

FIGURE 1.11 A screen dump of the user interface for the process planner along with a test part created in FBDS.

Tolerance synthesis is defined as the process of allocating tolerances to the size, form, and location of geometric entities of a component such that the cost incurred due to tolerances is minimized while satisfying functional and assembly requirements. Tolerance synthesis is a two-step procedure that involves the following tasks: (a) mapping part functions to equivalent functional tolerance limits and (b) constraining the functional tolerance limits with respect to manufacturing, assembly, and inspection constraints. Associated with tolerance synthesis is the issue of representation of tolerances such that it could be used effectively for the tasks of tolerance analysis and other downstream manufacturing activities. In this research work, we have demonstrated the application of ICAD techniques for performing the above two

TABLE 1.7 A View of the High-Level Process Plan along with the Various Machinability Parameters Generated by the BBPP System for Machining the Features in the Test Part

Process Plan Issue	Feature No.	Feature Type	Process Planner Output
Manufacturing-process	Feature-1	Square slot	Slab milling
	Feature-2,3	Square hole	Drilling and milling
	Feature-4,5,6,7,8,9,10	Cylindrical hole	Drilling
Manufacturing-setup	Feature -1	Square slot	1
(face no. in part)	Feature-2,3	Square hole	3
	Feature-4,5,6,7,8,9,10	Cylindrical hole	2

Manufacturing Operation Parameters			
Process plan details	Feature 1	Features 2 and 3	Features 4,5,6,7,8,9,10
Manufacturing operation	Slab milling	Drilling and milling	Drilling
Machine tool	Universal milling machine	Radial drilling and universal milling machine	Radial drilling machine
Cutting-tool type	HDPMC	PTSD, SSEM	PTSD
Cutting-tool specifications	ARCEARH	PLCH, PRP	PLCH
	Cps 1	Cps2	Cps3
Cutter diameter	1	3 (drill dia.),3	3/4
Cutter width	4	—	—
Feed (mm/tooth)	0.18	0.25 (mm/min) Drill bit and 0.018 (mm/tooth) Cutter	0.45 (mm/min)
Tool material grade	S2	S2	S2
Number of passes	4	7	—
Cutting fluid	EO	EO	EO
Speed (m/min)	29	21 (Drilling), 34	21
Depth of cut (mm)	0.65	1.0	—

Finishing Operation Parameters			
Finishing operation machine tool	No operation required	No operation required	Internal grinding VGM
Cutting-tool type	—	—	CyGW
Grit size	—	—	60
Infeed (m/min)	—	—	0.005
Traverse (mm/pass)	—	—	0.125
Wheel speed (m/min)	—	—	51
Work speed (m/min)	—	—	46
Wheel material	—	—	CBN
Wheel type	—	—	B180TV

Heat-Treatment Specifications

Carburizing at 870–950°C with $CaCO_3$ followed by water quenching.

tasks associated with tolerance synthesis, and we outline the architecture of a tolerance synthesis system (TSS) for automatic synthesis of tolerance specifications under the overall framework of an IPD system. The TSS infers the functional tolerance limits (task [b]) based on the qualitative descriptions of the component functions as specified by the user through the use of component function–functional tolerance limit knowledge bases. These knowledge bases relate functions associated with the component geometric entities (surfaces and features) with certain functional tolerance limits. The specified functional tolerance limits are further constrained (task [b]) by identifying appropriate constraints (manufacturing, assembly, and inspection criteria), which are applicable to the component geometric entities. This task (i.e.,

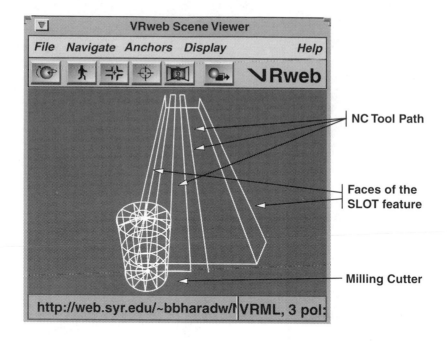

FIGURE 1.12 Display of the NC tool path generated by the NC tool path planning module for machining the square slot (Feature 1) feature in the test part.

task [b]) is performed by the TSS system through the use of tolerance optimization formulations,[19] process error-tolerance limit data bases, and geometric entity-process error relationship knowledge bases.[16] In this chapter, we further demonstrate the potential of applying ICAD techniques for the automatic synthesis of tolerance specifications for a test-case assembly, which is commonly encountered in various industrial applications.

ICAD Techniques in Tolerance Synthesis

The ability of TSS to synthesize tolerance specifications on components subjected to a variety of component functions and assembly requirements is dependent to a large extent on the availability, content, and accessibility of the required knowledge and data bases. The different knowledge and data bases that have been developed for the task of tolerance synthesis in TSS are

1. **Component function–functional tolerance limit** knowledge base: The qualitative and abstract descriptions of component functions and assembly requirements as specified by the user cannot be directly transformed into functional tolerance limits. In the past, various expert systems[20,21] have been developed for transforming abstract component function descriptions into functional tolerance limits. However, the domain of the expert systems was restricted to a few classes of components, which are provided in the standard fit and tolerance tables (ISO tables). This drawback limits the application of an expert system approach for the synthesis of tolerances. In order to overcome the above drawback for the synthesis of tolerances, we have proposed a two-step procedure for the mapping of abstract component functions into functional tolerance limits. The two steps consist of the following: (a) at first, the abstract component functions are transformed into the part function model (PFM) of the part,[16,22] and (b) in the next step, the functional tolerance limits are established by mapping the PFM model of the part with the appropriate entry in the component function-functional tolerance limit knowledge base. This knowledge base essentially

TABLE 1.8 Examples of Rules in the Component Function-Functional Tolerance Limit Knowledge Base

---

Rule 1:

IF interacting faces of two components have PFM model = [0,1,1,1,0,0, {Contact pressure = p_c}]

  AND have surface contact

THEN the two components have interference fit condition between them

at the interacting faces - | Interference | $\Rightarrow \Delta = \dfrac{\pi d_i \mu p_c}{42.2\left(\dfrac{d_o + 0.3d_i}{d_o + 6.33d_i}\right)}$

Alternative expression $\Rightarrow \Delta = \left(\dfrac{p_c d_c}{E_i}\right)\left(\dfrac{d_c^2 + d_i^2}{d_c^2 - d_i^2} - \nu_i\right) + \left(\dfrac{p_c d_c}{E_o}\right)\left(\dfrac{d_c^2 + d_i^2}{d_c^2 - d_i^2} - \nu_o\right)$

where Young's modulus of material of the two mating parts is given by E_i and E_o. d_o, d_i, and d_c are the external, inner, and contact surface diameters, respectively; ν_o, ν_i represent the Poisson ratio of the material of the two mating parts; Δ_i denotes the change in diameter of the inner contacting surface due to the contact pressure, p_c; Δ_o denotes the change in diameter of the outer surface due to the contact pressure, p_c; Δ represents the total change in both internal and external diameters of the contacting surfaces.

Application: The interference fit conditions between interacting faces are modeled as a contact between two cylinders. Depending on the interference and the dimensions of the surfaces, appropriate design tolerance limits (upper and lower size tolerance limits) are determined from the ANSI/ISO standard fit tables. In the case of planar faces, an equivalent cylindrical face is first established (using plane diagrams), and the corresponding tolerance limits are determined.

Rule 2:

IF the interacting face of a component has PFM model = [0,1,0,1,1,0, {Contact pressure = p_c}]

  AND maintains surface contact with mating face

  AND the surface type is cylindrical surface

THEN cylindricity form tolerance should be assigned on the face

---

consists of relationships among the following three entities: (a) the PFM, (b) the geometric nature of the component surfaces, and (c) the contact conditions that exist on the interacting surfaces of the components in the assembly. The relationships among these three entities have been encoded in the form of rules making them more amenable to the implementation of the knowledge base in a computer-based IPD system. In Table 1.8, two rules from the component function-functional tolerance limit knowledge base are shown. In Rule 1, the interacting faces of two mating components, which have a PFM model = [0,1,1,1,0,0,{Contact Pressure (Numerical value)}], would have an interference fit condition whenever the two interacting faces have surface contact between them. In this rule, the contact pressure term would be derived from the behavioral attribute that would be assigned to the PFM model of the face. Next, the necessary amount of interference (Δ) would be estimated using one of the equations provided in Rule 1. It should be noted that the interference between the two mating faces is directly influenced by the contact pressure (p_c) associated between the two faces as well as other geometric and design parameters. The appropriate size tolerance limits are determined based on the face dimensions and the required interference at the contact in conjunction with the ISO standard fit tables.[23] Similarly, two cylindrical surfaces, which maintain a surface contact between them and with PFM model = [0,1,1,1,0,0,{Contact Pressure (numeric term)}], would be assigned cylindricity form tolerance (Rule 2). It should be noted that the rules provided in the knowledge base can be applied to components that are subjected to a wide variety of component functions. This enhances the application and utility of the knowledge base in the synthesis of tolerance specifications.

2. **Geometric entity-process error relationship** knowledge base: The functional tolerance limits as established from the component function-functional tolerance limit knowledge base are further constrained with respect to various manufacturing, assembly, and inspection criteria associated with the different geometric entities of the components. The knowledge of the relationship between

TABLE 1.9 Process Capabilities for Straightness
Tolerance for Different Manufacturing Operations

Machining Process	Straightness Tolerance Zone (mm)
Drilling	0.001
Slotting	0.0005
Turning	0.0001
Boring	0.0001
Cylindrical grinding	0.00005
Reaming	0.00006
Surface grinding	0.00003
End milling	0.002

Note: The above data is available in References 23–25.

component geometric entities and various process errors that affect the accuracy of these geometric artifacts is stored in this knowledge base. The knowledge of the above relationships is stored in the form of rules that are applicable under the same specified conditions. For instance, a machined surface would be subjected to form errors arising from the deflection of cutting tools, vibration of tool posts, and cutting-tool wear. The knowledge of the above three parameters would be useful for determining the achievable form tolerances of surfaces manufactured using a variety of machining processes.

3. **Process-tolerance limit** data base: This data base provides information regarding the capabilities of the different manufacturing and assembly processes with respect to achieving certain size, form, and positional tolerances. This data base provides a general guideline of the range of process capabilities. A sample instance as shown in Table 1.9 depicts the process capabilities of different machining processes with regard to straightness tolerance.

Schema for Tolerance Synthesis within the TSS System

Given the knowledge regarding component functions, manufacturing processing information, and assembly plans, the global objective of a tolerance synthesis schema is to assign a set of size, form, and position tolerances along with datum reference planes to a component. This issue is further complicated by the fact that there exists an interdependency relationship between tolerances and the various inputs required for assigning tolerances.[27] This deadlock in assigning tolerances can be broken with a schema in which neither the tolerances nor the different tolerance inputs are affected by the breakdown in their interdependency relationships. In this section, we outline a tolerance synthesis schema that takes into consideration the above interdependency relationships and assigns tolerances on the basis of knowledge pertaining to component functions, manufacturing process considerations, and assembly process capabilities. With reference to Figure 1.13, the tolerance synthesis schema works in four distinct phases.

- **Phase #1 (Specification)**: In this phase, the tolerance synthesis schema requires the following inputs:
  Geometry description: The synthesis schema requires the organization of the part geometry along three hierarchical levels of representation (assembly, part, and feature level). The user is required to specify the part geometry description at the assembly level (i.e., spatial location of part in assembly), part level (i.e., spatial location of features in the part and their inter-relationships), and feature level (i.e., feature geometry).
  Part function specification: The user specifies the part behavior on the basis of a function vocabulary and describes the behavioral attributes associated with that function.
  Material and surface finish specification: The synthesis schema also requires that the user specify the material and surface finish characteristics of the part.

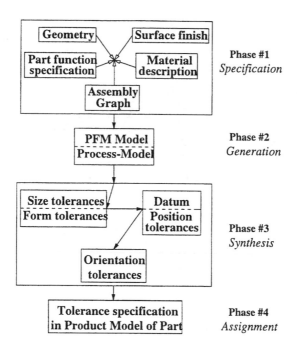

FIGURE 1.13   Tolerance synthesis schema.

Assembly graph: The assembly graph describes the procedure for assembling the different parts in the assembly without considering the effect of tolerances. At present, the assembly graph is specified by the user.

- **Phase #2 (Generation):** In this phase, the PFM model for the part surfaces, along with the process model for the part, is generated. The PFM model represents the equivalent part functions on the different surfaces of the part. The PFM model for a surface is essentially a collection of kinematic and force degrees of freedom (dof), along with other part behavior attributes. The kinematic and force dofs represent the absence/presence of motions and transmission of forces along a particular axis of the surface.[16] The process model represents the process plan for manufacturing the part without considering the effect of tolerances.[15]

- **Phase #3 (Synthesis):** The different tolerances for the part are synthesized at this stage. The synthesis procedure consists of two subphases: (a) transformation of the PFM model into functional tolerance limits, and (b) constraining the functional tolerance limits with respect to different constraints. The first and second subphases take extensive support of the component function-functional tolerance limit and geometric entity-process error relationship knowledge bases, and the process-tolerance limit data base. The size and form tolerances are assigned on the surfaces while the positional tolerances are assigned on the location of different geometric features in the part. The datum reference planes/features for the different geometric features of the part are also assigned at this phase. The data are assigned on the basis of the PFM assignments for the different surfaces of the part in conjunction with their manufacturability, assemblability, and inspectability criteria.

- **Phase #4 (Assignment):** After the assignment of data at the part level, different form tolerances have to be redefined with respect to their datum reference planes. In this stage, the form tolerances are substituted with different orientation tolerances using a form-to-orientation tolerance conversion chart.[16] After completion of this stage, the different tolerances are assigned to the product model of the part.

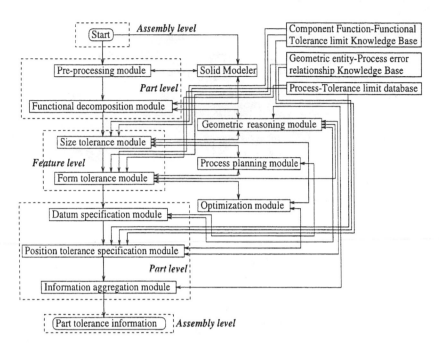

FIGURE 1.14   The system architecture of the tolerance synthesis system.

Implementation and Case Study

Implementation

The TSS for automatic allocation of tolerances on components has been implemented using the CM software package on a SUN SPARC workstation.[26] The IMSL math library and the CLIPS expert system shell are also used in the implementation of the TSS. The high-level system architecture of the system is shown in Figure 1.14. In addition to different computational modules (as shown in Figure 1.14), a number of other routines have been developed using the CLOS language for performing different utility functions. The TSS has been built in a modular fashion such that each one of the modules could be modified without affecting the rest of the system. At present, the TSS consists of about 11 modules that are built around the CM package. The 11 modules are

- *Pre-processing module*: In this module, the user specifications (such as geometry, part functions, material description, surface finish, and assembly graph) are represented as attribute values in the product model of the part. The information is distributed along three hierarchical levels of the part product model. These functions are performed in conjunction with the product model templates and the solid modeler in the CM system.
- *Solid modeler*: The solid modeler (SHAPES) is used for creating the geometric model (hybrid CSG/B-Rep based) of the part. The geometric attributes of the part (solid, surface, edge, and vertex information) can be attached as information attributes to the product model templates of the part.
- *Function decomposition module*: In this module, the component functions as specified by the user are decomposed into equivalent functional specifications (PFM) on the individual surfaces of the part. This is done by using the PFM algorithm[16] in conjunction with the geometric reasoning module.
- *Geometric reasoning module*: This module (AGM function-based) is an independent module that provides geometric reasoning support for a number of other modules within the TSM module. It

primarily uses the bounding-box, point-in-polyhedra query, polyhedral proximity, and polyhedral and surface intersection algorithms for the reasoning purposes.

- *Size tolerance module*: This module retrieves information from the function decomposition, process planning, and optimization modules in order to arrive at a set of size tolerance assignments on the surfaces of the part. This is done by transforming the PFM assignment on the part surface into equivalent design tolerance limits and determining the optimal size tolerance limits with regard to other constraints (such as manufacturing and quality considerations). The component function-functional tolerance limit knowledge base and the process-tolerance limit data base provide support to the size tolerance module.

- *Form tolerance module*: This module works in a similar fashion to the size tolerance module. In this module, the form tolerance type (such as flatness, circularity, etc.), along with the form tolerance zone, is assigned for the different surfaces of the part. During the synthesis phase, the component function-functional tolerance limit and geometric entity-process error relationship knowledge bases, and the process-tolerance limit data base assist in the assignment of tolerances.

- *Process planning module*: This module is used for creating the process model of the part using a blackboard architecture-based process planner.[15]

- *Optimization module*: In this module, the different optimization formulations that are encountered in tolerance synthesis are evaluated in order to assign the tolerances. This module offers an integrated platform for solving linear programming (LP) and nonlinear programming (NLP) optimization-related problems.

- *Datum specification module*: This module takes as its input the information from the geometric reasoning module, the PFM model of the surfaces, and the process model of the part in order to assign a set of data for referencing different features in the part. The datum assignment algorithm is provided in Reference 16.

- *Position tolerance specification module*: In this module, the position tolerances on the location of different features of the part are assigned on the basis of the component functions, the process model, and the inter-relationships between the location of different part features. The component function-functional tolerance limit and geometric entity-process error relationship knowledge bases along with the process-tolerance limit data base provide support to the position tolerance module.

- *Information aggregation module*: In this module, the size, form, and position tolerances assigned to the part are retrieved for presentation in a useful manner to the user. In addition, this module assists in transformation of some of the tolerances into other types as required (e.g., some of the form tolerance assignments are transformed into orientation tolerances, depending on the specified functional and assembly requirements).

Case Study: Tolerance Assignment for a Spindle Assembly

A spindle assembly consisting of three parts (shaft, hub, and key) as shown in Figure 1.16 is used for demonstrating the working of TSS. In Table 1.10, the user-specified inputs to TSS are shown. The input consists of (a) part geometry specification, part location in assembly with respect to a global coordinate reference frame, location, and type of feature in the part, and size dimensions of the features, (b) surface finish specifications (R_a values) for the faces of the features, (c) part function specification denoting the functions performed by the part, and (d) assembly-graph of the spindle assembly. Next, TSS generates the geometric models of the three parts and instantiates the respective product models (Figure 1.15). With reference to Figures 1.15 and 1.16, face interactions involving surface contact take place among the following faces: (a) face 3 (F6-3) of feature F6 is pressed against face 3 (F3-3) of feature F3, (b) face 4 (F6-4) of feature F6 is pressed against face 4 (F5-4) of feature F5, (c) face 2 (F4-2) of feature F4 is pressed against face 2 (F2-2) of feature F2, (d) face 5 (F6-5) of feature F6 is pressed against face (F3-5) and face (F5-5), and (e) face 6 (F6-6) of feature F6 is pressed against face (F3-6) and face (F5-6). In accordance with the proposed tolerance synthesis schema, the PFM model for the individual faces of the three parts is generated for the specified part functions and assembly configuration.

TABLE 1.10   User Input for the Tolerance Synthesis System

Part and Feature Geometry Description					
Part No.	F type (F-id)	F Geometry	F Dimension × 10 mm	F .Loc‡	Surface Finish
Part-1	*Bff*(F1)	Base-Cylinder	[2.75,1.5] (R × H)	[0,0,0]	0.01
	Nff(F2)	Hole-Cylinder	[1.75,1.5] (R × H)	[0,0,0]	0.004
	Nff(F3)	Slot-Block	[1.5,0.5,0.5] (L × B × H)	[0,0,15]	0.009
Part-2	*Bff*(F4)	Base-Cylinder	[1.75,7] (R × H)	[0,0,0]	0.01
	Nff(F5)	Slot-Block	[1.5,0.5,0.5] (L × B × H)	[0,0,15]	0.009
Part-3	*Bff*(F6)	Base-Block	[1,1,0.5] (L × B × H)	[0,0,15]	0.009

Part Function Specification			
Part No.	Part/Function	Function Description	Assembly Graph
Part 1†	Contact, self-rotary motion	Contact pressure: 30 kg/mm², Speed: 350 RPM, Load: NIL, Torque: 65 kg/mm	Pa*1* → (Part-2 ∅)
Part 2†*	Contact, self-rotary motion	Contact pressure: 30 kg/mm², Speed: 350 RPM, Load: 0.45 kg, Torque: 65 kg/mm	Pa2 → (Pa*1* Part-3)
Part 3†	Contact	Contact pressure: 30 kg/mm²	Pa3 → (Pa2 Part-1)

F = Feature, *Bff* = Base form feature, *Nff* = Negative form feature, R × H = Radius × Height, L × B × H = Length × Breadth × Height, Loc. = Locations, Assy. = Assembly, Pa*x* = Partial assembly state *x*.
‡ All locations are with respect to a global coordinate reference frame surface finish for each face of F is in μ.
† Part material is C-14 Steel and part has additional part function of nonpermanent assembly.
* Additional part function—House (for explanation refer to Reference 22).

In the next step, the appropriate size and form tolerances, and data and position tolerances for the features are identified in conjunction with component function-functional tolerance limit knowledge base (size, form, and position), geometric entity-process error relationship knowledge base, and process-tolerance limit data bases. In Table 1.11, the size, form, and orientation tolerances associated with all the different faces along with the data and position tolerances associated with the different form features of the three parts are shown.

For instance, based on the PFM specifications of Face 3 (F3-3) and in conjunction with rules similar to those shown in Table 1.8, size tolerance limits (upper limit = 7 mm, lower limit = 0 mm) and orientation tolerance (perpendicularity) of 0.001 mm are assigned (Table 1.11) on the face. The size tolerance limits are derived by first identifying the type of fit between the two mating faces. In this case, face F3-3 maintains an interference fit with the mating face. The amount of interference between the two mating faces is determined using Rule 1 provided in Table 1.8 by using the values of the function specifications associated with the mating faces and material properties of the part material (Table 1.10). Using the ISO standard fit tables,[23] the appropriate tolerance limits associated with the desired amount of interference are identified for the specified dimensions of the mating faces and are further restricted by manufacturing constraints in order to obtain the above size tolerance limits. A similar procedure was adopted for the assignment of orientation tolerance to face F3-3. Similarly, the feature (F3) containing face F3-3 was assigned a position tolerance of 0.002 mm with Face 1 (F1-1) of feature 1 (F1) and feature 2 (F2) serving as the data (Table 1.11). On a similar note, feature F2 is assigned a position tolerance of 0.0015 mm with Face 1 (F1-1) and feature 1 (F1) serving as the data based on the values of the part behavior specifications (N − 350 RPM, T − 65 kg/mm), which are assigned to the PFM model of face F2-2. Any change in the values of the part behavior specifications associated with face F2-2 (i.e., values of N or T) would affect the value of position tolerance (i.e., eccentricity associated with rotation of hub).

The tolerance values and required datum assignments are assigned to the product model of the individual faces and features at the end of this stage. In Figure 1.17, the technological information (including tolerance information) associated with a face of the key part is displayed. The TSS has been tested for a variety of cases and the tolerances synthesized by the system suggest the viability of our proposed tolerance model and tolerance synthesis schema.

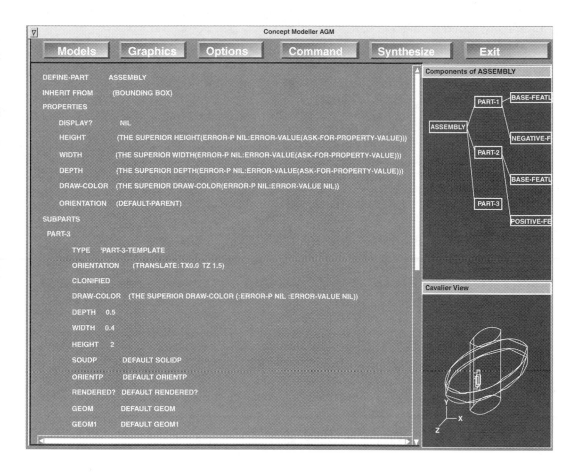

FIGURE 1.15   The instantiation of the initial part geometry and other attribute information in the product model of the assembly.

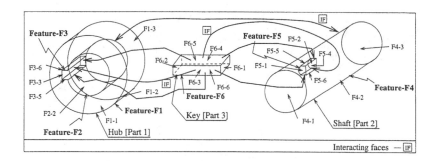

FIGURE 1.16   An illustration of the different faces of the parts in the test-case assembly.

1.6   Discussion and Conclusions

A comprehensive understanding of different issues involved in the application of ICAD techniques for the development of an IPD system has been discussed in this chapter. A fundamental understanding of these issues is the key toward the development of intelligent and integrated design systems with enhanced capabilities for product design. The development of prototype product design systems for design analysis, process planning, and tolerance assignment has also been reported in this chapter.

TABLE 1.11   Tolerance Specifications for the Different Parts in the Test-Case Assembly

				Size, Form and Orientation, and Position Tolerances					
				Surface Tolerances					
					Form and Orientation		Datum and Position Tol.		
Part No.	Feature Id	Face No.	Dimension × 10 (mm)	Size Tol. Limits × 10^{-3} mm	Type	Tol. Zone (mm), DRP	Datum Feature	Datum Planes	Tol. Value (mm)
Part-1	F1	1	0.5	137/115	□	0.001, —	—	—	—
		2	1.375	−90/−95	⊥	0.0015, F1-1			
		3	0.5	137/115	—	—			
	F2	2	0.875	11/0	⊥	0.001, F1-1	F1 *	F1-1◇	0.0015
	F3	3	0.25	7/0	⊥	0.009, F1-1	F2 *	F1-1◇	0.002
		5	0.25	7/0	⊥	0.001, F1-1			
		6	0.25	7/0	⊥	0.001, F1-1			
Part-2	F4	1	3.5	−320/−372	□	0.001, —	—	—	—
		2	0.875	−10/−15	⊥	0.001, F4-1			
		3	3.5	−320/−372	—	—			
	F5	1	0.5	45/0	—		F4 *	F4-1◇	0.002
		2	0.5	45/0	—				
		4	0.25	7/0	⊥	0.001, F4-1			
		5	0.25	7/0	⊥	0.001, F4-1			
		6	0.25	7/0	⊥	0.009, F4-1			
Part-3	F6	1	0.5	−15/−25	—	—	—	—	—
		2	0.5	−15/−25	—	—			
		3	0.25	16/10	⊥	0.0012, F6-6			
		4	0.25	16/10	⊥	0.0012, F6-6			
		5	0.25	16/10	∥	0.007, F6-6			
		6	0.25	16/10	□	0.001			

DRP = Datum Reference Plane, Tol. = Tolerance, ⊥ = Perpendicularity, ∥ = Parallelism, □ = Flatness, ◇ = Primary datum plane, * = Datum Feature (Medial axis used as secondary datum).

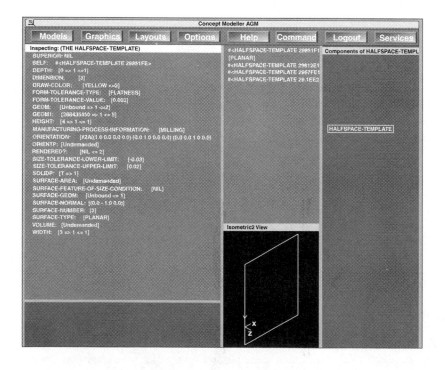

FIGURE 1.17   A view of the data in the slot values of a surface class.

The development of an IPD system is largely influenced by the product development models adopted for a specific application domain and the framework of the design process involving the synergic use of symbolic and numerical analysis techniques. From the three application domains considered in this chapter, it is evident that the practical utility of applying ICAD techniques for design automation is influenced by the knowledge encoded within the different knowledge bases, the decision making mechanisms, and the behavior of the integrated symbolic-numeric tools. Factors which influence the inadequacy of the models and the proposed framework are largely caused by limited domain knowledge of the design artifact in the various knowledge bases. Among subjective factors are mistakes and inaccuracies that are possible during the planning and construction of the knowledge bases. Our strong belief is that knowledge-based computational systems will revolutionize the entire task of product design.

For the future, research should be primarily focused on developing generic intelligent interfaces for a completely automated product design system. The capabilities of the IPD system should be enhanced to foster development of collaborative product design systems using the information infrastructure of the Internet and World Wide Web.

Acknowledgments

The authors would like to thank Concentra Inc. for providing the Concept Modeller software package for this research project on the development of an integrated product design system. The authors would also like to acknowledge the help provided by Prof. C. K. Mohan, Dr. Prakash Sarathy, Mr. Chris Ludden, Mr. Patrick Graham, and Mr. Anand Chavan during the different aspects of this research work.

Note: Concept Modeller, I-DEAS, ANSYS, CLIPS and IMSL are registered trademarks of Concentra Inc., Structural Dynamics Research Corporation, ANSYS Inc., NASA, and IMSL Inc., respectively.

Abbreviations

UMM—Universal milling machine; HDPMC—Heavy duty plain milling cutter; ARCEARH— [A—Axial rake angle, R—Radial rake angle, C—Corner angle, E—End cutting angle, A—Axial relief angle, H—Helix angle]; RDUMM—Radial drilling and universal milling machine; PTSD—Parallel taper shank drills; SSEM—Straight shank end mills; PLCH—[P—Point angle, L—Lip clearance angle, C—Chisel point angle, H—Helix angle]; PRP—[P—Primary land width, R—Radial primary relief angle, P—Primary clearance angle]; EO—Emulsifying Oil; VGM—Vertical grinding machine; CBN—Cubic boron nitride; S2—[Carbon = 0.95%, Chromium = 3.5%, Molybdenum = 8.2%; Tungsten = 1.5%, Vanadium = 1.7%]; Cps1—[15°, 15°, 30°, 10°, 7°, 25°]; Cps2—[118°, 15°, 135°, 32°] (Drilling), [13 mm, 0.4°, 28°] (Milling); Cps3—[118°, 15°, 135°, 32°]; CyGW—Cylindrical grinding wheel; Std. Tol.—+/−0.1 mm; All dimensions are in mm unless otherwise specified.

References

1. U. Roy and B. Bharadwaj. The role of product specifications in concurrent engineering based product design systems. In *Proc. of 5th Industrial Engineering Research Conference, Minneapolis, MN,* May 1996.
2. J. J. Cunningham and J. R. Dixon. Designing with features: the origin of features. In *Proc. of 1988 ASME Conference on Computers in Engineering,* 237–243, 1988.
3. F. Kimura, T. Kjellberg, F. Frause, and M. Wozny. Concurrent engineering research in review. Technical Report CERC-TR-RN-92-012, CERC, West Virginia University, 1992.
4. F. Kimura, T. Kjellberg, F. Frause, and M. Wozny. Concurrent engineering research in review. Technical Report CERC-TR-RN-92–013, CERC, West Virginia University, 1992.
5. M. S. Shephard and M. A. Yerry. Towards automated finite element modeling for the unification of engineering design and analysis. *Finite Elements in Analysis and Design,* vol. 2, 143–160, 1986.
6. M. S. Shephard and P. M. Finnigan. Integration of geometric modeling and advanced finite element preprocessing. *Finite Elements in Analysis and Design,* vol. 4, 147–161, 1988.

7. Z. Young and I. R. Groose. A rule-based computational system for automatic finite element modeling. *Computers in Engineering*, vol.14, 87–94, 1990.

8. J. E. Sneckenberger and S-H. Chung. An innovative design scheme for product development using AI-type engineering. In *Proc. of International Conference on CAD/CAM, Robotics and Factories of the Future*, 190–200, 1993.

9. P. H. Winston. *Artificial Intelligence.* Addison-Wesley, Reading, MA, 1993.

10. R. Engelmore and T. Morgan. *Blackboard Systems.* Addison-Wesley, Reading, MA, 1988.

11. U. Roy, B. Bharadwaj, and C. Ludden. Unification of CAD and FEM using knowledge engineering. *Concurrent Engineering: Research and Application*, 2:7–15, 1994.

12. U. Roy. Establishment of an intelligent interface between symbolic and numeric analysis tools for an intelligent computer-aided design system. In *IEEE International Workshop on Emerging Technologies and Factory Automation*, Melbourne, Australia, 317–322, 1992.

13. Swanson Analysis Systems, Inc., Houston, PA. *ANSYS Engineering Analysis System User's Manuals*, 1993.

14. T-C. Chang and R. A. Wysk. *An Introduction to Automated Process Planning Systems.* International Series in Industrial and Systems Engineering. Prentice-Hall, Englewood Cliffs, NJ, 1985.

15. U. Roy, B. Bharadwaj, A. Chavan, and C. K. Mohan. Development of a feature based expert manufacturing process planner. In *Proc. of the 7th IEEE International Conference on Tools with Artificial Intelligence.* IEEE, 1995.

16. B. Bharadwaj. A Framework for Tolerance Synthesis of Mechanical Components. Master's thesis, Syracuse University, NY, 1995.

17. U. Roy. Computer-Aided Representation and Analysis of Geometric Tolerances. Ph.D. thesis, Purdue University, West Lafayette, IN, 1989.

18. SDRC Inc. *CAEDS User's Guide*, OH, 1993

19. U. Roy and B. Bharadwaj. An enhanced framework for tolerance synthesis - Part B. *International Journal of Production Research*, 1997.

20. D. Janakiram, L. V. Prasad, and U. R. K. Rao. Tolerancing of parts using an expert system. *International Journal of Advanced Manufacturing Technology*, 4:157–167, 1989.

21. P. H. Gu and H. A. El Maraghy. Expert tolerancing consultant for geometric modelling. In *Symposium on Product and Process Design*, G. Chryssolouris, Ed., Manufacturing International 1988. ASME, 1988.

22. U. Roy and B. Bharadwaj. Design with part behaviors: behavior model, representation and applications. Submitted for publication, 1997.

23. K. Lingaiah. *Machine Design Data Handbook.* McGraw-Hill, NY, 1994.

24. I. S. Tarasevich and E. I. Iavoish. *Fits, Tolerances and Engineering Measurement.* Foreign Languages Pub. House, Moscow, 1963.

25. Society of Manufacturing Engineers, Dearborn, MI. *Tool and Manufacturing Engineers Handbook*, 4th edition, 1983.

26. Concentra Inc., Cambridge, MA. *Concept Modeller 1.3 Release Guide*, 1990.

27. U. Roy and B. Bharadwaj. Tolerance synthesis in a product design system. In *Proc. of the XXIV NAMRC/SME Conference*, MI. SME, May 1996.

2

Computer-Aided Design (CAD) for the Translation of Geometric Data to Numerical Control (NC) Programs for Manufacturing Systems

Han-Tong Loh
National University of Singapore

Ken-Soon Neo
National University of Singapore

Yoke-San Wong
National University of Singapore

2.1   Introduction

A numerical control (NC) program consists of a set of instructions or commands to the controller of a computer numerically controlled (CNC) machine to direct its operations on the part in the machine. Methods to generate the NC programs range from manual to highly automated:

- Manual part programming (e.g., in G and M codes for two to two-and-a-half-dimension parts).
- Computer-assisted part programming (e.g., using the APT programming language).
- Manual data input—Interactive graphic programming on the CNC monitor primarily in G and M codes and special function keys and commands.

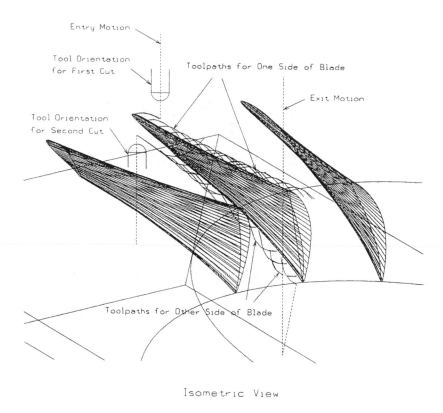

Isometric View

FIGURE 2.1   Tool paths generated for machining of turbine blade.

- CAD-based interactive NC programming —The raw and final parts are created on the CAD system and the desired set of processing operations to be performed is interactively selected or entered by the programmer from a given set of menu commands.[1-3] The processing operations (e.g., cutting tool paths) are then simulated on the high resolution graphics monitor for visual feedback or verification before the system is prompted to generate the required NC programs. Very complex shapes can be simulated and postprocessed for three-axis to five-axis machining, such as the turbine blade shown in Figure 2.1 for four-axis machining.[3]

- Computer-automated part programming (CAPP)—The computer decides the processing requirements on the part and subsequently generates the corresponding NC program required to execute the required processing tasks.

The last two approaches require the translation of geometric data from the CAD-represented part to extract geometric features to be generated, the corresponding processes required, and the tool paths. The most common processes under computer numerical control are machining operations. Except for the last approach, the NC programmer has to employ the expertise of a machinist in processing a part. In general, he must be familiar with the sequence of machining operations, the required cutting tools, feeds and speeds, and the basic programming methodology. Besides visualizing the entire process, the programmer must also be familiar with the particular machine being programmed. To automate the generation of NC programs from given geometry of a part defined in a CAD system, an effective computer-aided process planning system is required to determine the type of machining operations, corresponding machine tools, and fixtures for each type of machining operations (e.g., on a single machine): feasible machining sequences, cutter passes, and cutting conditions for both the roughing and finishing stages.

In this chapter, the focus of the CAPP is on the translation of a given part geometry to the required NC program for a single machine. The issues of tool-path generation are also presented. CNC machines typically can only direct a tool to move along linear and circular paths. However, geometric profiles of CAD-generated parts are usually defined by third- or higher-order curves and surfaces. Hence, it is necessary to fit lines and circular arcs, or their combination, to these CAD-defined curves to produce the CNC tool paths to some desired tolerance. To check and verify that the complex surfaces are generated to the desired accuracy, computer-controlled coordinate measurement machines (CMMs) are used. A logical approach is to use the same CAD-represented part to generate the NC program for the measurement of the part by the CMM. Setup, processing and data analysis requirements are quite different from those for machining operations. These are discussed in the last section.

2.2   Automation of Process Planning for Computer-Generated NC Programs

Process planning provides the necessary link between design and manufacture. Various computer-automated or computer-aided process planning (CAPP) systems have been developed and reported.[4–7] The recent trend is towards integrating CAPP with CAD and production planning and control activities.[6–7] Some systems aim to cater to the processing of a large variety of parts. Others are product- or geometry-specific, catering to a few specific families of parts, such as rotational parts. The applications of CAPP systems in industrial environment are somewhat limited and their full potentials have yet to be realized.[7] In general, a fully automated (generative) process planning system performs satisfactorily for well-defined families of parts with established process plans.[5] Even for a family of parts with pre-determined sequence of operations and finite variation in geometry, much development effort is still required to ensure that all possible permutations and combinations are considered and an efficient plan generated. The following two sections present CAPP for part families of more specific geometry where the parts are processed by machining and all processing requirements are completed on a single CNC machine:

- CAPP for NC turning[8,9]
- CAPP for NC milling of spherical parts: space frame nodes[10]

The outputs obtained from a CAPP system are used for job scheduling, NC programming and other manufacturing tasks. Prior to the generation of the NC programs, it is necessary to decide on the machining operations, select the cutting tools, and determine the cutter paths and cutting conditions. For a given machine tool, the activities involved in a typical machining operation planning include the following steps:

- Interpretation of part geometry or feature extraction
- Determination of cutting operations and boundaries
- Selection of setups and clampings
- Selection of tool types
- Determination of cut distribution and cut paths
- Determination of cutting parameters
- Determination of operation sequence and routing

CAPP for Turning

One approach to automating the decision making is to apply artificial intelligence (AI) techniques, such as expert system techniques. Using expert system techniques to provide a machining knowledge-based model, an integrated machining system (IMS) has been developed for the manufacture of rotational parts.[9] It aims to integrate and automate the feature extraction, operation planning, machinability data selection and NC program generation.

The architecture of this system is illustrated in Figure 2.2. This PC-based system comprises a user interface for interaction, a feature recognizer, a knowledge base of facts and rules. A software tool called GOLDWORKS (III) with COMMON LISP as the programming platform is employed in the knowledge building. With this frame-based knowledge building tool, the information being manipulated is repre-

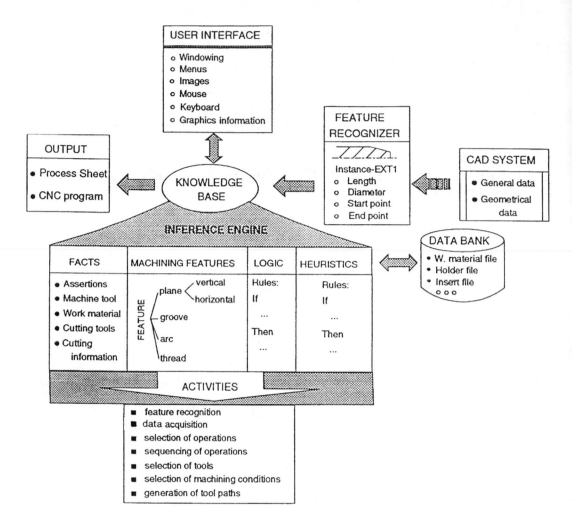

FIGURE 2.2   Architecture of the integrated machining system for turning.

sented by instances of frames, and the problem solving knowledge that manipulates the instances is represented by production rules. The software allows for easy modification, addition, or updating of information and production rules by an expert user. The system is designed to integrate knowledge in a modular form and incorporate heuristics so as to minimize human intervention when used.

The IMS first interprets a given CAD-represented part (AutoCAD DXF drawing file) and extracts the machining features. It assigns appropriate operations to all features identified, automatically selects the tools and determines the required machining conditions. Finally, tool paths are generated. Figure 2.3 shows the flow chart of the steps taken by the IMS to generate a process sheet and a set of CNC codes for a given setup. Fixture planning is not included in the IMS. It is assumed that component setups with suitable workholding are provided so that the solution generated by the IMS can be implemented. The inferential strategy, the feature extraction, and the knowledge base of facts and rules of the IMS are detailed in Reference 9.

A sample part drawing of a component created using the CAD software AutoCAD is shown in Figure 2.4. The first setup is used to machine to largest outer diameter feature, the left-most vertical feature and the two left-most internal features of the component. In the second setup the largest outer diameter of the component is clamped in the chuck. Based on this setup, the extracted features are shown in Figure 2.5. The process sheet and CNC program generated by the IMS are shown in Figures 2.6.

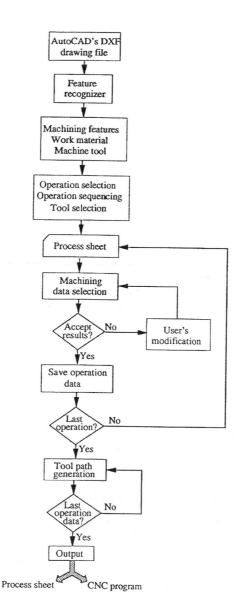

FIGURE 2.3 Process plan and NC program generation steps in the IMS.

CAPP for NC Milling of Spherical Parts (Space Frame Nodes)

Space-frame nodes are joints that interconnect linear members to form complex space frames, as shown in Figure 2.7. Each spherical node has a hole configuration that defines the orientation of the bolt holes in spherical coordinates according to its position in the space frame. The hole configurations are determined by a CAD software for space-frame design. For each hole configuration, the CAD software outputs the required number of nodes and a table giving the orientations of the holes with respect to the center of the node in spherical coordinates, such as the one given in Table 2.1. Nodes with the same hole configuration vary in number, from a few to tens and hundreds, depending on the design of the space frame.

Previously, according to the printed output of the tables, nodes were produced by drilling and tapping the holes into forged solid spheres using a manually adjusted two-axis indexing table mounted on a three-axis CNC machine. Because of the wide range of hole configurations, with only small and medium batches of nodes having the same configurations, frequent changes were required in the manually

TABLE 2.1   A CAD-Generated Hole Configuration for a
Space Frame Node

| Number of Nodes | Angle (normal) | | Bolt Size |
	Alpha	Beta	
70 (Number)	0.00	36.18	12
	44.98	0.00	12
100 (Diameter)	89.95	36.18	12
	134.98	3.90	12
	180.00	36.18	12
	224.98	0.00	12
	269.95	36.18	12
	314.98	3.90	12
	0.00	−90.00	12

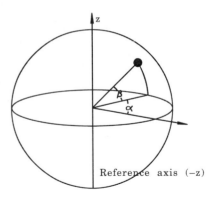

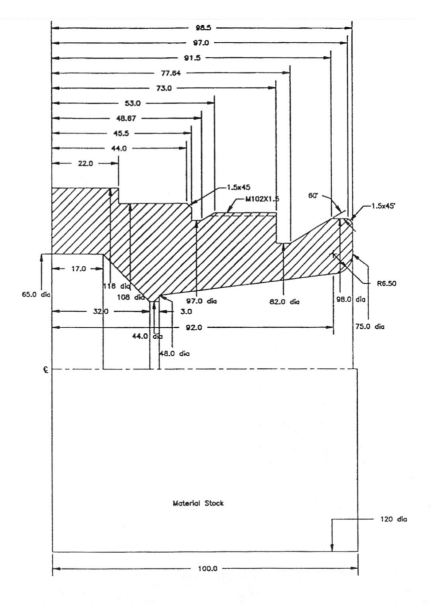

FIGURE 2.4   Part drawing of a turned part.

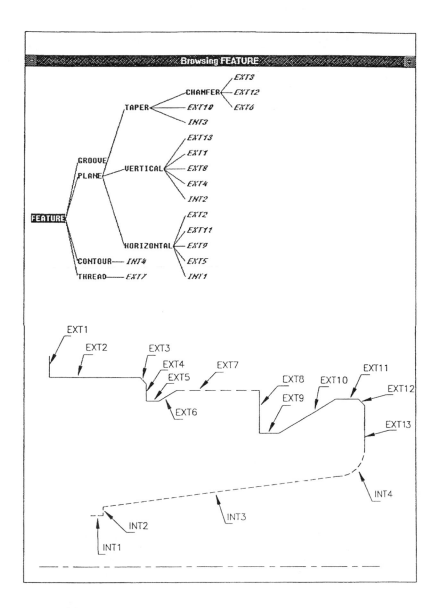

FIGURE 2.5 Extracted features from part drawing.

prepared process plans for tooling, fixturing, and NC programming, resulting in very slow production rate. Delivery was also uncertain because few job shops were able or prepared to machine the usually one-off orders.

To reduce the lead time for the manufacture of the space-frame nodes, a computer-automated machining cell for the manufacture of the nodes has been developed. The machining cell consists of a three-axis vertical machining center. With automatic tool-changing capability, the machining center can complete all required machining operations for a given setup without manual interruption. Machining centers are increasingly customized to automate machining of parts in batches that vary in lot size from as low as one for those with very complex profiles to several thousand.[11-13]

The machining center for the cell is linked to a two-axis CNC rotary table that is used to orient the spherical node (Figure 2.8). The cell is designed to machine spherical nodes of diameters 85, 100, 125 and 150 mm with bolt holes varying from M12 to M27. The nodes are machined in batches that have the same hole configurations. For each node, a mounting hole has to be machined so that it can be

NO	POS	OPERATION	FEATURES	TOOLHOLDER	INSERT
10	1	ROUGH_FACING_OP	(EXT13)	PCLNR_2525M_16	CNMG_12_04_16-MF_GC415
20	1	SEMI_FINISH_FACING_OP	(EXT13)	PCLNR_2525M_16	CNMG_12_04_16-MF_GC415
30	2	DRILL_OP	(INT1)	R416.1-0400-20-05	WCMX_06_03_08_R-53_GC-A
40	3	ROUGH_TURN_OP	(EXT12 EXT11 EXT10 EXT9 EXT8 EXT7 EXT6 EXT5 EXT4 EXT3 EXT2 EXT1)	SVJBR_2525M_16	VBMT_16_04_08-UM_GC415
50	3	SEMI_FINISH_TURN_OP	(EXT12 EXT11 EXT10 EXT9 EXT8 EXT7 EXT6 EXT5 EXT4 EXT3 EXT2 EXT1)	SVJBR_2525M 16	VBMT_16_04_08-UM_GC415
60	4	EXT_THREAD_OP	(EXT7)	R166.0FG-2525-16	R166.0G-6MM01-150_GC225
70	1	ROUGH_BORE_OP	(INT4 INT3 INT2 INT1)	S25T-PCLNR_12	CNMG_12_04_16-MF_GC415
80	1	SEMI_FINISH_BORE_OP	(INT4 INT3 INT2 INT1)	S25T-PCLNR_12	CNMG_12_04_16-MF_GC415

```
%
N5 G13
N10 G00 X200000 Z200000 F200 T010101 M03 M44
N15 G96 S397
N20 X334000 Z105000
N25 G85
N30 G82
N35 G00 X200000 Z200000 L1000 D2000
N40 X329000 Z182890 F150 S420 T010101 M03
N45 G01 X0 Z176320
N50 G80
N55 G00 X200000 Z200000 M05
N60 S70 T020202 M03 M44
N65 X0 Z110000
N70 G01 X0 Z110450 F100
N75 G01 X0 Z110000
N80 G00 X200000 Z200000 M05
N85 F180 S406 T030303 M03 M44
N90 X334000 Z105000
```

```
N220 G80
N225 G00 X200000 Z200000 M05
N230 F150 S420 T010101 M03 M44
N235 X254460 Z105000
N240 G85
N245 G00 X200000 Z200000 L1600 D3000
N250 G41 X285720 Z181320 F100 S443 T010101 M44 M03
N255 G01 Z176320
N260 G02 X273020 Z169970 I0 K-6350
N265 G01 X257620 Z113690
N270 G01 X257620 Z113690
N275 G01 X254460 Z110450
N280 G40
N285 G80
N290 G00 X200000 Z200000 M05
N295 M09
N300 MC2
```

FIGURE 2.6   Process sheet and partial listing of NC program.

mounted on the rotary table. Two parts are machined in the cell between each loading and unloading. One is a blank forged sphere that is suitably clamped in a fixed fixture for the machining of the mounting hole. The other node is the one that has just been machined with a mounting hole and transferred from the fixed fixture to the rotary table where it is bolted to the table using the mounting hole. A set of quick-change fixtures has been specially designed to secure and locate different diameters of nodes. For each bolt hole and mounting hole, a flat surface centered at the axis of the hole is machined off the spherical node at a depth that depends on the diameter of the node.

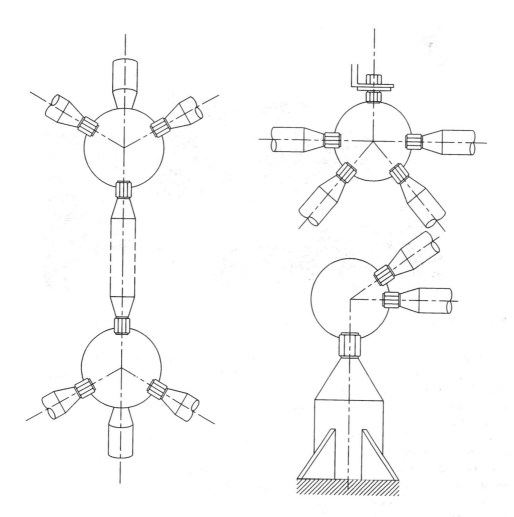

FIGURE 2.7 Space frame nodes.

For each batch of nodes to be machined, two NC programs are required: one for the CNC of the machining center and one for the CNC of the rotary table. To produce consistent and reliable NC programs quickly without specially employing a process planner/NC programmer, a PC-based, computer-automated process planning system has been developed to directly translate the aforementioned geometric data from the computer-aided, space-frame design software to NC programs for the two controllers, without intervention by the operator. The PC-based system is also connected by serial communication lines to the two CNC systems for direct transfer of the NC programs.

An important consideration in developing a suitable process plan for the machining of the space-frame nodes is the setup planning, in which the CAD-referenced hole configuration is transformed to a mounting-hole-referenced configuration that minimizes the required number of mounting holes and setups. The determination of the optimal mounting-hole-referenced configuration is detailed in Reference 10. Based on the mounting-hole-referenced hole configuration, NC programs are automatically generated for the following sequence of operations:

- Machining of the node in the fixed fixture
- Machining of the node mounted on the rotary table

Two machine-referenced coordinate systems (G54 and G55 in word address format for the FANUC CNC system) are set during the machining of the nodes. One is set with reference to the node in the fixed

FIGURE 2.8 Machining center for machining of space frame nodes.

fixture for the machining of the first mounting hole as well as the node code and the reference mark engraving. The other is set with reference to the node mounted on the rotary table. During the generation of the NC programs, necessary information, such as machinability data, tool-change codes, offset codes, etc., is automatically extracted from a data base that can be accessed and modified by authorized personnel with a given password.

The NC program begins with commands to the CNC machining center and rotary table for the machining of the first mounting hole in the node mounted in the fixed fixture. In addition, a node code of the node (e.g., T1 for the hole configuration shown in Table 2.1) is engraved on the flat of the mounting hole. If the mounting hole is a reference hole, a reference mark is also engraved on the flat. Both the node code and the reference mark are used in the assembly of the space frame. Once the mounting hole requirement has been determined, the orientation of every hole to be machined with respect to the mounting hole is appropriately transformed. If two mounting holes are required, two sets of hole configuration will be defined, appropriately transformed to the reference system based on each mounting hole. Figure 2.9 lists a sample section of an NC program for the machining center. The first section of the program includes NC instructions for the machining of the mounting holes. The corresponding NC program for the rotary table to rotate (in two axes) the holes to the correct vertical orientation for the drilling of the holes is partially listed in Figure 2.10.

2.3 Curve Fitting for NC Machining

In general, current CNC machines are capable only of linear and circular interpolations. General curves to be produced by CNC have to be approximated by fitting with lines, circular arcs, or a combination of both, to some given tolerance.[14-23] The aim is to obtain the least number of linear or circular segments required to approximate the curve to reduce the length of the NC program and the time to produce the approximate curve. Piecewise fitting using linear segments is very commonly used and reported,[24-28] as this approach is simplest to implement and fastest to compute. The resultant surface finish is, however, poor because of discontinuities between the piecewise-fitted segments. Linear approximation also tends to require a significantly larger number of segments than circular-arc approximation, unless the curvature of the curve to be fitted is small.[29,30] The circular-arc approximation may be applied in a piecewise manner to a predetermined set of points whereby the deviation of the arc from the points that it serves to replace

```
%
:1000
(NUMBER*TO*MACHINE*70)
G21G90G0G40G49G80T24
G91G30Y0Z0
T1
```
{*Machining of mounting hole of part in fixed fixture.*}
M6T5(FLAT) {*Start with machining a flat.*}

```
G54G90G0X46.378Y0M3S800
G43Z-3.5H1M8
G1X0F300
G0Z0M9
G91G30Y0Z0M5
```
M6H0T6(CENTER*M12)   {*Centre drill a shallow hole for*
```
G54G90G0X0Y0S1200M3
G43Z-2.5H5M8
G83G98Z-19.Q2.R-2.5F80
G80M9
G0Z0
G91G30Y0Z0M5
```
M6H0T7(DRILL*M12)   {*Drill the hole.*}
```
G54G90G0X0Y0S800M3
G43Z-2.5H6M8
G83G98Z-32.25Q2.R-2.5F80
G80M9
G0Z0
G91G30Y0Z0M5
```
M6H0T4(TAP*M12)   {*Tap the hole.*}
```
G54G90G0X0Y0S32M3
G43Z-2.5H7M8
G84G98Z-26.25R-2.5F56
G80M9
G0Z0
G91G30Y0Z0M5
```
M6H0T2(LOCATE)
```
G54G90G0X-15.Y0M3S1200
G43Z-2.5H4M8
G83G98Z-14.Q2.R-2.5F80
G80M9
G0Z0
G91G30Y0Z0M5
```
M6H0T1(ENGRAVE)   {*Engrave node code "T1."*}
```
G90G0X-2.25Y12.189M3S2200
G43Z0H2M8
G91G0Z-2.6F80
```
X0.5(CODE*T)   {*Engrave "T"*}
```
Y1.5
G1Z-1.
X1.95
X-.975
Y-3.
G0Z1.
X.975Y1.5
```
X0.5(#1)   {*Engrave "1"*}
```
Y.9
G1Z-1.
X.6Y.6
Y-3.
G0Z1.
Y1.5
G90G0Z0
X-1.5Y-12.189
G91G0Z-2.6F80
```
X1.5Y1.5(REF)   {*Drill reference hole.*}
```
G1Z-1.
X-1.5Y-3.
X3.
```

```
X-1.5Y3.
G0Z1.
X1.5Y-1.5
G90G0Z0M9
G91G30Y0Z0M5
H0
M0(OPTIONAL*STOP)
```
{*Machining of part mounted on rotary table.*}
M6T5(FLAT) {*Milling of all the flats for the bolt holes.*}

M75   {***}
```
G55G90G0X46.378Y153.328M3S800
G43Z-126.828H1M8
G1X0F300
G0Z0M9
M75
G55G90G0X46.378Y153.328
Z-126.828M8
G1X0
G0Z0M9
M75
G55G90G0X46.378Y151.987
Z-117.454M8
G1X0
G0Z0M9
M75
G55G90G0X46.378Y151.987
Z-117.454M8
G1X0
G0Z0M9
M75
G55G90G0X46.378Y117.799
Z-48.062M8
G1X0
G0Z0M9
M75
G55G90G0X46.378Y117.799
Z-48.062M8
G1X0
G0Z0M9
M75
G55G90G0X46.378Y117.799
Z-48.062M8
G1X0
G0Z0M9
M75
G55G90G0X46.378Y117.799
Z-48.062M8
G1X0
G0Z0M9
G91G30Y0Z0M5
```
M6H0T6(CENTER*M12) {*Centre drill all the bolt holes.*}

```
M75
G55G90G0X0Y153.328S1200M3
G43Z-125.828H5M8
G83G98Z-142.328Q2.R-125.828F80
G80M9
G0Z0
M75
```

*Note: This listing is only up to the center drilling of part on the rotary table. Comments between { } are not part of program listing. The rest of the NC program consists basically of repetitions of sub-sets of program blocks very similar to the sub-set for flat milling (as can be seen in the partial listing of the sub-set for center drilling). *M75 is signal to rotary table to position part (See Fig. 9 of NC program for the rotary table.).*

FIGURE 2.9   Sample NC program for machining center.

%
N001 XG91 XS44.98 XF10.00 XL001 YG91 YS90. YF5.00 YL001 YR01   *{Cycle for Flat Milling}*
N002 XG91 XS180. XF10.00 XL001 YG91 YS0 YF5.00 YL001 YR01   *{Each block positions the node to the*
N003 XG91 XS-90. XF10.00 XL001 YG91 YS-3.9 YF5.00 YL001 YR01   *required orientation and is invoked by M75 in the*
N004 XG91 XS-180. XF10.00 XL001 YG91 YS0 YF5.00 YL001 YR01   *NC program such as that listed in Figure 7.}*
N005 XG91 XS45.02 XF10.00 XL001 YG91 YS-32.28 YF5.00 YL001 YR01
N006 XG91 XS89.95 XF10.00 XL001 YG91 YS0 YF5.00 YL001 YR01
N007 XG91 XS90.05 XF10.00 XL001 YG91 YS0 YF5.00 YL001 YR01
N008 XG91 XS89.95 XF10.00 XL001 YG91 YS0 YF5.00 YL001 YR01
N009 XG91 XS135.03 XF10.00 XL001 YG91 YS36.18 YF5.00 YL001 YR01   *{Reset to initial position and repeat*
N010 XG91 XS180. XF10.00 XL001 YG91 YS0 YF5.00 YL001 YR01   *cycle for Center Drilling}*
N011 XG91 XS-90. XF10.00 XL001 YG91 YS-3.9 YF5.00 YL001 YR01
N012 XG91 XS-180. XF10.00 XL001 YG91 YS0 YF5.00 YL001 YR01
N013 XG91 XS45.02 XF10.00 XL001 YG91 YS-32.28 YF5.00 YL001 YR01
N014 XG91 XS89.95 XF10.00 XL001 YG91 YS0 YF5.00 YL001 YR01
N015 XG91 XS90.05 XF10.00 XL001 YG91 YS0 YF5.00 YL001 YR01
N016 XG91 XS89.95 XF10.00 XL001 YG91 YS0 YF5.00 YL001 YR01
N017 XG91 XS135.03 XF10.00 XL001 YG91 YS36.18 YF5.00 YL001 YR01   *{Reset and repeat cycle for M12 Drilling}*
Note: Listing is only up to the beginning of M12 drilling of part on the rotary table. The NC blocks for the positioning of the node by the rotary table are repeated between each operation as the tool performs all the operations at each position (per mounting) before it is changed.

FIGURE 2.10   NC program listing for CNC rotary table.

or best-fit, or the curve where the points are taken, is within the given tolerance.[20,28] Piecewise circular-arc-fitting algorithms suffer from a similar lack of smoothness in the machined surface, though to a lesser extent. One way to overcome this is to use biarc-curve fitting,[14,17,21] which ensures tangential (C^1) continuity between the circular segments.

The machining of objects with curved surfaces essentially reduces to machining a series of plane curves defined in a two-dimensional plane. Hence, without loss of generality, we can discuss the approximation of curves by linear or circular segments by considering only the case of curves defined in the xy plane. The following section will introduce two approaches in the generation of CNC cutter path data using linear or circular segments to approximate general curves that are used in the creation of curved objects in CAD systems. In the first method, we will illustrate the approach using the cubic Bezier curve as the general curve example and approximate it by a minimum number of linear or circular segment. In the second method, the general curve is given in the form of a large number of points that may have been digitized from a scaled model using CMM technology or laser-ranging technology. Biarcs are used to approximate the desired shape to within specified tolerances.

A C^0 Continuity Approach

Let $b_0 = (x_0, y_0)$, $b_1 = (x_1, y_1)$, $b_2 = (x_2, y_2)$ and $b_3 = (x_3, y_3)$ be the four control points in a two-dimensional space, and t be a parameter (from 0 to 1) that characterizes a cubic Bezier curve. The cubic Bezier curve can then be written as follows:

$$x(t) = \sum_{i=0}^{3} x_i B_i^3(t)$$

$$= x_0(1-t)^3 + 3x_1 t(1-t)^2 + 3x_2 t^2(1-t) + x_3 t^3 \tag{2.1a}$$

$$y(t) = \sum_{i=0}^{3} y_i B_i^3(t)$$

$$= y_0(1-t)^3 + 3y_1 t(1-t)^2 + 3y_2 t^2(1-t) + y_3 t^3 \tag{2.1b}$$

To approximate the curve described above by a minimum number of linear segments or circular arcs, the approach is to cast it as an optimization problem: *minimize the number of segments subject to the constraint that in each segment, the maximum deviation must not exceed the tolerance*, i.e.,

$$\text{minimize} \quad N \tag{2.2a}$$

$$\text{subject to} \quad (\text{max } d)_i \leq \varepsilon \quad i = 1, \dots, N \tag{2.2b}$$

where N is the number of segments, d the deviation and ε the tolerance.

The problem will involve finding this number N and the $N + 1$ break points on the curve. This falls under the class of a nonlinear mathematical programming problem having mixed integer-continuous variables, which is usually difficult to solve. The problem is further compounded by the fact that the number of constraints is not fixed because it depends on N, which is the objective itself.

Fortunately, in this case there is only one integer variable—the number of segments. Falling on a set of optimization principles known as the *Monotonicity Principles*,[31] the problem becomes tractable because the principles dictate that for the problem to be bounded, at least one of the constraints in (2.2b) must be active. One can argue that a solution to (2.2) is to have as many of the constraints active as possible. Hence, one would try to find successive segments with their deviations from the curve *equal to* the tolerance. The problem as stated in (2.2) can thus be turned into a series of iterative continuous problems: *for each segment, maximize the length of the segment subject to the constraint that the maximum deviation does not exceed the tolerance*, i.e.,

$$\text{maximize} \quad t \tag{2.3a}$$

$$\text{subject to} \quad \text{max } d \leq \varepsilon \tag{2.3b}$$

Monotonicity Principles again dictate that the constraint (2.3b) must be active for the problem in (2.3) to be bounded. Discussions on the determination of the deviations between the linear or circular segments and the Bezier curve will not be given here but can be found in Reference 29.

Algorithm for Linear Approximation

The algorithm for linear approximation may be summarized as

Initialization: The Bezier curve is first approximated by a single linear segment as shown in Figure 2.11(a). t_{low} and t_{high}, two variables which act as a control bracket, are initially set to **0** and **1**, respectively. Initially, t_0 is set to be equal to t_{low} and t_1 to t_{high}. k is a counter that tracks the number of segments required. Set $k = 1$ initially. The maximum deviation d between the line and the Bezier curve is then calculated and compared with the given tolerance ε.

Step 1: When $d > \varepsilon$, set $t_{high} = t_1$, $t_1 = 0.5 \, (t_{low} + t_1)$, as shown in Figure 2.11(b). This procedure is repeated as long as $d > \varepsilon$, as shown in Figure 2.11(c).

Step 2: When $d < \varepsilon$, set $t_{low} = t_1$, $t_1 = 0.5 \, (t_1 + t_{high})$, as shown in Figure 2.11(d). This procedure is repeated as long as $d < \varepsilon$.

Steps 1 *and* 2 are repeated until:

    $d < \varepsilon$ by a specified convergence criterion (e.g., fixed at 10% of ε) or,

    $d < \varepsilon$ and t_1 is equal to 1.

Then, the break point is considered found. t_1 is then assigned to an array element $t(k)$. Increment k.

Step 3: The value of t_1 is assigned to t_0 and t_{low}. t_1 and t_{high} are then set to the value of **1** as shown in Figure 2.11(e). The maximum deviation d is again calculated and compared with ε.

Steps 1 *to* 3 are repeated until $t(k) = 1$.

End.

Figures 2.12a and 2.12b show two typical Bezier curves approximated by linear segments.

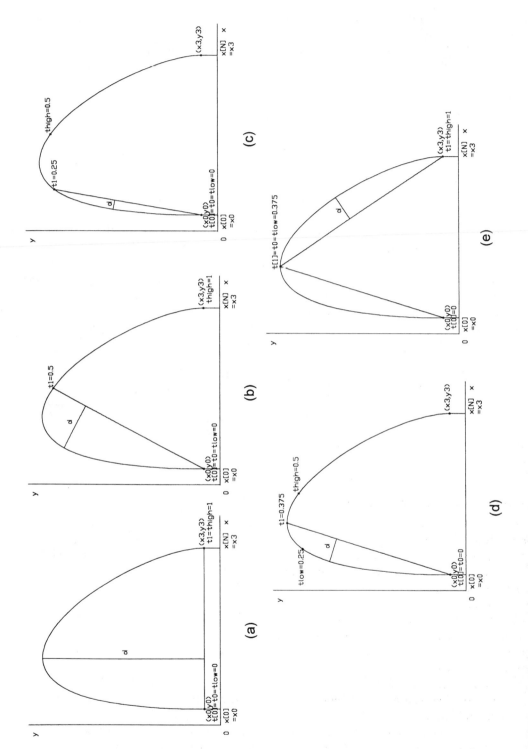

FIGURE 2.11 Linear approximation of a cubic Bezier curve.

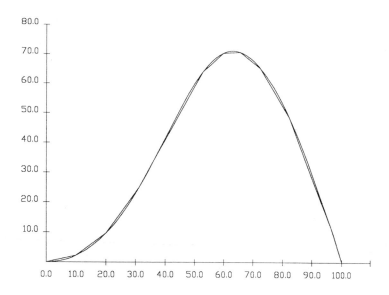

FIGURE 2.12(a) Linear Approximation for Bezier curve A. Tolerance : 0.6, convergence criteria : 0.06, number of straight lines: 10, processor time: 0.06.

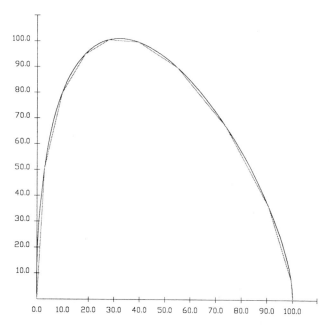

FIGURE 2.12(b) Linear Approximation of Bezier curve B. Tolerance : 0.9, convergence criteria : 0.09, number of straight lines: 10, processor time: 0.06.

Algorithm for Circular Approximation

Only two points are required to define a linear segment. Three points, however, are required to define a circular arc. Hence the algorithm for circular approximation is correspondingly more complex than that for the linear approximation, although the same basic idea is employed. Figure 2.13 illustrates the procedure for circular approximation.

Initialization: The Bezier curve is first approximated by a single circular arc which passes through three points, defined by $t_0 = 0.0$, $t_1 = 0.5$ and $t_2 = 1.0$, as shown in Figure 2.13(a). Two control brackets (t_{1l}, t_{1h}) and (t_{2l}, t_{2h}) are used. t_{1l} and t_{2l} are initially set to **0**, while t_{1h} and t_{2h} are set to **1**. d_0 is the maximum deviation

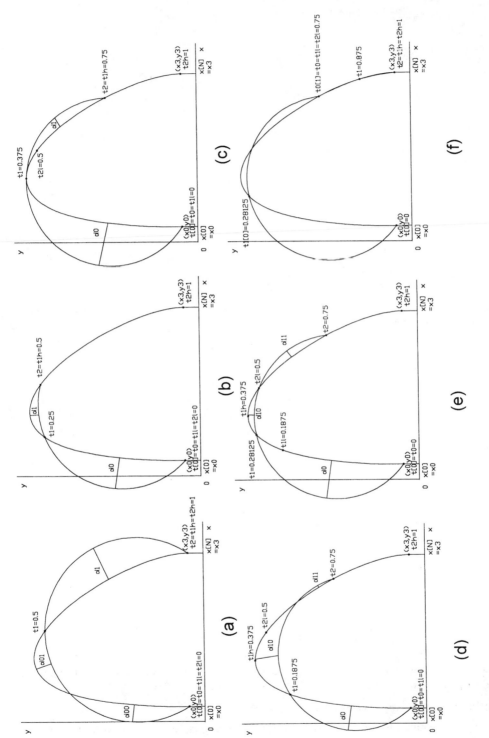

FIGURE 2.13 Circular approximation of a cubic Bezier curve.

between t_0 and t_1, while d_1 is the maximum deviation between t_1 and t_2. These maximum deviations are obtained by using the Golden Section Search[32] and compared with the given tolerance ε.

Step 1: If both d_0 and $d_1 > \varepsilon$, set $t_{2h} = t_2$, $t_2 = 0.5 \, (t_2 + t_{2l})$, $t_{1h} = t_2$ and $t_1 = 0.5 \, (t_0 + t_2)$, as shown in Figure 2.13(b). This procedure is repeated as long as both d_0 and $d_1 > \varepsilon$.

Step 2: If both d_0 and $d_1 < \varepsilon$, set $t_{2l} = t_2$, $t_2 = 0.5 \, (t_2 + t_{2h})$, $t_{1h} = t_2$ and $t_1 = 0.5 \, (t_0 + t_2)$, as shown in Figure 2.13(c). This procedure is repeated as long as both d_0 and $d_1 < \varepsilon$.

Step 3: If $d_0 > \varepsilon$ and $d_1 > \varepsilon$, set $t_{1h} = t_1$ and $t_1 = 0.5 \, (t_1 + t_{1l})$, as shown in Figure 2.13(d). This procedure is repeated as long as $d_1 > \varepsilon$ and $d_0 < \varepsilon$.

Step 4: If $d_0 > \varepsilon$ and $d_1 < \varepsilon$, set $t_{1l} = t_1$ and $t_1 = 0.5 \, (t_1 + t_{1h})$, as shown in Figure 2.13(e). This procedure is repeated as long as $d_0 > \varepsilon$ and $d_1 < \varepsilon$.

Steps 1 to 4 are repeated until:

both d_0 and $d_1 < \varepsilon$ by the specified convergence criterion, or,

both d_0 and $d_1 < \varepsilon$ and t_2 is equal to 1.

Then, the break point is considered to be found. t_2 is then assigned to an array element $t(k)$. The center of the circular arc is stored in $x_c(k)$ and $y_c(k)$, and its radius in $r(k)$. Increment k.

Step 5: Set $t_0 = t_2$, $t_{1l} = t_0$, $t_{2l} = t_0$, $t_2 = 1$, $t_{1h} = 1$ and $t_{2h} = 1$, as shown in Figure 2.14(f). The maximum deviation d_0 and d_1 are again calculated and compared with ε.

Steps 1 to 5 are repeated until $t(k) = 1$.

End.

Figures 2.14a and 2.14b show the circular approximation for the same two Bezier curves as in Figures 2.12a and 2.12b. With the same tolerances and convergence criteria, the number of segments needed are much fewer.

A C^1 Continuity Approach

In the second approach, biarcs are used as the interpolating curve. A biarc is defined by two consecutive circular arcs joined together in a C^1 continuous manner and was first proposed by Bolton[14] as an interpolating curve to overcome the C^0 discontinuity of using just simple arcs. By imposing suitable end conditions, chains of biarcs can be built up that maintain the C^1 continuity throughout. Depending on the topological relationship of the two circular arcs, biarcs can be classified into *C-shaped* and *S-shaped*

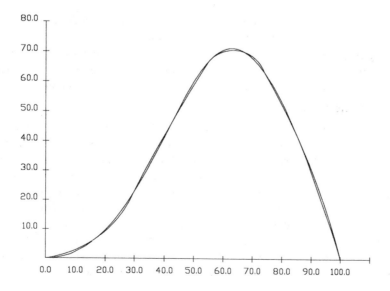

FIGURE 2.14(a)   Circular approximation of Bezier curve A. Tolerance : 0.6, convergence criteria : 0.06, number of circular arcs: 4, processor time: 0.5.

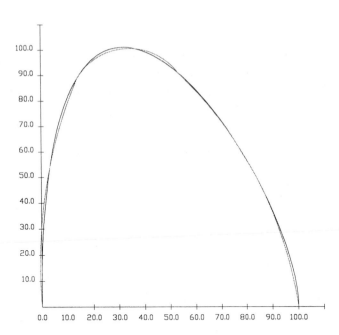

FIGURE 2.14(b) Circular approximation of Bezier curve B. Tolerance : 0.9, convergence criteria : 0.09, number of circular arcs: 3, processor time: 0.33.

biarcs. A C-shaped biarc has both the centers of the circular arcs on the same side of the curve while an S-shaped biarc has them on opposite sides.

A few methods have been proposed to fit biarcs to approximate or interpolate a set of two-dimensional data points. Some of these methods assume that the biarcs have to pass through all the data points while the rest only require that the data points are within a specified tolerance of the fitted biarcs. Because defining a biarc given only the end-point positions and tangent directions is an under-constrained problem, researchers have proposed methods to introduce the additional constraint required to uniquely define these biarcs by considering minimal total curve length,[23,33] minimal spline strain energy,[17,34] minimal radii difference,[14,15] minimal radii ratio,[15] or minimal curvature difference.[17,21,23] Ong et al.[35] had proposed minimization of total undercut area as an optimizing criterion, but the method was applied to approximating B-spline curves—though it can be easily modified to a set of discrete data points as well. In all of the methods described above, except for those of Meek and Walton[21] and Ong et al.,[35] the number of biarcs has to be predetermined beforehand and is usually overspecified.

Formulation of Biarc Curve

Before we describe our approach,[36] we will develop a formulation of the biarc curve that is necessary for the rest of the analysis. For practical purposes, we can assume that the spanned angles of the two circular arcs of a biarc do not exceed 2π. Without loss of generality, we can transform the two end points P_1 and P_2 to lie on the horizontal line with P_1 as the origin and P_2 on the positive axis. The two end tangents t_1 and t_2 form angles α and β, respectively, to the vector from P_1 to P_2, the angle being positive if it is in the counterclockwise direction and negative in the clockwise direction. An additional variable θ which denotes the angle from P_1 to Q, the joining point of the two circular arcs, is introduced to assist in the formulation (see Figure 2.15) From the figure, it can be seen that to maintain C^1 continuity at Q, the two arc centers O_1 and O_2 must be collinear with Q. This condition, together with the C^0 continuity requirement, implies that the two following equality constraints must be satisfied:

$$R_1(\sin(-\alpha) + \sin(\theta + \alpha)) + R_2(-\sin(\theta + \alpha) + \sin(\beta)) = L$$
$$R_1(\cos(-\alpha) - \cos(\theta + \alpha)) + R_2(\cos(\theta + \alpha) - \cos(\beta)) = 0 \tag{2.4}$$

where L is the distance from P_1 to P_2.

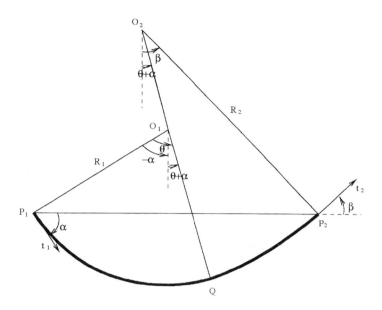

FIGURE 2.15 Formulation of biarc.

Solving Equation (2.4) above yields the radii of the two circular arcs, R_1 and R_2:

$$R_1 = -\frac{\sin\left(\frac{\theta + \alpha + \beta}{2}\right)}{2\sin\left(\frac{\theta}{2}\right)\sin\left(\frac{\alpha - \beta}{2}\right)}L$$

$$R_2 = -\frac{\sin\left(\alpha + \frac{\theta}{2}\right)}{2\sin\left(\frac{\alpha - \beta + \theta}{2}\right)\sin\left(\frac{\alpha - \beta}{2}\right)}L$$

(2.5)

It should be noted that R_1 and R_2 are signed quantities; positive radius indicates that the arc center lies to the left of the tangent and negative radius to the right. For θ, positive value is counterclockwise and negative value clockwise. For the special case of $\theta = 0$, it means that the segment from P_1 to Q is in fact a straight line. Figure 2.16 shows the different types of biarcs formed by various combinations of R_1 and R_2.

It can be observed that the angles of the circular arcs must have the same sign as the radii; in other words, the relationships $R_1\theta > 0$ and $R_2(\beta - \alpha - \theta) > 0$ hold. Furthermore, restricting θ to the range $[-\pi, \pi]$ for practical purposes leads to the following results:

$$
\begin{aligned}
-2\alpha < \theta < -(\alpha + \beta), &\quad \text{if } \alpha > \beta \\
-(\alpha + \beta) < \theta < -2\alpha, &\quad \text{if } \alpha < \beta
\end{aligned}
$$

(2.6)

Specifying the variable θ determines uniquely the biarc parameters which can be obtained by solving Equation (2.5). It is possible for the system of Equation (2.4) to produce a degenerate result. This occurs when α equals β and an infinite number of solutions exists. In this case, it is necessary to specify an additional variable, e.g., R_1, to constrain the solution.

Therefore, given the end points $P_1(x_1, y_1)$, $P_2 (x_2, y_2)$, and the tangent angles α, β, after obtaining R_1 and R_2, the arc centers O_1 and O_2 can be obtained.

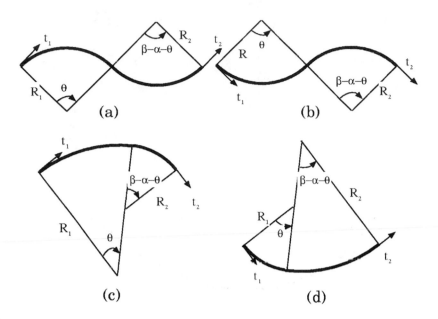

FIGURE 2.16 Types of biarcs.

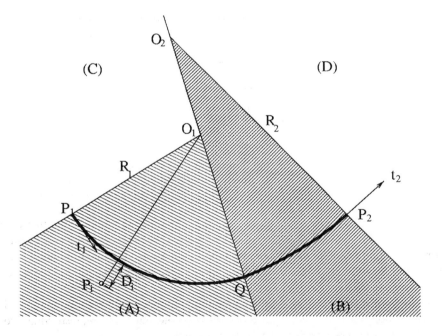

FIGURE 2.17 Distance between a point and biarc b.

The Optimal Biarc Sub-problem

Let P_i be an arbitrary data point lying between the two end points P_1 and P_2 (see Figure 2.17). Let D_i be the deviation between P_i and the biarc passing through P_1 and P_2. This distance can be easily calculated. The sector in which point P_i lies needs to be first identified. If P_i is in the P_1O_1Q sector, the distance D_i can be obtained by

$$D_i = \sqrt{(x_i - x_{O_1})^2 + (y_i - y_{O_1})^2} - R_1 \tag{2.7}$$

Similarly, if P_i is in the sector P_2O_2Q, the distance D_i can be obtained. The only exception is when P_i lies outside either of these sectors, in which case D_i is defined as the minimum distance between P_i and either of the end points P_1 and P_2.

Having the deviation formula available, we can therefore find the optimal biarc which passes through two end points P_1 and P_2 by formulating it as a least square problem:

$$\text{Minimize} \ \sum_i (D_i(\theta))^2 \tag{2.8a}$$

$$\text{subject to} \quad (\theta_{min} < \theta < \theta_{max}) \tag{2.8b}$$

where θ_{min} and θ_{max} are constrained by the Equation (2.6)

This is a nonlinear optimization problem of one variable with simple-bounded constraints and can be easily solved by any of the commonly available optimization algorithms such as the sequential quadratic programming method.

Biarc Curve Fitting to a Large Set of Discrete Data Points

Having developed the formulation of the biarc curve and the optimal biarc sub-problem, we are now in the position to tackle the approximation of a large set of data points so that the resulting biarc curves are C^1 continuous and within a specified tolerance of the data points. We assume the data points are given in the sequence that traces out the desired profile. The approach combines both a search process and an optimization process. In the search process, the algorithm starts from an initial point and searches forward to find the largest number of consecutive data points that can be approximated by a single biarc and still not violate the tolerance constraint. The search process uses a binary search approach, whereas finding the optimal biarc within the search steps solves an optimal biarc sub-problem as formulated by the model in the previous section.

The algorithm can be summarized as follows:

```
if (max_deviation_opt_biarc(1,n) < tolerance)
    add optimal_biarc(start,end);
else
    start = 1
    while (start < n)
        end = longest_end(start, start+1, n);
        add optimal_biarc(start,end);
        start = end;
    endwhile
endif
```

where *optimal_biarc* returns the optimal biarc between two end points and *longest_end* is the binary search to find the furthest point that does not violate the tolerance constraint.

The routine longest_end (Integer Routine longest_end(start, low, high)) can be summarized as follows:

```
deviation = max_deviation_opt_biarc(start, high)
while (deviation > tolerance)
        high = (low + high)/2
        if (high = low) return high;
        else deviation = max_deviation_opt_biarc(start, high)
        endif
endwhile
return high
```

where *max_deviation_opt_biarc* solves the optimal biarc subproblem and returns the maximum deviation of the optimal biarc.

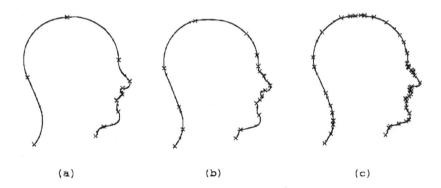

       (a)   (b)   (c)

FIGURE 2.18   Approximation of scanned head.

This algorithm will provide a minimum or near-minimum number of biarcs to fit a given set of data points. Figure 2.18 shows a cross-sectional profile of a scanned head approximated 12 inches by 12 inches in size. The initial data set consists of 760 points. Using 0.1 inch, 0.05 inch and 0.01 inch tolerances, respectively, the number of biarcs needed are only 10, 17 and 47, respectively.

2.4   Design Data Extraction for Computer-Aided Measurement

In terms of measurement and inspection equipment, the computer controlled coordinate measuring machine (CMM) is a highly flexible piece of equipment for inspection of different parts that effectively replaces costly and specialized gauges. There are essentially three programming methods for the CMM:

- *Self-teaching*, whereby the inspection program is created by manually operating the CMM to inspect a part but at the same time recording the procedure as a CMM program. The drawbacks of this method include reduction of the machine productive time and requirement that the physical part must already be available.[37]

- *Off-line Manual Programming*, whereby the operator has to manually program the CMM using programming languages such as Dimensional Measuring Interface Specification (DMIS)[38] or the native language of the specific CMM. The off-line programming method has the advantage of not incurring CMM downtime and not requiring the physical part to be available and set up in the CMM. However, the programming process is very tedious to the operator compared to the self-teaching approach.

- *Computer-aided Programming* through interaction with a CAD/CAM system.[39,40] Although this is the most suitable approach for CMM programming in terms of having a standard platform for design and manufacture, the commercially available CMM programming option for CAD/CAM systems still requires interactive input from the programmers. Automation level in terms of path planning, selection of inspection points, and other operation parameters is still limited.

All the aforementioned methods of CMM programming currently suffer from the following drawbacks:

- CMM programming requires trained personnel who must be familiar with the programming system (especially for CAD/CAM system), as well as the process knowledge for CMM inspection.
- The programming process is still very tedious, leading to high labor turnover—especially for highly skilled personnel.
- The flexibility in programming will leave room for inconsistencies in measurement approaches by different programmers.

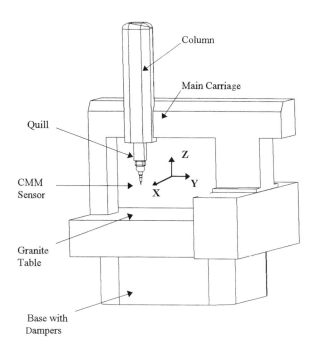

FIGURE 2.19 A coordinate measuring machine.

- The cycle time on the CMM is a bottleneck in the manufacturing process because the sampling technique is inherently slow. Optimization of the measurement parameters, which is important to reduce cycle time on the CMM, is no trivial task, based on current programming methods.

Computer-aided inspection planning for CMM (CAIPS-CMM) to automate the generation of CMM part programs can minimize problems associated with CMM programming mentioned above.

A central issue in the automated CMM part programming is the extraction of design data from a CAD model for the various planning tasks. Information extracted from the CAD model coupled with input from planning decisions has to undergo further re-organization that is suitable for the automatic generation of the CMM part programs. The following subsections deal with aspects of the coordinate measuring machine, the computer-aided inspection planning systems, and a suitable data structure for the measurement of a prismatic part based on extracted design data from a CAD model.

Coordinate Measuring Machine (CMM)

A CMM is a highly accurate three-dimensional spatial digitizer for any physical object. It carries a sensor that is mounted in the quill of a rigidly constructed structure that allows for three degrees of movement in a Cartesian frame, as shown in Figure 2.19. The axes of a computer controlled CMM are normally driven by servo motors and may have accurate scale of up to 0.5 micron resolution. The motion control of the CMM, and the capture of data point by activation of the sensor is carried out by the system controller that receives instructions from the host computer. The function of the host computer is to interpret CMM commands, and process measurement data from the CMM for evaluation purposes. The other function of the host computer is to provide the physical link to the operator interface, a CAD/CAM system where CMM programs may be created, or other computers in a computer integrated manufacturing system.

CMM sensors that are used to detect surface points on an object can be broadly divided into tactile and non-tactile types. Non-tactile sensors using camera and laser are currently limited in accuracy and their operation is highly dependent on conditions such as ambient lighting and the surface texture of

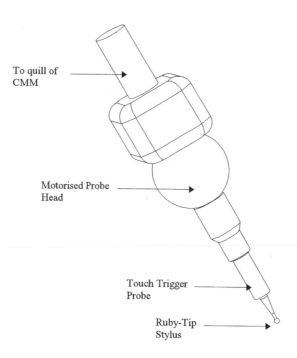

To quill of
CMM

Motorised Probe
Head

Touch Trigger
Probe

Ruby-Tip
Stylus

FIGURE 2.20   A touch trigger probe configuration.

the specimen. The electro-mechanical touch trigger probe, a tectile sensor, is of better accuracy and the most widely used sensor system for the CMM in the industry today. An example of a touch trigger probe system suitable for automatic part inspection is shown in Figure 2.20. Such a probe system is flexible because it includes a motorized probe head for aligning the probe in various orientations and is capable of automatic probe change. The touch trigger probe can be attached directly or indirectly via a probe extension to the motorized probe head. The touch trigger probe gives a trigger signal from the deflection of the stylus attached to it when the ruby tip at the end of the stylus is in contact with the surface to be measured. The stylus and ruby can be of different dimensions, depending on the applications. In using a touch-trigger type of probing system for the CMM, the geometric data of the object to be inspected is of importance because

- The geometry of the probe system will affect the accessibility of the measurement region on the workpiece. Collision with the workpiece and the working volume of the CMM have to be considered when selecting a probe configuration for measurement.
- The data points to be sampled by the CMM must have both spatial coordinate values as well as the surface normals. The surface normals are used for radius compensation due to the spherical contact tip.

The digitized points obtained from measurement on the CMM have to undergo further processing on the host computer depending on applications. The primary application of CMM is for evaluating the geometric and dimensional tolerance (GD&T) specifications for a part. GD&T is used to define the acceptable variation of a part geometry and the essential functional relationship between geometric elements on the part. The chief motivation of GD&T is to ensure that usable sizes and locations are created for assembly of parts; hence ensuring interchangeability of parts even before they are made. Size and location are the main tolerance specifications for assembly of parts; in order to allow for greater amount of size and location tolerances, surface quality control tolerances such as profile, form, orientation, and runouts are also defined. GD&T specifications have to conform to internationally accepted standards such as ANSI Y14.5M[41] for meaningful applications. A general data structure based on the ANSI Y14.5M standard has been defined for each tolerance specification as shown in Tables 2.2 and 2.3.

TABLE 2.2 Data Structure for Each Tolerance Specification

S/No	Identifier	Description
1	Tol_ID	Identifier for tolerance specification
2	Tol_Type	Tolerance type (size, form, profile, location, runout, orientation)
3	Tol_zone_val	Tolerance zone value
4	Tol_zone_type	Tolerance zone type
5	Matl_Cond	Material condition
6	Tol_Ref	Size tolerance reference for material condition specified
7	Basic_Element	Basic element reference for tolerance
8	Basic_Dim	Basic dimensions
9	U_tol	Upper tolerance
10	L_tol	Lower tolerance
11	Primary_Datum	Primary datum reference
12	Primary_Matl_cond	Primary datum reference material condition
13	Pri_Dat_MC_Tol_Ref	Primary datum material condition/size tolerance reference
14	Secondary_Datum	Secondary datum reference
15	Secondary_Datum_Matl	Secondary datum material condition
16	Sec_Dat_MC_Tol_Ref	Secondary datum material condition/size tolerance reference
17	Tertiary_Datum	Tertiary datum material condition
18	Tertiary_Datum_Matl	Tertiary datum material condition
19	Ter_Dat_MC_Tol_Ref	Tertiary datum material condition/size tolerance reference

TABLE 2.3 Tolerance Information

Characteristics	1	2	3	4	5	6	7	8	9	10	11	12	13	14	15	16	17	18	19
Tolerance																			
Size																			
Linear	Y	11	N	N	N	N	Y	Y	Y	Y	N	N	N	N	N	N	N	N	N
Diameter	Y	12	N	N	N	N	Y	Y	Y	Y	N	N	N	N	N	N	N	N	N
Angle	Y	13	N	N	N	N	Y	Y	Y	Y	N	N	N	N	N	N	N	N	N
Location																			
Position	Y	51	Y	Y	Y	Y	Y	Y	N	N	Y	Y	Y	Y	Y	Y	Y	Y	Y
Concentricity	Y	52	Y	Y	N	N	Y	N	N	N	Y	N	N	N	N	N	N	N	N
Composite	Y	53	Y	Y	Y	Y	Y	Y	N	N	Y	Y	Y	Y	Y	Y	Y	Y	Y
Form																			
Circularity	Y	21	Y	Y	*1	N	Y	N	N	N	N	N	N	N	N	N	N	N	N
Cylindricity	Y	22	Y	Y	*1	N	Y	N	N	N	N	N	N	N	N	N	N	N	N
Flatness	Y	23	Y	Y	*1	N	Y	N	N	N	N	N	N	N	N	N	N	N	N
Straightness	Y	24	Y	Y	Y	Y	Y	N	N	N	N	N	N	N	N	N	N	N	N
Profile																			
Curve	Y	31	Y	*2	N	N	Y	Y	Y	Y	Y	N	N	Y	N	N	Y	N	N
Surface	Y	32	Y	*2	N	N	Y	Y	Y	Y	Y	N	N	Y	N	N	Y	N	N
Runout																			
Circular	Y	41	Y	Y	*1	N	Y	*3	N	N	Y	N	N	Y	N	N	Y	N	N
Total	Y	42	Y	Y	*1	N	Y	*3	N	N	Y	N	N	Y	N	N	Y	N	N
Orientation																			
Parallelism	Y	61	Y	*4	S	*5	Y	N	N	N	Y	Y	Y	Y	Y	Y	Y	Y	Y
Perpendicularity	Y	62	Y	*4	S	*5	Y	N	N	N	Y	Y	Y	Y	Y	Y	Y	Y	Y
Angularity	Y	63	Y	*4	S	*5	Y	Y	N	N	Y	Y	Y	Y	Y	Y	Y	Y	Y

*1 Modifier: Perfect form not required at MMC. (It is always assumed that perfect form exists at MMC for form tolerance specification.)

*2 Indicate that the tolerance zone is all around the profile or surface.

*3 Conditional.

*4 Always rectangular zone.

*5 Yes, if feature of size is referenced.

TABLE 2.4 Substitute Elements in CMM Software Evaluation

Substitute Element	Parameters			
	Location	Orientation	Size	Angle
Line	x, y and z coordinates of a point on the line	cx, cy, cz cosine vectors of the direction of the line	nil	nil
Plane	x, y and z coordinates of a point on the plane	cx, cy, cz cosine vectors of the normal to the plane away from the material	nil	nil
Circle	x, y and z Center of the circle	cx, cy, cz cosine vectors of the normal to the plane that contains the circle	r radius of circle	nil
Sphere	x, y and z center of the sphere	nil	r radius of sphere	nil
Cylinder	x, y and z coordinates of a point on the axis of the substitute cylinder	cx, cy, cz cosine vectors of the axis of the cylinder	r radius of cylinder	
Cone	x, y and z coordinates of a point on the axis of the substitute cone	cx, cy, cz cosine vectors of the axis of the cone which point in direction of increasing diameter	r radius of cone at the point (x, y, z)	phi apex angle of the substitute cone
Torus	x, y and z coordinates of a point at the centre of the substitute torus	cx, cy, cz cosine vectors of the axis of the torus	r1 radius of circular section of the tube of the torus r2 mean radius of ring of the torus	

For application in evaluating GD&T specifications for a part, sampled points from the CMM are used to compute the substitute geometric elements by the CMM software. A representative set of substitute geometric elements for CMM metrology is shown in Table 2.4. The locations, orientations, sizes, and angles of the part features required to be checked are calculated from the relevant parameters of the substitute elements and compared with the GD&T specifications in the design representation. The method for computation of substitute elements will have significant impact on the final evaluation results. The default computation approach is the Gaussian least square method which minimizes the sum of the squares of the perpendicular distances of the points in the measured data set to the substitute element. The Chebyshev approach, which minimizes the maximum perpendicular distance from the points in the measured data to the element, is also being used. For circular elements like the circle, cone, sphere, cylinder, and torus, the minimum circumscribed approach or the maximum inscribed approach may also be used. Regardless of the computation approach, invariably the goodness of the substitute element is highly dependent on the sample size of the measured data. Generally with larger sample size, the goodness of the substitute element is increased, but at the expense of a longer measurement cycle time. Determination of the ideal sample size is a rather complicated process as it depends on many factors such as manufacturing process capability, tolerance specifications, measurement confidence level, etc. Attempts have been made to address this issue using a statistical approach[42] and feature-based approach using artificial neural networks.[43] The assumption that manufacturing process is constant in both approaches, however, poses some limitations in real-life applications. Developers of CMMs are moving towards the use of analogue contact probe, which can scan a larger sample size at a faster rate.

CMMs are usually supplied with high-level programming languages that vary with different vendors. More recently, a neutral CMM language called DMIS (Dimensional Measuring Interface Specifications)

has been developed as a neutral interface for the bi-directional communication of data between CAD systems and computer-controlled CMMs in a multi-vendor, computer-integrated environment. DMIS was developed by Computer-Aided Manufacturing Incorporated (CAM-I) and it was approved in 1990 by the American National Standards Institute (ANSI) as an American National Standard. The main advantages of DMIS is that it is APT-like and it allows measurement results to be carried back in the same neutral format. This is particularly important in computer integration to both quality and management information systems in a CIM architecture and is fast evolving as the main industry standard.

DMIS carries two main types of statements: process-oriented commands and geometry-oriented statements. Process-oriented commands consist of motion statements, machine parameters statements, part alignment statements, probe calibration statements, and other statements related to the inspection process. Geometry-oriented statements are used to describe the geometrical elements, geometric dimension and tolerances, coordinate systems, and other types of data which may be included in a CAD database. DMIS has been designed to be compatible with complete part model definitions supporting simple geometrical features like point, circle, ellipse, arc, sphere, cylinder, cone, and plane—as well as more complicated ones like the general curve element, general surface element, pattern elements, rectangle element, parallel plane element, and general raw data element. It also supports GD&T specifications according to ANSI Y14.5M which is widely adopted in the industry. Another powerful feature of DMIS is its high level language capability. This allows for variable definitions as well as program control, providing for high flexibility in automated inspection planning. Although it has been reported that there are interfacing problems between the CAD system and the CMM,[44] DMIS will evolve as the main interface standard as further refinements are made.

Computer-Aided Inspection Planning System for the Coordinate Measuring Machine (CAIPS-CMM)

The development of a computer-aided inspection planning system for the coordinate measuring machine (CAIPS-CMM) is characterized by the generative approach with few reported instances of the variant approach.[45] In the generative approach, applications are broadly divided into:

- Inspection of turned parts[46]
- Inspection of prismatic parts[47–49]
- Inspection of complex surfaces for molds.[50–52]

For turned and prismatic parts, the application of a feature-based technique is very apparent. Expert systems are also mainly used for the reasoning process in planning, coupled with techniques like object-oriented programming. For the inspection of complex surfaces, the focus of researchers is on minimization of inspection points and comparative analysis.

Regardless of the applications, a CAIPS-CMM should generally cover the following areas in order to automate the task of CMM part programming in a computer-integrated environment:

- An interface to a CAD product model for interpretation of measurement requirements. Inspection requirements for the CMM are driven by the GD&T specifications which usually refer to geometrical surfaces or datum on the product model. Hence a product model should ideally encompass complete geometric information as well as GD&T information. Higher level information like features can be incorporated which could help in the planning process. The structure of the product model representation is outlined in the next section.
- A framework for inspection planning based on the knowledge of the operational procedures of the CMM is required.
- Knowledge of the machine capabilities in terms of the machine configuration, the available sensor configurations, the accuracy, the operating speed, the supporting evaluation functions, etc., is essential for decisions such as selection of probes for the CMM.

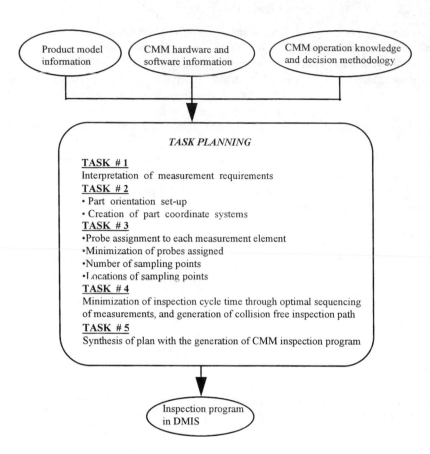

FIGURE 2.21   CMM planning requirements and tasks.

- A methodology for the assignment of set-ups, part coordinate systems, probe configurations, sampling size, and minimization of probe travel is important for the optimization of CMM inspection in terms of cycle time.
- A methodology for collision-free path planning in order to realize feasible inspection programs
- An interface to the CMM in the form of a neutral programming language such as DMIS which will carry the final inspection commands for any CMM.

To realize the final objectives of a computer-aided inspection planning system for the CMM, there are generally five inspection tasks to be carried out, as outlined in Figure 2.21.

1. The first inspection task is to interpret the measurements required for the part based on the product model representation. Interpretation would include recognizing geometric elements on the part to be measured for the evaluation based on their associated GD&T specifications, and determining the evaluation functions to be used and the output to be produced by the CMM.
2. The second inspection task is to determine the setup of the part on the CMM for inspection. The major consideration is the minimum setup orientation for the part. The task plan for the CMM will include the part coordinate systems to be used for the orientation, the elements to be measured or constructed to create the part reference, and the elements that are to be measured in the setup.
3. The third inspection task is to assign the probe configurations and operation parameters for the measurement elements. Major considerations would include assignment of suitable probes, minimization of the probes assigned, and the sampling points required for the measurement.

4. The fourth task is to optimize the entire operation by selecting optimal sequences for the measurements in each setup and to generate optimal and collision-free paths for the probe.
5. The final task is to synthesize the plan and generate the inspection program in a neutral programming language such as DMIS for the CMM.

A framework for the computer-aided inspection planning system for the CMM (CAIPS-CMM) is shown in Figure 2.22. The input to the CAIPS-CMM is a product model that carries features, solids and GD&T information. The output of the system is the CMM inspection program in DMIS format.

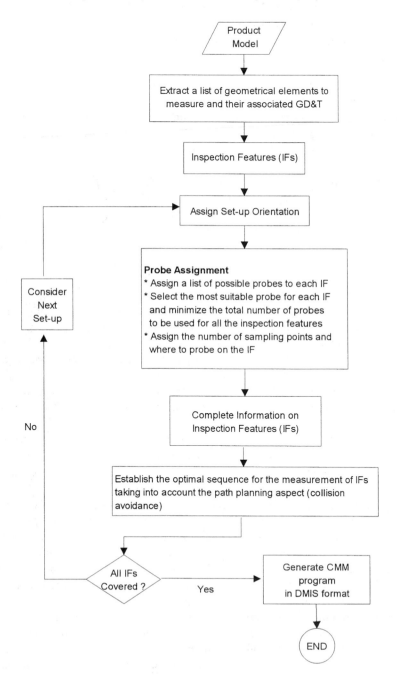

FIGURE 2.22 Flowchart showing the framework of the CAIPS-CMM.

Product Model Representation for Computer-Aided Measurement

The product design representation for a part must carry enough information and be structured such that it is amenable for computer interpretation for automated inspection planning. Three levels of information are required:

- Geometrical information pertaining to shape, volume, and surfaces. Such information is required for visualization of the part model, extraction of surface data for measurement, accessibility analysis of contact probes, and path planning for inspection on the CMM.
- Measurement evaluation as defined by the GD&T specifications.
- High-level engineering information pertaining to shape or region on the part. Usually such regions are linked to specific manufacturing or measurement processes.

It is quite established at this stage that a solid modeling scheme is capable of providing complete and unambiguous representation for the above purpose. There are many solid representation schemes available such as constructive solid geometry (CSG) and boundary representation (BREP). In CSG, solids are represented by a collection of geometric primitives, such as box, cylinder, wedges, etc., associated together through a binary tree. The nodes of the tree represent Boolean operations such as union, intersection, difference, and rigid transformation, while the leaves of the tree represent the primitives. CSG models represent the volume of the objects very well but do not explicitly store the edge information that most CAD applications require. Representation and evaluation of complex surfaces are the problem areas for the CSG model. BREP models store information about the surfaces, curves, and points that form the boundary of the object. Since surface information is well known, direct manipulation of the surfaces is possible for applications in path planning and digitizing surface points in CMM planning. BREP models tend to be very difficult to create compared to CSG and their data files tend to be very big. Hence, it is common to find that many commercial packages have CSG as a means to create the model, but the actual model is evaluated in terms of BREP.

The second level of information required for CAIPS-CMM is the GD&T specifications. The specifications will help to determine the CMM evaluation functions to be applied to geometric elements that are measured. For example, a circularity tolerance specification for a cylinder element will only require a circle evaluation function, whereas a cylindricity tolerance specification will require a cylinder evaluation function. Most CAD/CAM systems do not have a consistent approach for linking the GD&T information to the product model. Some systems allow for the GD&T specifications to be added as descriptive notes associated with certain geometry, while others allow for the GD&T specifications to be integral to the CAD model.

A product model that is created based purely on geometrical information has little engineering significance in terms of engineering analysis and planning for manufacturing and inspection. A feature-based modeling technique has been developed to provide a more natural interface between the designer and other engineering applications such as process planning. A feature can essentially be defined as a frame of information organized such that tasks like planning for manufacture and inspection can be carried out more easily. From this definition, it can be seen that many different features can be defined based on the different applications as different shape and engineering knowledge are required. For example, a machinable feature is mappable to a generic shape that has engineering significance in terms of machining, while a technological feature contains non-geometric attributes related to function and performance of an entity in the feature-based model. Features are likened to the basic building blocks for a product model and it is quite apparent that more than one type of feature is required. A more extensive review on features for design and manufacture can be found in Shah.[53] It should be emphasized that feature-based modeling alone is not sufficient for engineering applications as geometric reasoning and processing still have to be carried out using solid model information. Hence it is quite important to have a hybrid system that has both solid (usually BREP) and feature information.

The basic building block for the inspection model in a CAIPS-CMM is the measurement associated with a geometrical surface element on a part. The measured element can then be evaluated for downstream operations such as evaluation of GD&T and construction of other geometrical elements, and for establishing part coordinate systems. For convenience, it is appropriate to call this building block for the CAIPS-CMM an **inspection feature (IF)**. An inspection feature can be defined as a frame of information necessary for planning measurement of a geometrical element on a part with the objective of evaluating all GD&T associated with it. The inspection feature defined above essentially consists of the following classes of information:

- *Measurement element*: This is the basic surface entity on the part model where measurement points have to be taken for evaluation of GD&T or for construction of other elements. The surface entity is intended to conform to the basic geometrical elements supported by DMIS. The list of elements is shown in Table 2.2. The data structure for such an element will include information such as geometric element type, element identity, reference to the construction feature, number of points to sample, where to sample, and CMM operation parameters such as the start point, end point, speed, approach, and retract distance.
- *Evaluated element*: This is computed from the sample points of a measurement element based on mathematical function. The function to be used is dependent on the GD&T or construction elements to be evaluated.
- *GD&T specifications*: This is a list of GD&T specifications associated with a measurement element. The data structure for the GD&T specification is outlined in Tables 2.2 and 2.3. It should be noted that there may be more than one GD&T specification for a particular measurement element. For example, a cylindrical surface of a hole may be evaluated as a cylinder or a circle depending on the GD&T specifications. Moreover, the measurement element could have been used as a reference for other elements.
- *Setup and part coordinate system (PCS)*: The selection of the setup to carry out the measurement. This will include the orientation of the part as well as the selection of a suitable PCS. The data structure will include the identifier for the setup, identifier for the PCS, evaluated elements, and parameters for establishing the PCS.
- *Probe selection*: The probe to be used for the measurement element will depend on considerations such as accessibility and tolerance requirements. The data structure for the probe selection will include information such as the identifier for the probe selection, parameters for the probe orientation, and the physical configurations of the probe.

It is clear that it may not be possible to assign all information based on the product model itself. Information such as the assignment setup, PCS, and probe system for each inspection feature has to undergo further reasoning processes in the CAIP-CMM system as shown by the flow chart of Figure 2.22.

Figure 2.23 shows a prismatic model created using ProEngineer, which is a CAD system that uses feature and parametric technology for product model creation. The basic module supports the design-by-features concept for machining operations with features like hole, cut, slot, etc. and supporting features like datum and PCS. The feature-based model for ProEngineer is evaluated in BREP for solid representation. A listing of some relevant information for the model created is shown in Table 2.5. GD&T can also be specified for the model; this information can be tied to the feature or the geometry. Figures 2.24 and 2.25 show some of the geometric and dimensional tolerances specified for the model, while Tables 2.6 and 2.7 show partial listings of the GD&T information.

A module has been written in ProDevelop to extract information from the CAD model to organize it into inspection features for computer aided measurement. Since inspection is directed by GD&T specifications, the CAD data extraction module first searches through the data base for all GD&T specifications and then obtains a listing of geometrical elements required for evaluation. This geometrical element may be direct

TABLE 2.5 Partial Listing of Construction Features Used for CAD Model

Feature ID	Feature Type	Axis ID	Surface IDs
1	FT_FIRST_FEAT	Nul	[2] [7] [12] [14] [16] [18]
20	FT_DATUM	Nul	[21]
22	FT_DATUM	Nul	[23]
24	FT_DATUM	Nul	[24]
26	FT_CSYS	Nul	[2] [18] [12]
28	FT_CUT	Nul	[36] [41] [43] [45]
67	FT_HOLE	84	[79] [81]
95	FT_HOLE	108	[100] [103] [105]
117	FT_SLOT	Nul	[125] [130] [132] [134] [136]
174	FT_HOLE	189	[181] [184] [186]
194	FT_CUT	Nul	[200] [203]
242	FT_HOLE	257	[249] [252] [254]
262	FT_CUT	Nul	[268] [271]

TABLE 2.6 Partial Listing of Geometric Tolerances for CAD Model

GTOL ID	GTOL Type	Value	Mat'l Cond.	GTOL Reference	Feature ID	Datum Reference
0	LOC/POS	0.1	MMC	AXIS [84]	67	[20] [22] [24]
1	LOC/CONC	0.2	MMC	AXIS [108]	95	A_1 [67]
2	FORM/FLAT	0.1	RFS	SRF [7]	1	NUL
4	FORM/STRGHT	0.4	RFS	AXIS [84]	67	NUL
5	FORM/CIR	0.01	RFS	SRF [105]	95	NUL
6	FORM/CYL	0.02	RFS	SRF [186]	174	NUL
7	ORIENT/ANG	0.1	RFS	SRF [203]	194	[20]
9	ORIENT/PERP	0.1	RFS	AXIS [189]	174	[22]
10	ORIENT/PAR	0.01	RFS	SRF [7]	1	[20]

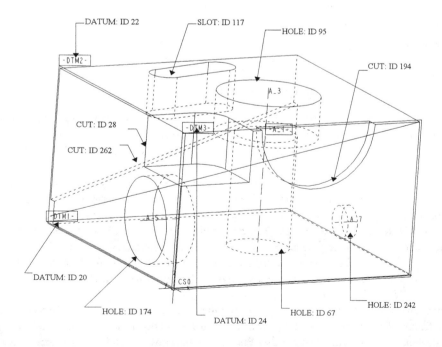

FIGURE 2.23 Representative construction features from a CAD model.

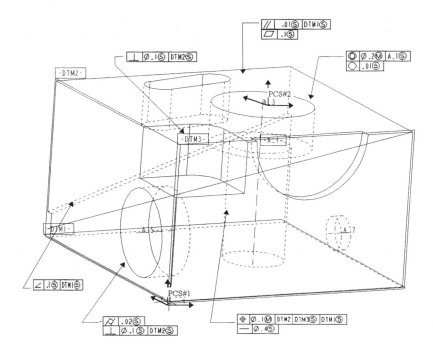

FIGURE 2.24 Representative geometric tolerances from a CAD model.

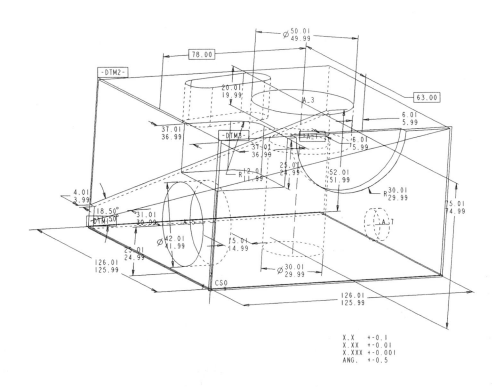

FIGURE 2.25 Representative dimensional tolerances from a CAD model.

TABLE 2.7   Partial Listing of Dimensional Tolerances for a CAD Model

DTOL ID	DTOL Type	Upper Value	Lower Value	DTOL Element	Feature ID	Reference Element
0	LINEAR	126.01	125.99	SRF [16]	1	[SRF 12]
1	LINEAR	126.01	125.99	SRF [14]	1	SRF [18]
2	LINEA R	75.01	74.99	SRF [7]	1	SRF [2]
3	LINEAR	25.01	24.99	SRF [36]	28	SRF [31]
6	RADIUS	12.01	11.99	SRF [43]	28	NUL
7	LINEAR	37.01	36.99	SRF [41]	28	SRF [18]
8	LINEAR	37.01	36.99	SRF [45]	28	SRF [12]
9	DIAMETER	30.01	29.99	FEAT [67]	67	NUL
13	DIAMETER	50.01	49.99	FEAT [95]	95	NUL
14	LINEAR	20.01	19.99	SRF [125]	117	SRF [120]
15	LINEAR	50.01	49.99	SRF [136]	117	AXIS [108]
17	LINEAR	44.01	43.99	SRF [132]	117	AXIS [108]
18	LINEAR	32.01	31.99	SRF [132]	117	AXIS [108]
19	RADIUS	8.01	7.99	SRF [136]	117	NUL
25	DIAMETER	42.01	41.99	FEAT [174]	174	NUL
26	LINEAR	25.01	24.99	FEAT [174]	174	SRF [2]
27	LINEAR	31.01	30.99	FEAT [174]	174	FEAT [84]
28	LINEAR	4.01	3.99	SRF [200]	194	SRF [197]
31	ANGULAR	18.5	17.5	SRF [203]	194	SRF [2]
38	DIAMETER	16.01	15.99	FEAT [242]	242	NUL
39	LINEAR	10.01	9.99	FEAT [242]	242	SRF [2]
41	LINEAR	6.01	5.99	SRF [268]	262	SRF [265]

TABLE 2.8   Partial Listing of Initial Inspection Features Generated from a CAD Model

Inspection Feature ID	Element Type	Element Surfaces	Element Feature	Datum Plane	Datum Axis	Dimensional Tolerances	Geometric Tolerances
0	PLANE	[2]	1	20	NUL	NUL	NUL
1	PLANE	[18]	1	22	NUL	NUL	NUL
2	PLANE	[12]	1	24	NUL	NUL	NUL
3	CYLIN	[79] [81]	67	NUL	84	[9]	[0] [4]
4	PLANE	[7]	1	NUL	NUL	[2]	[2] [10]
5	PLANE	[16]	1	NUL	NUL	[0]	NUL
6	PLANE	[14]	1	NUL	NUL	[1]	NUL
7	CYLIN	[43]	28	NUL	NUL	[6]	NUL
8	PLANE	[41]	28	NUL	NUL	[7]	NUL
9	PLANE	[45]	28	NUL	NUL	[8]	NUL
10	PLANE	[36]	28	NUL	NUL	[3]	NUL
11	CYLIN	[103] [105]	95	NUL	NUL	[13]	[1] [5]
12	PLANE	[136]	117	NUL	NUL	[15]	NUL
13	PLANE	[132]	117	NUL	NUL	[17] [18]	NUL
14	CYLIN	[134]	117	NUL	NUL	[19]	NUL
15	PLANE	[125]	117	NUL	NUL	[14]	NUL
16	CYLIN	[186] [184]	174	NUL	NUL	[25] [26] [27]	[6] [9]
17	PLANE	[203]	194	NUL	NUL	[31] [32]	[7]
18	PLANE	[200]	194	NUL	NUL	[28]	NUL
19	CYLIN	[252] [254]	242	NUL	NUL	[39] [38]	NUL
20	PLANE	[268]	262	NUL	NUL	[41]	NUL

reference or datum reference from the GD&T, and it forms the measurement element of the inspection feature. Since a geometrical element may have more than one GD&T specification, a measurement element may be split into subelements depending on the GD&T specifications. Table 2.8 shows a partial listing of the initial inspection features extracted from the CAD model based on measurement elements and GD&T information.

Further information pertaining to the inspection features has to be built up based on planning methodologies for setup planning, PCS selection, and surface data points extraction based on the number of sampling points and accessibility analysis of the probe.

This example shows an approach of extracting information from a product model for computer-aided measurement and organizing this information into inspection features. The inspection features will form the basic building blocks which will help to facilitate the generation of the CMM part program.

References

1. Chia, A. P. H., An Integrated CAD/CAM Approach for the Manufacture of Wind-Tunnel Models, B. Eng. Thesis, National University of Singapore, 1994.
2. Giam, K. Y., Manufacture of Wind-Tunnel Model Using CAD/CAM Technique, B.Eng. Thesis, National University of Singapore, 1993.
3. Tan, L. B., Manufacture of Models Employing Integrated CAD/CAM Techniques, B.Eng. Thesis, National University of Singapore, 1995.
4. Alting and Zhang, H. C., Computer-aided process planning: the state-of-the-art survey, *International Journal of Production Research*, Vol. 27, No. 4, 1989, 553–585.
5. Nolen, *Computer-automated Process Planning for World-class Manufacturing*, Marcel Dekker, Inc., NY, 1989, 65–70.
6. Eversheim W. and Schneewind, J., Computer-aided process planning—state of the art and future development, *Robotics and Computer Integrated Manufacturing*, Vol. 10, No. 1/2, 1993, 65–70.
7. Elmaraghy, H. A., Evolution and future perspectives of CAPP, *Annals of the CIRP*, Vol. 42, No. 2, 1993, 1–13.
8. Yeo, S. H., Wong, Y. S., and Rahman, M., Integrated knowledge-based machining system for rotational parts, *International Journal of Production Research*, Vol. 29, No. 7, 1991, 1325–1337.
9. Yeo, S. H., An integrated knowledge-based machining system for rotationally symmetric parts, Ph.D. Thesis, National University of Singapore, 1992.
10. Wong, Y. S. and Wang, Z., Automated process planning for CNC machining of spherical space-frame nodes, *Journal of Manufacturing Systems*, Vol. 14, 1995, 369–377.
11. Eade, R., For flexibility, you can't beat them, *Manufacturing Engineering*, May 1989, 49–52.
12. Coleman, J. R., Machining centers cut it in job shops, *Manufacturing Engineering*, March 1990, 41–45.
13. Hines, F., Specialized CAM: a success story, *Manufacturing Engineering*, May 1989, 88–89.
14. Bolton, K. M., Biarc curves, *Computer-Aided Design*, Vol. 7, No. 2, 1975, 89–92.
15. Su, B. Q. and Liu, D. Y., *Computational Geometry—Curve and Surface Modeling*, Academic Press, NY, 1989.
16. Moreton, D. N. and Parkinson, D. B., The application of a biarc technique in CNC machining, *Computer-Aided Engineering Design*, 8, 1991, 54–60.
17. Schonherr, J., Smooth biarc curves, *Computer-Aided Design*, Vol. 25, No. 6, 1993, 365–370.
18. Sharrock, N., Biarcs in three dimensions, in Martin, R.R., Ed., *Mathematics of Surfaces* II, Oxford University Press, UK, 1986.
19. Parkinson, D. B., Optimized biarc curves with tension, *Computer-Aided Geometric Design*, Vol. 9, 1992, 207–218.
20. Piegl, L., Curve fitting algorithm for rough cutting, *Computer-Aided Design*, Vol. 18, No. 6, 1986, 79–82.
21. Meek, D. S. and Walton, D. J., Approximation of discrete data by G^1 arc splines, *Computer-Aided Design*, Vol. 23, No. 6, 1991, 411–419.
22. Meek, D. S. and Walton, D. J., Approximating quadratic NURBS curves by arc splines, *Computer-Aided Design*, Vol. 25, No. 6, 1993, 371–376.
23. Nutbourne, A. W. and Martin, R. R., *Differential Geometry Applied to Curve and Surface Design*, Volume I: Foundations, Ellis Horwood, NY 1988.

24. Makinouchi, S., Okamoto, M., and Yamagata, K., Optimal curve fitting for NC machining by dynamic programming technology, *Reports of the Osaka University*, Vol. 16, No. 2, 1976.

25. Cantoni, A., Optimal curve fitting with piecewise linear functions, *IEEE Transactions on Computers*, C-20, No.1, 1971, 59–67.

26. Faux, I. D. and Pratt M. J., *Computational Geometry for Design and Manufacture*, Ellis Horwood Limited, N.Y., 1985.

27. Ding, Q. and Davies, B. J., *Surface Engineering for Computer-Aided Design*, Halsted Press, Chichester, England, 1987.

28. Vickers, G. W., Ly, M. H., and Oetter, *Numerically Controlled Machine Tools*, Ellis Horwood, N.Y.

29. Teng, C. S., Loh, H. T., and Wong, Y. S., A study of the deviation of best-fitted lower-order curves for the cubic Bezier curve, Internal Report, Mechanical and Production Engineering Department, National University of Singapore, 1993.

30. Wong, Y. S., Loh, H. T., and C. S. Teng, An optimal piecewise curve-fitting approach for CNC machining, *Transactions of NAMRC*, Vol. 23, 1995, 157–162.

31. Papalambros, P. Y. and Wilde, J. W., *Principles of Optimal Design: Modeling and Computation*, Cambridge University Press, 1988.

32. Jasbir, S. A., *Introduction to Optimum Design*, McGraw Hill, NY, 1989.

33. Hoschek, J., Circular splines, *Computer-Aided Design*, Vol. 24, 1992, 611–618.

34. Parkinson, D. B. and Moreton, D. N., Optimal biarc curve fitting, *Computer-Aided Design*, Vol. 23, 1991, 411–419.

35. Ong, C. J., Wong, Y. S., Loh, H. T., and Hong, X. G., An optimization approach for biarc curver-fitting of B-spline, *Computer-Aided Design*, Vol. 28, No. 12, 1996.

36. Kao, J. H., Loh, H. T., and Prinz, F., Least-square biarc curve fitting for CNC machining, *Proceedings of 17th Computers in Engineering Conference*, 1997.

37. Evershiem, W. and Auge, J., Automatic generation of part program for CNC-coordinate measuring machine linked to CAD/CAM systems, *Annals of the CIRP*, Vol. 35, No.1, 1986.

38. DMIS, "Dimensional Measuring Interface Specification," Version 2.1, Computer-Aided Manufacturing International Inc., 1989.

39. Intergraph/Mechanical Probe CMM Option, Intergraph Corporation, Huntsville, Alabama.

40. Pro/CMM, ProEngineer CMM Option, Parametric Technology Corporation, 128 Technology Drive, Waltham, MA 02154, USA.

41. ANSI Y14.5M-1982, Dimensioning and Tolerancing, American Society of Mechanical Engineers, USA, 1983.

42. Yau, H. T. and Menq, C. H., An automated dimensional inspection environment for manufactured parts using coordinate measuring machines, *International Journal of Production Research*, Vol. 30, No. 7, 1992, 1517–1536.

43. Zhang, Y. F., Nee, A. Y. C., Fuh, J. Y. H., Neo, K. S., and Loy, H. K., A neural network approach to determining optimal inspection sampling size for CMM, *Journal of Computer Integrated Manufacturing Systems*, Vol. 9, No. 3, 1996, 161–169.

44. Neo, K. S. and Wong, Y. S., Link between a CAD system and a computer-controlled CMM via the dimensional measuring interface specifications (DMIS), *Proceedings of the Industrial Automation '92 Conference*, 20–23 May 1992, World Trade Centre, Singapore, 179–188.

45. Pao, Y. H., Komeyli, K., Shei, D., Leclair, S., and Winn, A., The episodal associative memory: managing manufacturing information on the basis of similarity and associativity, *Journal of Intelligent Manfacturing.*, Vol. 4, 1993, 23–32.

46. Elmaraghy, H. A. and Gu, P. H., Expert system for inspection planning, *Annals of CIRP*, Vol. 36, No.1, 1987, 85–89.

47. Merat, F. L., Radack, G. M., Roumina, K., and Ruegsegger, S., Automated inspection planning within the rapid design system, *IEEE International Conference on Systems Engineering*, 1991, 42–48.

48. Merat, F. L. and Radack, G. M., Automatic inspection planning with a feature based CAD system, *Robotics & Computer Integrated Manufacturing*, Vol. 9, No. 1, 1992, 61–69.

49. Tao, L. G. and Davies, B. J., Knowledge based 2 1/2 D prismatic component inspection planning, *International Journal of Manufacturing Technology,* Vol. 7, 1992, 339–347.

50. Wong, C. L., Menq, C. H., and Bailey, R., Computer integrated dimensional inspection of manufactured objects having sculptured surfaces, *Advanced Manufacturing Engineering,* Vol. 3, 1991, 37–44.

51. Yau, Y. T. and Menq, C. H., Computer-aided coordinate metrology, *Computers in Engineering,* ASME, 1993.

52. Pahk, H. J., Kim, Y. H., and Hong, Y. S., Development of Computer-aided inspection system with CMM for integrated mold manufacturing, *Annals of the CIRP,* Vol. 42, No. 1, 1993, 557–560.

53. Shah, J. J., *Features in Design and Manufacture, Intelligent Design and Manufacture,* John Wiley & Sons, Inc., NY, 1992, 39–71.

3

CAD-Based Task Planning and Analysis for Reconfigurable Workholding for Workpiece Positioning and Constraining

Bijan Shirinzadeh
Monash University

Workholding is one of the most basic and necessary processes required for all manufacturing operations, such as machining, assembly, and joining. Workholding, commonly referred to as fixturing, is performed to position and constrain the workpiece for presentation to the manufacturing or assembly devices. If flexible manufacturing systems (FMS) are to be truly flexible, then the fixturing process must also be flexible. This chapter briefly describes the economic justifications for flexible fixturing within computer-integrated manufacturing (CIM) environment. Various approaches to flexible fixturing are presented. The reconfigurable workholding (or fixturing) is one of the most appropriate flexible fixturing techniques for CIM environment. Reconfigurable fixturing employs a number of fixture modules that are set up, adjusted and changed automatically by robotic devices without human intervention. The requirements for locating and constraining workpieces using such fixture modules are presented. Furthermore, the fundamental issues for planning and analysis of fixture setup are described. A technique for computer-aided planning and analysis of reconfigurable fixture is also presented in this chapter.

3.1 Introduction

Workpiece positioning and constraining, commonly referred to as workholding or fixturing, is an important issue in manufacturing processes.[1] In general, for any manufacturing operation to be successful, the workpiece must be located and held to remain in a desired position and orientation when subjected to external forces during manufacturing operations. Jigs and fixtures are the devices employed to position and constrain the workpiece for presentation to the manufacturing device. The traditional approach to fixturing involves designing and manufacturing special-purpose fixtures for specific workpieces and manufacturing operations.[2] In addition to this expensive process, a significant amount of intervention by skilled operators is needed in setting up and changing the fixtures when the manufacturing operation is completed or modified. Therefore the traditional approach is generally time consuming and thus costly. This is especially true for current and future manufacturing requirements characterized by rapid change of product design and reduced production cycle time.

To reduce the dependence on dedicated fixtures, recent research efforts have been directed toward developing alternative approaches. Several flexible fixturing strategies have been proposed and developed.[3] These strategies include sensory-based techniques, modular fixturing, reconfigurable fixturing, conformable clamps, and many others. It is believed that reconfigurable fixturing is one of the most appropriate strategies in a computer-integrated manufacturing (CIM) environment. However, for any fixturing technique to be effective, it must be supported by a computer-based planning and analysis system. This chapter will briefly describe the economic justifications for flexible fixturing within a computer-integrated manufacturing (CIM) environment, and, more importantly, within flexible manufacturing systems. Flexible fixturing strategies will be briefly described and fixture design requirements will be presented. However, the main focus will be on requirements, guidelines, structure, and formulation for the CAD-based planning and analysis of fixture configuration. These will be described in detail.

3.2 Economic Justification

A great deal of attention has been directed towards the development of advanced manufacturing systems. Manufacturing systems can be classified into two distinct categories: high volume and low- to medium-volume production systems. It is quite apparent that investments in manufacturing technology are more easily justified in high-volume production environments than in low- to medium-volume environments. High-volume production systems typically produce a large number of similar parts with similar processes. Therefore, factors justifying such investments in automated manufacturing technology include:

- Large production batches of same parts with similar processes
- Very few design changes over a long period of time
- Reduction of labor costs for support functions (such as materials handling and part positioning and constraining) by the deployment of dedicated and hard-tool fixturing
- Simplified scheduling in conjunction with less need for program development due to the repetitive nature of tasks and large batches.

On the other hand, low- to medium-volume production systems produce a larger number of diverse parts in much smaller batch sizes.[4] Therefore, in low- to medium-volume production systems, factors that prevent investments in automated manufacturing technology are basically the reverse of those in the high-volume production systems:

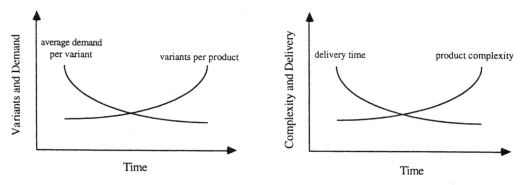

FIGURE 3.1 Trends in flexible manufacturing environment. (Reprinted from Reference 5, with permission from Elsevier Science.)

- Small production batches of diverse parts with dissimilar processes
- Frequent design changes
- Little reduction of labor costs for support functions, such as materials handling and part presentation using traditional or dedicated fixtures
- Sophisticated scheduling and control functions, together with a greater need for computer-aided program development to accommodate the variety.

Therefore, as products become more and more customer-oriented, smaller batch production volume will be required and faster changes between component manufacture becomes an important issue. The environmental conditions of production systems characterized by increasing changes over time leading to the increase of variants per product as shown in Figure 3.1.[5] Some of the major design and development challenges presented by low- to medium-volume production systems include:

1. Flexible fixturing and tooling concepts
2. Automation of the fixturing process and tool delivery
3. CAD-based planning, analysis, automated program development
4. Development of flexible information systems.

Having considered the practical and economical aspects and design challenges of low- to medium-volume production systems, one can define a flexible manufacturing system as a production unit capable of manufacturing a family of products with a minimum amount of manual intervention. Such a system would consist of work stations equipped with multi-axis robotic systems to perform manufacturing operations. Furthermore, it would operate as an integrated system under full programmable control. The traditional approach to workholding is costly due to the long lead time and effort required to design and manufacture dedicated fixtures. In addition, these fixtures require manual setup and change of fixtures when a manufacturing operation is completed or modified. There is also an overhead cost associated with storing and retrieving a multiplicity of fixtures.

Thus, the traditional approach is not suited to a modern and flexible manufacturing environment where production volume is low and product variation is high. If flexible manufacturing systems are to be truly flexible, therefore, the fixturing systems must also be automated and flexible. In order to replace or reduce the requirements for dedicated fixtures, recent research efforts have been directed towards developing alternative approaches to traditional fixturing methods.

3.3 Flexible Workholding Strategies

Flexible fixturing involves employing a single fixturing system to hold workpieces of various shapes and sizes within a family of manufacturing operations. The implementation of such systems is strongly dependent upon development of advanced fixturing techniques, together with integration of automation/robotics

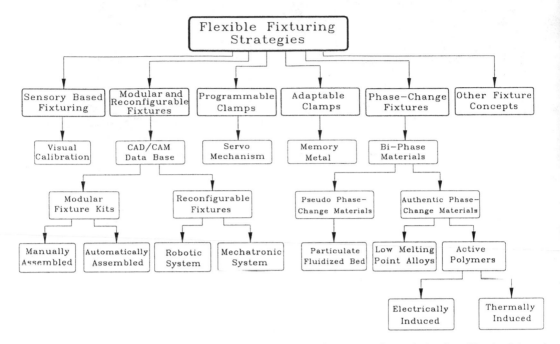

FIGURE 3.2 Flexible workholding strategies. (Reprinted from Reference 5, with permission from Elsevier Science.)

and CAD/CAM, which are essential ingredients of modern manufacturing. Figure 3.2 shows a schematic diagram highlighting the approaches that have received significant attention in the past decade.[6] These approaches are briefly described in this section.

Sensory-Based Workholding Techniques

In order to reduce the accuracy requirements for workholding, sensory-based information may be used to compensate for positioning errors in an assembly setup. Machine vision techniques have been applied to manufacturing processes to create a "visual fixture." Visual fixturing utilizes triangulation-based formulations and computer control to generate a workpiece "offset matrix" for use by a robot controller. The matrix effectively defines the location of the workpiece in six degrees of freedom. This technique has been applied to the fixturing problem encountered for insertion of windshields onto automobile bodies. It must be noted that this is a method to correct for location errors on parts for which traditional fixtures have already provided adequate clamping.

Modular and Reconfigurable Fixtures

Modular fixturing employs a number of elements such as V-blocks, rectangular blocks, and clamps to construct a fixture layout.[7] These pieces are bolted down to a T-slotted plate or a base plate with plain and tapped holes. The plate is usually part of the machine bed. The elements are aligned precisely with the aid of measuring devices (e.g., height gauge). Toggle clamps are often used to allow for rapid changing of the workpiece. The clamps push the workpiece against the V-blocks and rectangular blocks, thus constraining the workpiece. The accurate alignment of the workpiece requires great care.

Modular fixturing kits are based on the traditional machinist's approach of developing a variety of fixtures by assembling primitive elements.[8] Research efforts in modular fixturing have focused on the development of strategies for automatic setup of these elements using robots. Efforts have also been aimed at developing computer-assisted design techniques and geometric reasoning for assembly of the elements.[9–14]

The concept of reconfigurable fixtures has received considerable attention.[6] This approach uses a number of modules or mechanisms that are reconfigured or rearranged to form layouts that position

FIGURE 3.3 Reconfigurable fixture.

and constrain different workpieces within a family.[15] The reconfiguration or arrangement is generally performed by a robot or dedicated device built into the system. The fixture modules are retrieved from the storage magazines and placed on a fixture platform by a robot which then proceeds to perform adjustment operations on these modules.[15–17] The workpiece is placed on this configuration and the modules will be in contact with the workpiece at the desired locations (see Figure 3.3). The fixture modules are generally suitable for applications with low-to-medium force, and torque conditions such as light flexible assembly and finishing operations.

Adaptable Fixtures

A set of clamps has been designed to use shape memory alloy (SMA) wires to exert motive force.[3] The clamps consist of a 4 × 4 array of fingers. The fingers are normally locked by the squeezing action of heavy duty belleville springs. When current is applied to a SMA wire, the wire is heated; the temperature increase induces a martensite phase transformation in the material, which then attempts to assume its "trained" position. The resulting force releases the fingers so they conform to irregular workpiece surfaces. A second set of SMA wires is then used to apply the horizontal clamping force. This class of fixtures was originally developed to be used as end-effectors on robot manipulators, but they may also be utilized in other applications.

Programmable Conformable Clamps

Programmable conformable clamps are workholding devices employed specifically for the machining of titanium forgings for the turbine blades of large power generating equipment.[1] The clamps consist of octagonal frames that are hinged so that they may be opened to accommodate a blade and then be closed. The clamps may be located at almost any position along the blade; generally, two or three clamps will hold a blade rigidly enough for machining purposes. These clamps are light and compact enough to travel with the blade and they can accommodate a large number of different blade configurations. However, the initial accurate positioning of the workpiece within the clamps requires a great deal of care, and a different reference surface for each workpiece at the clamping station must be built. Other researchers have developed a numerically controlled clamping system integrated with a horizontal milling center for machining of workpieces. The system positions supports, locators, and clamps on servo-controlled turntables, providing a variety of support configurations. The clamping is performed by repositionable hydraulic cylinders. The system can be used for fixturing workpieces such as castings for machining purposes.

Phase-Change Fixtures

Phase-change fixtures can be divided into two categories: (1) fixtures incorporating materials which undergo an authentic phase change, and (2) fixtures incorporating materials which undergo a pseudo phase change. There have been numerous reports on this class of fixtures, which utilizes materials that undergo a phase change between liquid and solid states.[3,18] Authentic phase-change fixtures are classified according to how the bed material undergoes phase change, which may be temperature induced, electrically induced, or a combination of both. An example of authentic phase-change fixtures is the encapsulation which traditionally has been employed for special purpose machining, such as the milling of turbine blades.

The pseudo phase-change fixtures utilize the two-phase nature of materials that undergo a phase change between liquid and solid states. An example of this approach is the fluidized-bed fixture depicted in Figure 3.4. The fluidized-bed fixture consists of a container which retains the bed materials.[7] The container has a porous base through which air passes at a controlled rate. Under these conditions the bed materials behave as fluid into which the workpiece is partially immersed. When the air supply is cut off, the bed materials fall onto the base due to gravitational loading, and are compacted to form a solid mass and thus secure the workpiece. The removal of the workpiece from the fixture is achieved by switching on the air supply to create fluid phase again.

The main advantage of this approach is the flexibility of accommodating workpieces of various geometrical configurations by exploiting the fundamental properties of fluids, which are the ultimate conformable surface. However, the approach may generally be employed for low force and torque applications. Furthermore, the workpieces must generally be of prismatic shapes providing surfaces for frictional holding capacity. They must also be placed within the container using an accurate device and be held in the desired position and orientation until the vacuum compaction process is complete. Research efforts have focused on modeling the torsional displacement.[18] Studies have also been carried out on the effect of the workpiece and bed geometries on workpiece lateral displacement.[19]

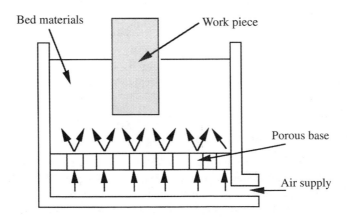

FIGURE 3.4   Fluidized bed fixture.

Other Fixturing Techniques

There are several other techniques proposed for workholding systems. Examples of these include the multi-leaf vice, petal collet, and stackable washer fixtures.[1] These have received less attention due to the lack of robustness and the small range of workpieces which they can accommodate. Generally, these techniques do not provide the flexibility, accuracy, and integration capability required in a world-class manufacturing environment. The most suitable approaches for flexible and computer-integrated manufacturing environments are deemed to be modular and reconfigurable fixturing, together with programmable clamps.

3.4   Reconfigurable Fixture Modules

In this chapter, attention is mainly focused on reconfigurable workholding. However, most of the planning and integration techniques can readily be applied to other fixturing approaches such as modular or programmable. In fact, for any fixturing system to be acceptable in a manufacturing environment, it must be capable of integration with the CAD/CAM systems and also automation devices such as robots and machine tools.[20] The reconfigurable workholding approach employs a number of adjustable fixture modules to locate and constrain the workpiece in the desired location. For the purpose of establishing the planning and analysis strategies, four types of fixture modules have been developed.[21] These modules include vertical support, horizontal support, horizontal clamp, and vertical clamp, as shown in Figure 3.5. These modules are mainly single axis and have been developed for fixturing workpieces for robotic assembly; however, the concept can be extended for the design of multi-axis modules for panel type workpieces (see Figure 3.6).

The fixture setup procedure involves a robot retrieving a number of vertical supports from the fixture magazine and placing these on a fixture platform such as an electromagnetic chuck. The robot proceeds to adjust the height of these modules, which support the workpiece at the required height. The electromagnetic chuck is used to secure the fixture modules once they are placed at the assembly station before any adjustments are to take place. Next, a number of horizontal supports are placed on the chuck where they would be in contact with the workpiece reference faces or locating tabs. The robot then places the workpiece within this fixture layout (referred to as the "initial" fixture layout); a number of horizontal clamps and vertical clamps are then placed on the chuck. The robot adjusts the height of these modules and activates the clamping mechanisms on them, and thereby fixturing the workpiece into the desired position and orientation (pose). The robot may now perform the subsequent assembly operations on the workpiece. It must be noted that other fixture modules have also been developed by other researchers for operations such as assembly and drilling.[22–23]

3.5   Fixture Task Planning and Analysis

An important objective of any flexible workholding system is to provide for rapid design of fixture configuration and analysis, and automated generation of fixture setup programs.[13,17,24] The fixture planning and analysis system must reside within the overall process planning of the manufacturing system. An approach is now presented describing an interactive computer-aided fixture planning and analysis (CAFPA) system for reconfigurable fixtures. In the proposed approach, the CAFPA consists of six submodules comprising pre-processor, planning, analysis, verification, interference checking, and post-processor (Figure 3.7). The following sections describe the individual submodules and their functionalities.

Pre-Processor Phase

The first operation in the pre-processor segment is to access the CAD data base and retrieve the model data of the workpiece. The boundary representation (B-Rep) scheme provides an excellent structure for manipulation of appropriate data and has a format as shown in Figure 3.8. There are several CAD systems

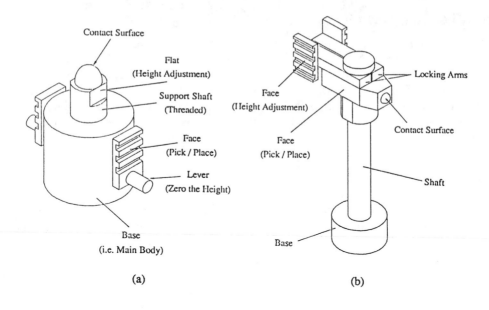

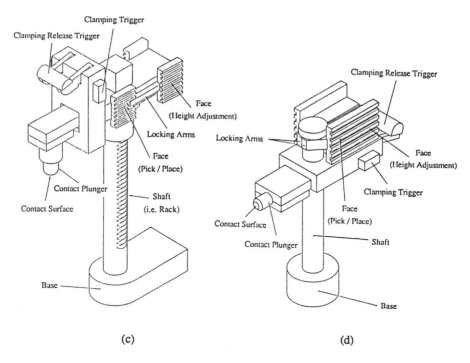

FIGURE 3.5   Schematic diagram of (a) vertical support, (b) horizontal support, (c) vertical clamp, and (d) horizontal clamp. (Reprinted from Reference 37, with permission from Elsevier Science.)

such as AutoCAD, CATIA, Intergraph, and Medusa which provide for data exchange facilities. The geometrical data retrieved from CAD is transformed into the B-Rep format for internal use by the planning and analysis system. Thus, the geometric information of the object is represented by its bounding faces. Each face is described by its bounding edges, and, in turn, each edge is defined by its vertices. Furthermore, an internal boundary loop, f_{1i} (i.e., hole in a face), is transversed in a clockwise direction opposite to the external loop, f_{1e} (anticlockwise), and is stored as a forbidden region for possible fixture

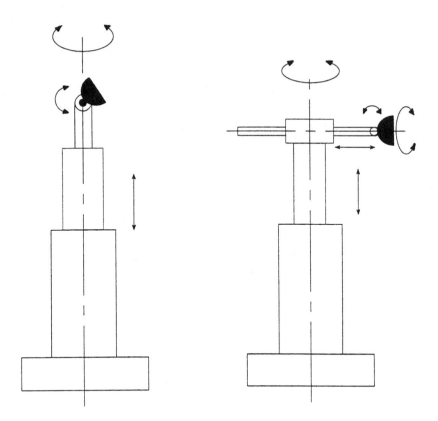

FIGURE 3.6   Schematic diagram of multi-axis fixture modules.

contact point. The geometrical information is then used to determine the components of the normals for all polygons (i.e., faces) and also to generate views of the workpiece for the purpose of display and verification. Let the equation of the polygon (i.e., face) be given by:

$$AX + BY + CZ + D = 0 \qquad (3.1)$$

where A, B, and C are the components of the normal, and D is a constant. The coefficients A, B and C describing the equation of a polygon (i.e., face) are proportional to the areas of the projections of the polygon onto the YZ, XZ, and XY planes, respectively. Thus, the coefficients A, B, and C may be easily determined using the Newell formulation. The next operation in the preprocessor phase is to retrieve inertial properties (e.g., center of mass, etc.) and store this information for future use. The preprocessor software must also provide facility for retrieval of the previously generated data, if any.

Fixture Configuration Planning

The planning phase involves specification of the location of the fixture modules by determination of the contact positions on the workpiece in the CAD environment. There is a total of 12 degrees of freedom, six linear movements ($+x$, $+y$, $+z$, $-x$, $-y$, $-z$), and six rotational movements (clockwise or counterclockwise around each of the three axes) associated with a body that is totally unconstrained as shown in Figure 3.9(a). The positive location of a workpiece is generally guaranteed if it is in contact with three points in the first plane, two points in the second plane, and a single point in the third plane, as seen in Figure 3.9(b). This is the traditional three-two-one rule (i.e., 3-2-1 location rule).[25]

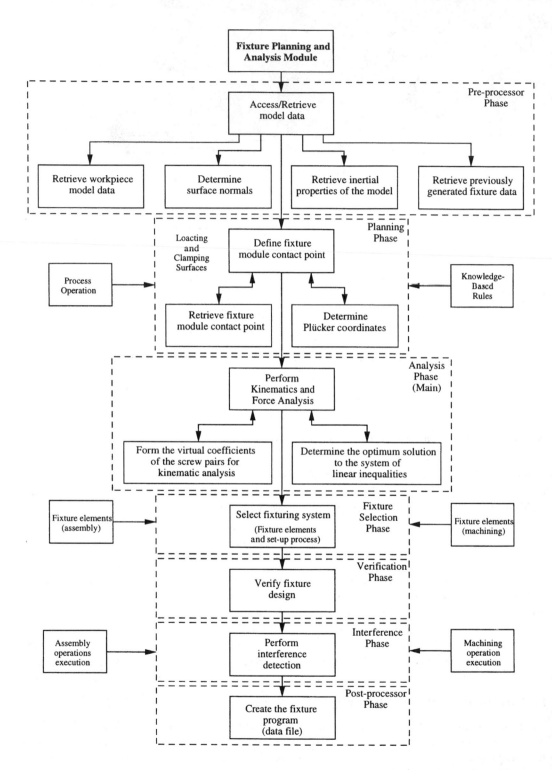

FIGURE 3.7 Structure of computer-aided fixture planning and analysis (CAFPA). (Reprinted from Reference 5, with permission from Elsevier Science.)

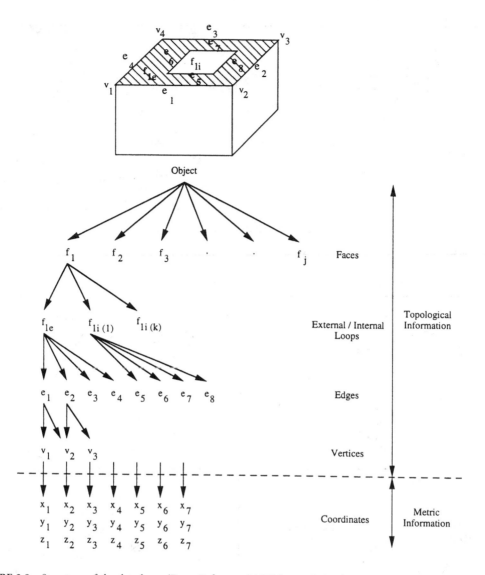

FIGURE 3.8 Structure of the data base. (From Reference 24. With permission.)

The vertical support fixture modules are employed to provide supports in the first plane, upward z-direction. In practice, at least three vertical support points are required to provide a stable positioning (Figure 3.10[a]). The vertical support contact points are generally selected to be on the reference planes. In addition, it may be required to provide additional supports in order to allow a uniform weight distribution and prevent object deflection. The horizontal supports are placed against machined surfaces or locating holes. These fixture modules, in general, provide supports in the second plane perpendicular to the first, and in the third plane perpendicular to the first and second planes as shown in Figure 3.10(b). This satisfies the positive location criterion for fixturing the workpiece when the subsequent clamping operation is carried out.

The horizontal and vertical clamps are employed to constrain (i.e., clamp) the workpiece in the horizontal and vertical planes, respectively. These fixture modules may be placed on any surface that the geometry of the workpiece allows. However, it is recommended that the point of contact for the first horizontal clamp should be chosen such that it would locate, and to an extent constrain, the workpiece in the X-Y plane (Figure 3.10[c]), thus satisfying the positive location criterion.[26] Whenever possible,

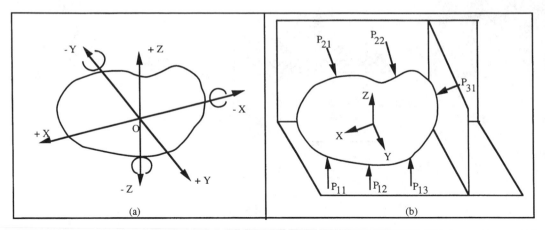

FIGURE 3.9   (a) Totally unconstrained rigid body, and (b) located rigid body. (From Reference 21. With permission.)

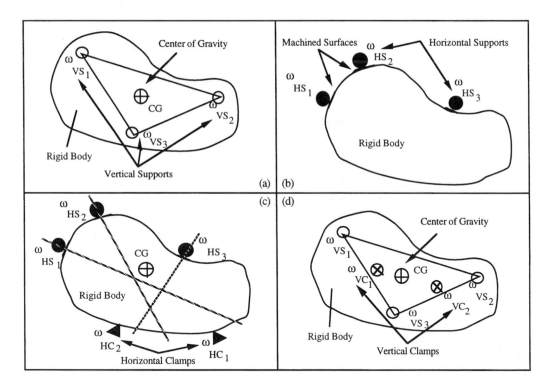

FIGURE 3.10   Configuration planning principles.

vertical clamps are generally placed within the region bounded by the vertical supports such that no rotational movement is generated due to activation of the clamps (see Figure 3.10[d]). This allows any force exerted by the vertical clamp fixture modules to be generally supported by the vertical supports underneath the workpiece in that region.

As the contact points are chosen, the three necessary coordinates, P_{C_x}, P_{C_y}, and P_{C_z}, of the chosen contact point are determined by transforming the CAD model coordinates into real world coordinates. The screw theory may be used to perform the kinematic analysis (described in the next section) of the fixture layout. The approach requires generating the wrenches acting on the workpiece. The wrenches (i.e., Plücker coordinates) for each contact point are equivalent to the components of the face normals and moments about the contact point. For every contact position that the designer chooses, the software module checks the position against the equations of the surface polygons and determines on which face the contact point lies. This is performed using the plane equation rewritten as[27]

$$d = \frac{AP_{C_X} + BP_{C_Y} + CP_{C_Z} + D}{\left(A^2 + B^2 + C^2\right)^{1/2}} \tag{3.2}$$

where P_{C_X}, P_{C_Y} and P_{C_Z} are the position of the contact point with respect to the assembly station coordinate frame, and A, B and C are the direction numbers for a given face stored in memory. The distance, d, of the chosen position from the plane of the polygon. If $d < 0$, then the point is outside, or on the negative side of the plane, since the normal inwards are taken as positive. If $d > 0$, then the chosen point is inside, or on the positive side of the plane. If $d = 0$, then the point is on the plane and this is the face that the chosen point lies on. It must be noted that this test is performed for every polygon describing the surface of the workpiece. If more than one polygon satisfies the test, then the coordinates of the chosen point are tested against the boundaries of the polygons, and the correct polygon will be identified. This identification is performed because there may be two planes perpendicular to each other and the chosen point of contact may be on the surface (e.g., the base of the workpiece) of one polygon while also satisfying the equation of the other plane. Rewriting the equation of the plane, in a normalized form, using more specific terminology (for future use):

$$A_{X_j}P_{C_X} + A_{Y_j}P_{C_Y} + A_{Z_j}P_{C_Z} + A_{C_j} = 0 \qquad j = 1, n \tag{3.3}$$

where A_{X_j}, A_{Y_j} and A_{Z_j} are the direction cosines, and j signifies the identification number.

Kinematic Analysis

Mathematical tools are required to describe the instantaneous and spatial motions of the workpiece when it is in contact with a set of fixture modules. The kinematic analysis can be performed using the conventional force and displacement vectors. Several researchers have used the conventional force approach for the analysis of modular fixtures for applications in machining.[25,28] In general, the three locating planes established using the 3:2:1 rule must provide a force field that opposes external forces and torques imposed on the workpiece such as the cutting force. The effects can be modeled as a resultant force R and a resultant moment M about the center of mass of the workpiece. The formulation is well suited to rectangular workpieces or workpieces with quadrangular prisms. The conditions for static equilibrium are as follow:

$$\sum_{i=1}^{3} F_i = 0 \tag{3.4}$$

and

$$\sum_{i=1}^{3} M_i = 0 \tag{3.5}$$

where F_i and M_i are the Cartesian components of the resultant force and moment vectors. Therefore, the following relationships describing the resultant forces and moments can be written:

$$R = R_x + R_y + R_z \tag{3.6}$$

where

$$R_x = F_{31} \tag{3.7}$$

$$R_y = -(F_{21} + F_{22}) \tag{3.8}$$

$$R_z = F_{11} + F_{12} + F_{13} \tag{3.9}$$

where reaction forces at each locating contact point are denoted by F_{ij}, and

$$M = \sum_{i=1}^{3} \sum_{j=1}^{4-i} r_{ij} \times F_{ij} \tag{3.10}$$

or

$$M = M_1 + M_2 + M_3 \tag{3.11}$$

where r_{ij} are the moment arm of F_{ij} and

$$M_i = \sum_{j=1}^{3} F_{ij} a_{i+3,j} \quad \forall i \in \{1, 2, 3\} \tag{3.12}$$

Combining equations (3.10) and (3.12) provides a matrix equation in the form of $AX = B$. In this formulation the positions of locating fixture modules may have to be reconsidered if the coefficients of the matrix A are singular.

The analysis can also be carried out using screw pairs. This approach is chosen for this task because screw representation of displacement and forces emphasizes the underlying representation upon line geometry and the surface model of the workpiece. Detailed accounts of the screw theory and its development may be found in Reference 29.

A set of fixture modules in contact with a workpiece can be modeled as a set of zero pitch wrenches acting on the workpiece. This stems from the fact that any system of forces and couples can be reduced to a wrench of a given intensity acting on a given screw. Therefore, we are interested in the total freedom of the workpiece when it is acted upon by a set of wrenches (see Figure 3.11). Once the contact face has

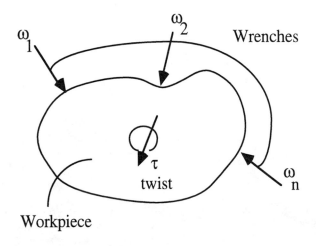

FIGURE 3.11   Schematic diagram of wrenches acting on a workpiece. (From Reference 24. With permission.)

been identified, the normals are used to determine the moment components and thus the wrench coordinates. Therefore, the Plücker coordinates for a given contact position with respect to a fixed point (e.g., center of gravity) on any surface, S, of the workpiece are determined using the following:

$$\omega_{1,i} = \left(\frac{\partial S}{\partial x}\right)_j$$

$$\omega_{2,i} = \left(\frac{\partial S}{\partial y}\right)_j$$

$$\omega_{3,i} = \left(\frac{\partial S}{\partial z}\right)_j$$

$$\omega_{4,i} = (P_{C_Y} - Y_{cg})\left(\frac{\partial S}{\partial z}\right)_j - (P_{C_Z} - Z_{cg})\left(\frac{\partial S}{\partial y}\right)_j$$

$$\omega_{5,i} = (P_{C_Z} - Z_{cg})\left(\frac{\partial S}{\partial x}\right)_j - (P_{C_X} - X_{cg})\left(\frac{\partial S}{\partial z}\right)_j$$

$$\omega_{6,i} = (P_{C_X} - X_{cg})\left(\frac{\partial S}{\partial y}\right)_j - (P_{C_X} - Y_{cg})\left(\frac{\partial S}{\partial x}\right)_j$$

(3.13)

where $\left(\frac{\partial S}{\partial x}\right)_j$, $\left(\frac{\partial S}{\partial y}\right)_j$, and $\left(\frac{\partial S}{\partial z}\right)_j$ are the components of the surface normal (i.e., A_{X_j}, A_{Y_j}, A_{Z_j}), i defines the fixture module number (i.e., number of wrenches), and j is the face number on the workpiece. It must be noted that the components of the moment about the assembly station coordinate frame may also be found using the above formulation by deleting the expressions for the center of gravity.

The components of the wrenches are used to perform the kinematic analysis. The generalized screw system is used to determine the possible movement that the workpiece may have when it is partially or fully constrained in a fixture layout. The governing equation describing the virtual coefficient may be written in general form as[24]

$$w_{\tau\omega} = \omega_{n,1}\tau_4 + \omega_{n,2}\tau_5 + \omega_{n,3}\tau_6 + \omega_{n,4}\tau_1 + \omega_{n,5}\tau_2 + \omega_{n,6}\tau_3 \qquad (3.14)$$

where $w_{\tau\omega}$ is the virtual coefficient for the twist, τ, and the wrench, ω, and n is the number of wrenches (i.e., fixture modules) in contact with the workpiece. The above formulation allows us to deal with the uni-directional nature of the constraints.[28] It also allows the identification of the specific contacts that may be broken for a given twist. The equations governing the virtual coefficients are formed for various stages of the fixturing process. These relationships include the following:

$$w_{\tau\omega} > 0 \qquad \text{repelling screw pair} \qquad (3.15)$$

$$w_{\tau\omega} = (0) \qquad \text{reciprocal screw pair} \qquad (3.16)$$

$$w_{\tau\omega} > 0 \qquad \text{contrary screw pair} \qquad (3.17)$$

An important design criteria is that the fixture layout must provide for ease of loading and unloading. The robot or the manipulation arm is required to place the workpiece into the fixture layout from an outside location in the "free-space" region without jamming the workpiece in the layout. The fixture layout design must therefore accommodate an initial stage where the workpiece can be moved into the fixture layout location from an outside (i.e., free-space) location. This fixture layout will be referred to as the "initial" fixture layout, and it must guide and locate the workpiece into the desired position

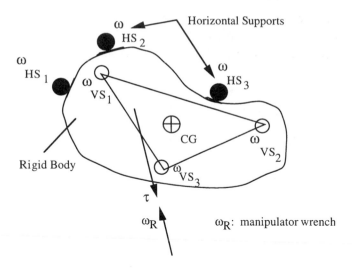

FIGURE 3.12   Initial fixture layout and approach/departure vector.

and orientation. Equation (3.15) is used to form the set of inequalities describing the initial fixture layout. Therefore, the vertical and horizontal supports will be included in the formulation as these are the only modules with which the workpiece will be in contact. This fixture layout requires that there be at least one non-zero twist repelling to all the wrenches exerted by support fixture modules. In other words, this set of inequalities must generally have a solution where τ_1, τ_2, and τ_3 (i.e., rotational twist) are zero. Thus, there will be no rotational twist about the axes X, Y, and Z. However, there must be a twist repelling to all the wrenches such that τ_4, τ_5, and τ_6 (i.e., linear twist) are non-zero, and the robot may approach the fixture layout from a free-space region to place the workpiece into the initial fixture layout. The approach and departure vector may be derived from this twist, as illustrated in Figure 3.12.

Another important requirement of a fixture is that the workpiece must be located in a specific pose for presentation to the manufacturing device. In other words, during the clamping operation the workpiece must be guided and constrained at the desired location where it would be in contact with all the fixture modules. Equation (3.16) may be used to form the set of equations describing the "final" fixture layout. This set of equations must have only a homogeneous solution where τ_1, τ_2, τ_3, τ_4, τ_5, and τ_6 must all be zero. This condition will provide for a null displacement. This means that there is no rotational twist about the axes X, Y, and Z, and no linear twist along the X, Y, and Z axes.

Equation (3.17) must be satisfied when it is desirable to add additional wrenches to the set for constraining the workpiece, if it is not already constrained. This simply means that at any stage additional wrenches may be added to the set. However, these wrenches must form a contrary screw pair with the twist that is obtained for the initial fixture layout. This condition also implies that the positioning twist applied to the workpiece by the robot or the clamping fixture modules must be contrary to the wrenches exerted by the initial fixture layout. Therefore, the contrary screw pair allows the formulation of wrenches from the initial to final fixture layouts. The system of linear inequalities that was formed by equations (3.15) through (3.17) may be solved using linear programming techniques.

3.6 Fixture Layout Verification

The solution of the linear inequalities can be obtained by a number of linear programming (LP) techniques. The simplex method is a good candidate for this. The solution obtained is simply the twist possible when the wrenches are acting on the workpiece because that is when the given set of fixture modules are in contact with the workpiece. The general linear programming problem can be stated as one of choosing the variables:

$$\{\tau\}^T = \{\tau_1 \tau_2, \ldots, \tau_n\} \tag{3.18}$$

such that

$$Z = \sum_{j=1}^{n} C_j \tau_j \rightarrow \min \tag{3.19}$$

subject to

$$\sum_{j=1}^{n} \omega_{ij} \tau_j \langle \leq, =, \geq \rangle_{b_i} \qquad i = 1, \ldots, m \tag{3.20}$$

and

$$\tau_j \geq 0 \qquad j = 1, \ldots, n \tag{3.21}$$

where ω_{ij}, b_i and C_j are constant coefficients. The notation $\langle \leq, =, \geq \rangle$ means that the constraints might be either equalities or inequalities ($\leq$ or $\geq$). The verification segment allows viewing the workpiece model with the wrenches at the points of contact. This submodule also verifies the virtual coefficients by substituting the components of the twist back into the governing equations and displaying the results in a tabulated form.

Figure 3.13 shows a fixture verification template illustrating the initial fixture layout for a rectangular workpiece. This fixture layout comprises four vertical supports and three horizontal supports in contact

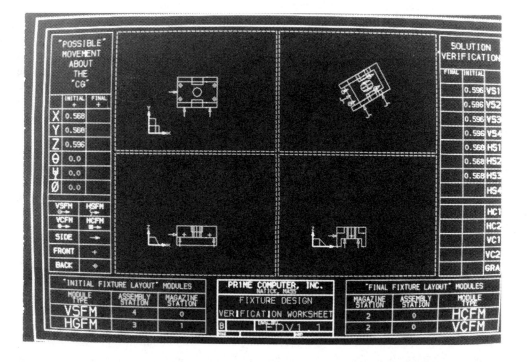

FIGURE 3.13 CAD analysis for an initial fixture configuration. (From Reference 24. With permission.)

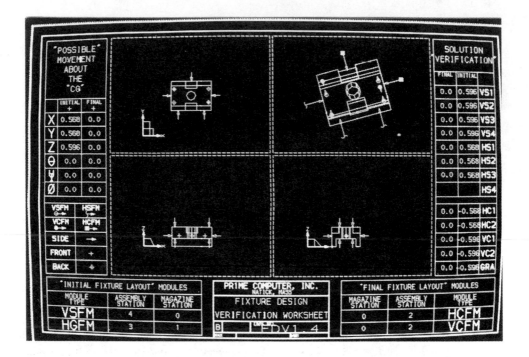

FIGURE 3.14   CAD analysis for a final fixture configuration. (From Reference 24. With permission.)

with the workpiece. The left hand column shows the normalized values of the twist. The twist, or the scalar multiple of the twist, would break the contact with all the fixture module. Therefore, the virtual coefficients formed by this twist and wrenches applied by the fixture modules are greater than zero (i.e., repelling).

Further, this also means that the wrench exerted by the robot manipulator, in placing the workpiece onto the initial fixture layout, is required to be opposite and thus contrary to the twist obtained for the initial fixture configuration. In fact, the components of the twist for the initial layout are used to determine the approach vector with which the robot moves to place the workpiece onto the fixture layout. Figure 3.14 shows the final fixture configuration with two horizontal and vertical clamps added to the initial fixture layout.

3.7   Interference Detection and Post Processing

The next important phase in any computer-assisted process planning and programming is the interference detection and automated generation of the machine programs. A number of sophisticated robot programming systems have been developed in recent years.[30-32] These systems combine the capabilities of modeling, task description, manipulator-level programming, simulation, and collision avoidance.[33] In general, these systems rely on interactive user specification of the objects in the workcell, and path planning is performed manually. The collision detection is usually carried out during simulation.[34-35] Such simulation and programming packages include ROBCAD, Deneb's Envision, CATIA, GRASP, and CIMstation. These generally utilize the "ideal" model parameters and robot controller algorithms to simulate the programmed motion. However, the ideal model generally differs from the actual model of the robot parameters.[36] Therefore, the programmed locations may require on-line correction using teach-mode programming.

In addition, computer-aided design (CAD) packages provide facilities to check for interference among the objects in a scene. However, when using CAD packages, each assembly must be modeled, and, depending on the details of the models, the interference detection would be time consuming and inefficient. Another shortcoming of such systems is their inability to take into account clearances as required in the physical world for any robotic assembly operation to succeed. Such clearances between fixture modules are necessary due to positioning errors which may be introduced by the robot manipulator.

Fixture Module Location

An important objective of a reconfigurable fixturing system is to provide means for a rapidly validating fixture construction program.[37] The fixture construction program, consisting of the placement, adjustment, and retrieval operations, is too complex and must be generated automatically. The information required by the interference detection module consists of the locations of the reference frames of the fixture modules with respect to the assembly station coordinate frame. All tool center point (TCP) locations for robot movements are determined by translation and rotation transformations of the reference frame position on the fixture modules. The positioning and orienting of the fixture module are performed in four-dimensional space: three to position and one to orient. Therefore, only rotation about the *z*-axis is required to orient the fixture module on the fixture bed, in the *X-Y* plane. The following expression describes the nomenclature for the location description of fixture modules:

$$\left[HC_{F_{i_j}}\right] \rightarrow \left[\text{Module Type}_{\text{Position Frame Designation}_{i\,=\,(X,\,Y,\,Z,\,\Psi,\,\Theta,\,\Phi)_j}}\right]_{\text{Station}} \quad (3.22)$$

As an example, the location of reference frame for a horizontal clamp, retrieved from the planning and verification software modules, is provided to the interference detection software module and is specified by a six parameter field as below:

$$\left[HC_{F_{i_j}}\right]_A = \left[HC_{F_{X_j}},\, HC_{F_{Y_j}},\, HC_{F_{Z_j}},\, HC_{F_{\Psi_j}},\, HC_{F_{\Theta_j}},\, HC_{F_{\Phi_j}}\right] \quad (3.23)$$

The first three parameters define the position (*x*, *y*, and *z*) and the last three parameters describe the orientation (roll Ψ, pitch Θ, and yaw Φ) of the parent reference frame (*F*) of the fixture module. Furthermore, *j* is the number of fixture modules in any particular category (e.g., HC: horizontal clamp). It must be noted that the roll and pitch parameters are zero but included only to be consistent for off-line programming purposes. Equation (3.24) describes the pose of the robot TCP during the placement of the horizontal clamp on the fixture bed.

$$\left[HC_{P_{i_j}}\right]_A = \left[HC_{P_{X_j}},\, HC_{P_{Y_j}},\, \Delta Z_{HC_P},\, HC_{F_{\Psi_j}},\, HC_{F_{\Theta_j}},\, HC_{F_{\Phi_j}}\right] \quad (3.24)$$

where ΔZ_{HC_P} denotes the height of the robot end-effector with respect to the fixture bed during the placement of the fixture module on the bed. The transformation from the reference datum on the fixture module to various critical locations at which the robot is required to interact with the fixture modules may be given as follows:

$$[P] = [T][\delta r] \quad (3.25)$$

where *P* is the pose of the robot TCP, *T* is the transformation containing the position and orientation of the fixture module's reference datum on the fixture bed, and δr is the appropriate distance from the functional locations on the fixture modules. Thus, the above formulation may be expressed as follows:

$$\begin{bmatrix} \cos\Phi & -\sin\Phi & 0 & X \\ \sin\Phi & \cos\Phi & 0 & Y \\ 0 & 0 & 1 & Z \\ 0 & 0 & 1 & 1 \end{bmatrix} \begin{bmatrix} 1 & 0 & 0 & \Delta R \\ 0 & 1 & 0 & 0 \\ 0 & 0 & 1 & \Delta Z \\ 0 & 0 & 0 & 1 \end{bmatrix} \quad (3.26)$$

Using the above equation, the following expressions may be written describing the robot TCP position during the placement of horizontal clamps with respect to the fixture bed coordinate frame.

$$HC_{P_{X_j}} = HC_{F_{X_j}} + (\Delta R_{HC_P} + \Delta R_{HC_{CF}})\cos(HC_{F_{\Phi_j}}) \tag{3.27}$$

$$HC_{P_{Y_j}} = HC_{F_{Y_j}} + (\Delta R_{HC_P} + \Delta R_{HC_{CF}})\sin(HC_{F_{\Phi_j}}) \tag{3.28}$$

$$HC_{P_{Z_j}} = Z + \Delta Z_{HC_P} \qquad \text{where } Z = 0 \text{ @ fixture bed coordinate frame} \tag{3.29}$$

where the ΔR_{HC_P} is the translation in the X-direction from the coordinate frame describing the reference fixture position to that of the pick/place position, and $\Delta R_{HC_{CF}}$ is the initial (i.e., unclamped) clearance between the contact surfaces of the horizontal clamp and the workpiece. As an example, Figure 3.15 shows various locations at which the manipulator is required to interact with the horizontal clamp in order to perform pick/place and height adjustment operations. Therefore, functional locations on the fixture module and robot TCP poses may be obtained using the above formulation in a similar manner.

Calculation of Interference

Collision avoidance requires magnitude and direction information; it is generally used for path planning of robots providing alternative robot position. Thus, collision avoidance is not appropriate for application in fixture planning. This is mainly due to the fact that the fixture modules must be at locations where they would be in contact with the workpiece. The technique appropriate for fixture construction is interference detection requiring only a true or false result. Therefore, we are only interested in interference detection at critical locations during fixture construction.

The proposed approach utilizes three tests to predict if any interference occurs between the fixture modules during the construction of the fixture. These tests include minimum separation, maximum separation, and model boundary. The underlying reason for minimum and maximum separation tests is to create a minimum region in which interference will occur, and a maximum region in which interference will not occur, regardless of the orientation of the fixture modules (Figure 3.16). The model boundary test is performed if the condition for minimum separation is met and the condition for maximum separation is not, thus taking into account the orientation of the fixture modules. The strategy and mathematical formulations for each test will be outlined using examples of cases involving interference between horizontal supports. Cases of interference detection involving other fixture modules follow similar hierarchical structure and formulations.

Minimum Separation

The test for minimum separation determines if the distance between the axes of the pick/place/adjustment coordinate frames of the two horizontal supports is smaller than the minimum distance required such that the bases of the two horizontal supports would not interfere with each other. Thus, if this condition is not satisfied, regardless of the orientation of the fixture modules, the two horizontal supports will collide during fixture setup. Figure 3.17 shows the schematic diagram of two horizontal supports located so that the shortest distance between the base centers is obtained. The geometrical constraint may be formulated mathematically as follows:

$$\Delta_{CC} \quad \beta_{HS} + \beta_{HS} + \mu \tag{3.30}$$

or

$$\chi_{\min} = \Delta_{CC} - 2\beta_{HS} - \mu \quad 0 \tag{3.31}$$

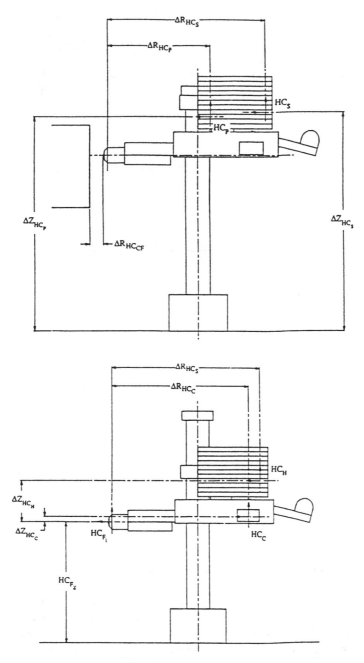

FIGURE 3.15 Illustration of functional locations on the horizontal clamp. (Reprinted from Reference 37, with permission from Elsevier Science.)

where β_{HS} is the minimum separation radius for the base signature of the horizontal support, μ is the clearance between the two supports, and Δ_{CC} is the distance between the base reference frames on the fixture modules and can be obtained using the following expression:

$$\Delta_{CC} = \sqrt{(HS_{P_{X_i}} - HS_{P_{X_j}})^2 + (HS_{P_{Y_i}} - HS_{P_{Y_j}})^2} \tag{3.32}$$

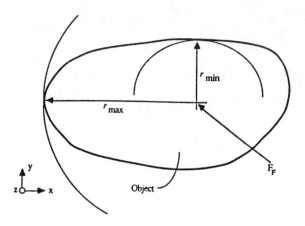

FIGURE 3.16 Schematic diagram of minimum and maximum separation conditions. (Reprinted from Reference 37, with permission from Elsevier Science.)

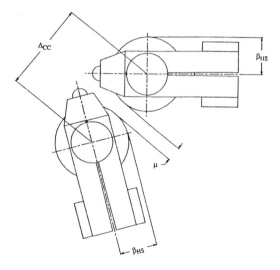

FIGURE 3.17 Minimum separation test between two horizontal supports.

where HS_p signifies the pick/place reference location on the horizontal support with respect to the fixture bed coordinate frame for i and j horizontal supports. If the above condition is not met, it can be concluded that the two supports will interfere with each other regardless of their orientation on the fixture bed. Therefore, there is no need to continue the interference analysis for the horizontal supports in question. However, if the geometrical constraint for minimum separation is met then the test for maximum separation must be performed.

Maximum Separation

The test for maximum separation determines if the distance between two fixture modules is greater than a distance such that the orientation of the fixture modules starts to play a role in the interference detection. If the test for maximum separation yields true, it can be concluded that the two fixture modules will not collide during fixture setup, and the analysis may be terminated. In this case, the maximum radius of the jaws from the appropriate datum will be chosen to formulate the geometrical constraint. Figure 3.18 shows a schematic diagram of the location of the two horizontal supports including the gripper signature

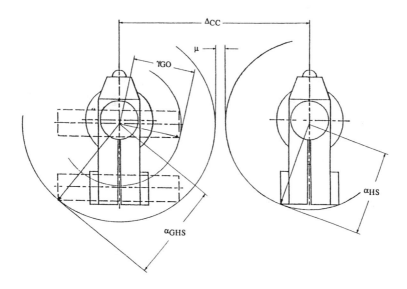

FIGURE 3.18   Minimum separation test between the horizontal supports.

(view from the above); one of the performing height adjustment operation on one of the horizontal supports. The following mathematical formulation may be formed as a general constraint for this case.

$$\Delta_{CC} \quad \alpha_{GHS} + \alpha_{HS} + \mu \tag{3.33}$$

or

$$\chi_{max} = \Delta_{CC} - \alpha_{GHS} - \alpha_{HS} - \mu \quad 0 \tag{3.34}$$

where Δ_{CC} is again the distance between the coordinate frames of the tool application on the two horizontal supports, α_{GHS} is the diagonal distance from the axis of the end-effector to the corner of the jaws when the jaws are open, and α_{HS} is the distance from the datum to the corner of the adjustment lever on the other horizontal support. Again, μ is the clearance specified.

However, if the above geometrical constraint is not met, then a model boundary test must be performed. This is due to the fact that the two horizontal supports may be positioned such that the interference-free fixture construction may or may not succeed, depending on the orientation of the fixture modules with respect to each other. It must be emphasized that the procedure is similar to those for all other fixture modules.

Interference Formulation Between Convex Polyhedral Objects

This phase of the interference detection again requires the geometrical data of the objects. The models of the various fixture modules and workstation have been created and are resident in the CAD data base. The major components of the workstation include the workpiece, the robot, the fixture bed, and the gripper. However, for the purpose of interference detection of the final fixture configuration, only the fixture modules and the gripper are used in the analysis. The objects are represented by clusters of convex polyhedral models. A boundary representation (B-rep) scheme may again be employed to describe each polyhedral model providing a description of the topology and a specification of dimensions. The topological representation is hierarchical in structure as shown in Figure 3.19. The representation consists of a list of:

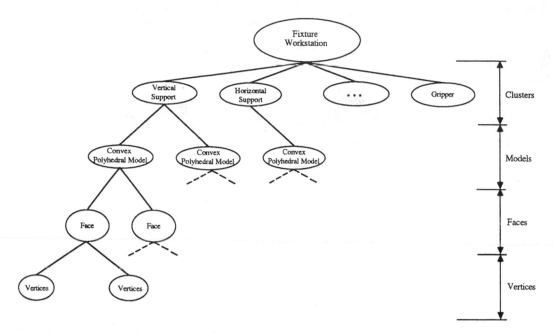

FIGURE 3.19 Topological representation for the convex polyhedral models. (Reprinted from Reference 37, with permission from Elsevier Science.)

- Clusters of objects in the fixture workstation
- Convex polyhedral models describing each cluster
- Faces on each model
- Vertices bounding each face
- Dimensional information for each vertex.

The necessary information required to perform the interference detection analysis can be derived from the above topology. This information includes the following:

- A list of faces on a cluster
- A list of vertices on a cluster
- A table of face normals for each cluster.

The approach for this phase of the interference detection consists of checking the vertices of the convex polyhedral models of one fixture module against the face normals of the convex polyhedral models for the second fixture module cluster. The reverse operation is also carried out, where the vertices of the second fixture module cluster are checked against the face equations of the first fixture module cluster.

Let A and B be objects represented by convex sets $\Omega_q \in R^u$, $q = 1, \ldots , g$. For two-dimensional space, $u = 2$, and for three-dimensional space, $u = 3$. If we define the set of unit outward normals for the faces bounding the model as $N = \{n : n \in R^u, \|n\| = 1\}$, the set of constants satisfying the equations for the faces as C, and the set of vertices attached to each face as $V = \{v : v \in R^u\}$, then the general formulation may be written as follows:

$$\Pi_{i,j} = \sum_{k=1}^{3} n_{A_{i_k}} v_{B_{j_k}} + C_{A_i} \geq \mu \qquad (3.35)$$

where i represents the identification number of the face on object A (i.e., $i = 1, \ldots, m$), j represents the identification number of vertex attached to the object B (i.e., $j = 1, \ldots, p$), k signifies the coordinates x, y, and z, and μ is the specified clearance. In the above formulation, the solutions $\Pi_{i,j} >, =$, and < 0 signify that the vertex is outside (i.e., positive direction) the face, on the face, and inside (i.e., negative

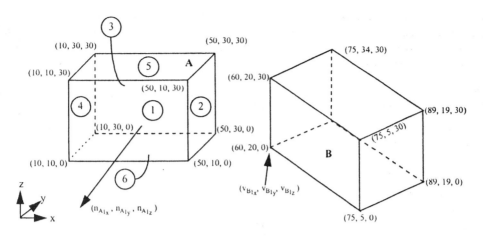

Faces on object A	General constraints c_i on object A	Constrains c_i for face i on object A	Vertex v_j ($j = 1$: 60, 20, 0) on object B	Resultant $\Pi_{i,\,i}$
1	$-y + c_1 = 0$	10	$-20 + 10$	-10
2	$x + c_2 = 0$	-50	$60 - 50$	10^* terminate
3	$y + c_3 = 0$	-30	$20 - 30$	-10
4	$-x + c_4 = 0$	$+10$	$-60 + 10$	-50
5	$z + c_5 = 0$	-30	$0 - 30$	-30
6	$-z + c_6 = 0$	0	$0 - 0$	0

FIGURE 3.20 Illustration of two convex polyhedral models. (Reprinted from Reference 37, with permission from Elsevier Science.)

direction) the face, respectively. The only acceptable solutions for the condition of disjointness between the two objects are $\Pi_{i,j} >, = \mu$. It must be emphasized that the value for clearance, μ, is selected based on the accuracy of the robot and the application.

As an example, consider two polyhedral objects A and B (Figure 3.20). If n_{A_i} denotes the "outward" normal for faces $i = 1, \ldots, 6$, on object A, and v_{B_j} denotes the vertices $j = 1,\ldots, 8$ for the object B. The above formulation simply stems from the substitution of components of a vertex of object B into general equation of the face on object A. This formulation determines upon which side of the face the vertex in question resides. It can be seen that for vertex (1) belonging to object B to be inside the region bounded by the faces (1–6) of object A, the solution to the set of equations resulting from the formulation must all be less than zero. In other words, the following relationships must hold:

$$\Pi_{1,1} = n_{A_{1_x}} v_{B_{1_x}} + n_{A_{1_y}} v_{B_{1_y}} + n_{A_{1_z}} v_{B_{1_z}} + C_{A_1} < 0$$

$$\Pi_{2,1} = n_{A_{2_x}} v_{B_{1_x}} + n_{A_{2_y}} v_{B_{1_y}} + n_{A_{2_z}} v_{B_{1_z}} + C_{A_2} < 0$$

$$\cdot$$

$$\cdot \qquad\qquad (3.36)$$

$$\cdot$$

$$\Pi_{6,1} = n_{A_{6_x}} v_{B_{1_x}} + n_{A_{6_y}} v_{B_{1_y}} + n_{A_{6_z}} v_{B_{1_z}} + C_{A_6} < 0$$

If the results of the above equations are all less than zero, then the vertex of object B is within the region bounded by the faces of object A. In other words, there will be interference between the two objects. It must be emphasized that all the vertices belonging to the object B must be checked, in a similar manner, against all the faces of object A. The reverse must also be performed: the vertices of object A are tested against the faces of object B. The fixture modules and the end-effector are modelled as convex polyhedral objects with respect to a datum coordinate frame, and these are stored in the data base.

It must be emphasized that this phase of interference detection is carried out based on the constraint of the general equation.[35] Therefore, the analysis for a given vertex of the first object against faces of the second object is terminated if the solution to any of the equations in the set, $\Pi_{i,j}$, yields = or > μ. In other words, the vertex is not bounded by the faces. However, if these constraints are not satisfied (i.e., $\Pi_{i,j} < \mu$), then the identification numbers and the interference location are reported to the user, and the analysis proceeds to the next objects.

It must also be emphasized that, in most cases, the process of interference detection is concluded by the minimum and maximum separation tests. There also exists the possibility of the objects crossing each other with the vertices located on the outside of the faces. For such cases, the object may be modeled as smaller objects effectively generating a mesh where the nodes are treated as vertices.

Sequence of Interference Detection

The sequence for the interference detection follows the sequence of fixture assembly. Vertical supports are tested, followed by horizontal supports, horizontal clamps, and finally the vertical clamps. Therefore, a four-phase sequence for interference detection is followed, as below:

Phase I: Do perform interference detection for vertical support against
  - other vertical supports
  - all horizontal supports
  - all horizontal clamps
  - all vertical clamps

  Enddo

Phase II: Do perform interference detection for horizontal support against
  - other horizontal support
  - all horizontal clamps
  - all vertical clamps

  Enddo

Phase III: Do perform interference detection for horizontal clamp against
  - other horizontal clamps
  - all vertical clamps

  Enddo

Phase IV: Do perform interference detection for vertical clamp against
  - other vertical clamps

  Enddo

The above sequence for interference detection is followed in conjunction with the procedure and mathematical formulations described in the previous sections. Therefore, the minimum separation test is performed first, followed by the maximum separation and the model boundary analysis. Figure 3.21 shows a schematic diagram of the software structure. Upon completion of the interference analysis, the results are provided to the design engineer. The locations of the modules are displayed in a CAD template by the software module (Figure 3.22). This intereference detection technique may be used in other reconfigurable and modular fixturing approaches reported in the literature.[22,23,38] The final operation, generally referred to as "post processing," will generate a fixture configuration data file, or configuration

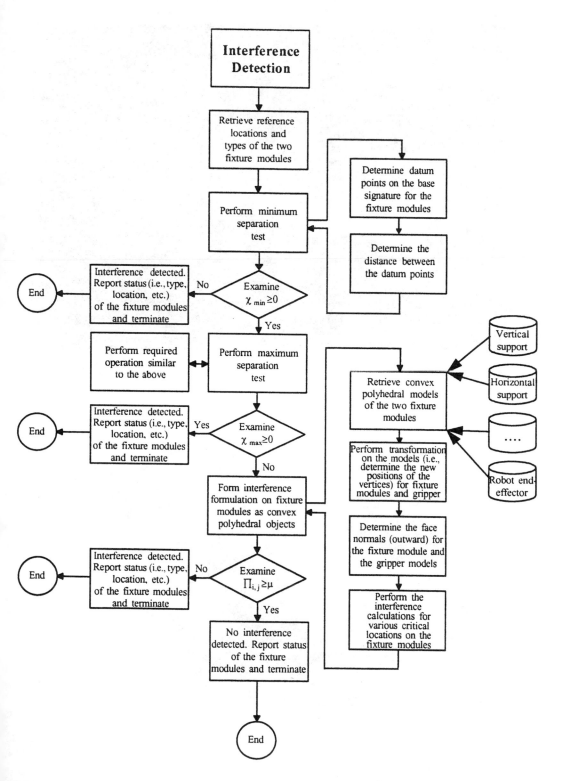

FIGURE 3.21 Software structure for the interference analysis. (Reprinted from Reference 37, with permission from Elsevier Science.)

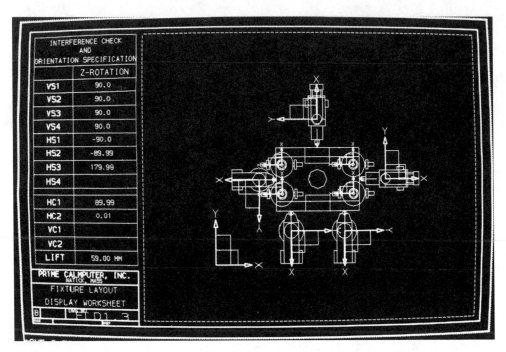

INTERFERENCE CHECK AND ORIENTATION SPECIFICATION	Z-ROTATION
VS1	90.0
VS2	90.0
VS3	90.0
VS4	90.0
HS1	-90.0
HS2	-89.99
HS3	179.99
HS4	
HC1	89.99
HC2	0.01
VC1	
VC2	
LIFT	59.00 MM

PRIME CALMPUTER, INC.
NATICK, MASS
FIXTURE LAYOUT
DISPLAY WORKSHEET

FIGURE 3.22   Example of the graphics display screen. (Reprinted from Reference 37, with permission from Elsevier Science.)

FIGURE 3.23   Example of a fixtured workpiece.

planning file. This file will contain all the necessary information regarding the position and orientation, the height adjustments, and the number of fixture modules required to locate and constrain the workpiece. The robot program to build the fixture, perform the assembly operation,[11] and dismantle the fixture layout can now be generated automatically by a robot task planner based on the configuration data file. Figure 3.23 shows an example of workpiece fixtured using the approach described.

3.8   Conclusion

A reconfigurable fixturing technique must provide for accurate planning, analysis, and automated fixture construction program. A CAD-based planning and analysis approach for reconfigurable fixturing in CIM and FMS has been described in this chapter. The approach employs a number of adjustable fixture modules to locate and hold workpieces of various shapes and sizes. The planning and selection of the fixture configuration are performed within a CAD environment. This is followed by the kinematic analysis and verification using the generalized screw pairs carried out by a dedicated software program. The interference detection is performed next. Finally, the post processing is carried out to generate a configuration file and robot program. These will consist of all the appropriate information to set up and change the fixture layout and to perform the assembly operations. Some of the advantages of employing such a system include:

1. Reduction of the lead time and effort associated with the design and manufacture
2. Reduction of the non-productive time between batches in a manufacturing plant
3. Reduction of the overhead cost associated with storing and retrieving a multiplicity of traditional fixtures
4. Full exploitation of robotic and automated assembly for low- to medium-volume production batches in FMSs and CIMSs.

Acknowledgments

This research is partially supported by an Australian Research Council (ARC) Small Grant, a Monash Research Fund (MRF) grant, and a Harold Armstrong Research Fund. The author wishes to thank his research assistants at Robotics & Mechatronics Research Laboratory, Department of Mechanical Engineering, Monash University.

References

1. Thompson, B. S. Flexible fixturing—a current frontier in the evolution of flexible manufacturing cells. *ASME Transactions* No. 84–WA/Prod–16, pp. 1–7, 1984.
2. Gandhi, M. V., Thompson, B. S., and Maas, D. J., Adaptable fixture design: an analytical and experimental study of fluidized-bed fixturing. *Journal of Mechanics, Transmissions, and Automation in Design,* Vol. 108, 15–21, 1986.
3. Hazen, F. B. and Wright, P. K., Workholding automation: innovations in analysis, design, and planning. *Manufacturing Review,* Vol. 3, No. 3, 224–237, 1990.
4. Krauskopf, B., Fixtures for small batch production. *Manufacturing Engineering,* 41–43, 1984.
5. Shirinzadeh, B., Strategies for planning and implementation of flexible fixturing systems in a computer integrated manufacturing environment. *Computers in Industry,* Vol. 30, No. 3, 175–183, 1996.
6. Shirinzadeh, B., Flexible and automated workholding systems. *Industrial Robot,* Vol. 22, No. 2, 29–34, 1995.
7. Gandhi, M. V. and Thompson, B. S., Automated design of modular fixtures for flexible manufacturing systems. *Journal of Manufacturing Systems,* Vol. 5, No. 4, 243–252, 1986.
8. Grippo, P. M., Gandhi, B. S., and Thompson, B. S., The computer-Aided design of modular fixturing systems. *The International Journal of Advanced Manufacturing Technology,* Vol. 2, No. 2, 75–88, 1987.
9. Beerma, J. R. and Kals, H. J. J., FIXES, a system for automatic selection of set-ups and design of fixtures. *Annals of the CIRP,* Vol. 37, 443–446, 1988.
10. Markus, A., Ruttkay, Zs. and Vancza, J., Automating fixture design—from imitating practice to understanding principles. *Computers in Industry,* Vol. 14, 99–108, 1990.
11. Pham, D. T. and Lazaro, A., Autofix—an expert CAD system for jigs and fixtures. *Journal of Machine Tools Manufacturing,* Vol. 30, No. 3, 403–411, 1990.
12. Cabadaj, J., Theory of computer-Aided fixture design. *Journal of Computers in Industry,* Vol. 15, 141–147, 1990.
13. King, L. S. and Hutter, I., Theoretical approach for generating optimal fixturing locations for prismatic workparts in automated assembly. *Journal of Manufacturing Systems,* Vol. 12, No. 5, 409–416, 1993.

14. Chou, Y. C., Geometric reasoning for layout design of machining fixtures, *Int. J. of Computer Integrated Manufacturing,* Vol. 7, No. 3, 175–185, 1994.
15. Asada, H. and By, A. B., Kinematic analysis and design for automatic workpart fixturing in flexible assembly. *Proc. of the 2nd Int'l. Symp. of Robotics Research,* Kyoto, Japan, 50–56, 1984.
16. Shirinzadeh, B., A flexible automatic fixturing system. *Third National Conference on Robotics,* 317–328, Melbourne, Australia, June 1990.
17. Bausch, J. J. and Youcef-Toumi, K., Kinematic methods for automated fixture reconfiguration planning. *Proc. of IEEE Robotics and Automation Conference,* Cincinatti, Ohio, May 1990.
18. Gandhi, M. V. and Thompson, B. S., Phase-change fixturing for FMS. *Manufacturing Engineering Systems,* Vol. 93, No. 6, 79–80, 1984.
19. Abou-Hanna, J. and Okamura, K., Finite element approach to modelling particulate bed fixtures. *Journal of Manufacturing Systems,* Vol. 11, No. 1, pp. 1–12, 1992.
20. Trappey, J. C. and Liu, C. R., A literature survey of fixture-design automation. *International Journal of Advanced Manufacturing Technology,* Vol. 5, 240–255, 1990.
21. Shirinzadeh, B., Issues in the design of the reconfigurable fixture modules for robotic assembly. *Journal of Manufacturing Systems,* Vol. 12, No. 1, 1–14, 1993.
22. Bausch, J. J. and Youcef-Toumi, K., Automated reconfiguration planning for sheet metal fixturing systems. *Proc. of Japan-USA Symposium on Flexible Automation,* Kyoto, Japan, July 1990.
23. Chan, K. C., Benhabib, B., and Dai, M. Q., A reconfigurable fixturing system for robotic assembly. *Journal of Manufacturing Systems,* Vol. 9, No. 3, 206–221, 1990.
24. Shirinzadeh, B., A CAD-based design and analysis system for reconfigurable fixtures in robotic assembly. *Computing & Control Engineering Journal,* Vol. 5, No. 1, 41–46, 1994.
25. Nnaji, B. O. and Lyu, P., Rules for an expert fixturing system on a CAD screen using flexible fixtures. *Journal of Intelligent Manufacturing,* Vol. 1, 31–48, 1990.
26. Trappey, A. J. C. and Matrubhutam, S., Fixture configuration using projective geometry. *Journal of Manufacturing Systems,* Vol. 12, No. 6, pp. 486–495, 1993.
27. Foley, J. D. and Dam, S. K., *Computer Graphics, Principles and Practice.* Addison-Wesley, Reading, MA, 1989.
28. Trappey, A. J. C. and Liu, C. R., An automatic workholding verification system. *Journal of Robotics & Computer-Integrated Manufacturing,* Vol. 9, No. 4/5, 321–326, 1992.
29. Ohwovoriole, M. S. and Roth, B., An extension of screw theory. *Journal of Mechanical Design,* Vol. 103, 725–734, 1981.
30. Fanghella, P., Galleti, C., and Giannotti, E., Computer-aided modelling and simulation of mechanics and manipulators, *Computer-Aided Design,* Vol. 21, No. 9, pp. 577–583, 1989.
31. Shiller, Z. and Dubowsky, S., Robot path planning with obstacles, actuators, grippers, and payload. *The International Journal of Robotics Research,* Vol. 8, No. 6, 3–18, 1989.
32. Owens, J., Robot simulation—seeing the whole picture. *Industrial Robot,* Vol. 18, No. 4, 10–12, 1991.
33. Thangaraj, A. R. and Doelfs, M., Reduce downtime with off-line programming. *Robotics Today,* Society of Manufacturing Engineers, Vol. 4, No. 2, pp. 1–3, 1991.
34. Steiner, K. V., Keefe, M., and Wolff, A., Interactive graphics simulation with multi-level collision algorithm. *Journal of Manufacturing Systems,* Vol 11, No. 6, 462–469, 1992.
35. Sorenti, P., GRASP for simulation and off-line programming of robots in industrial applications. *Proc. of Welding Engineering Software,* DVS 156, 55–58, Essex, 1993.
36. Beaumont, R. G. and Crowder, R. M., Real-time collision avoidance in two-armed robotic systems. *Computer-Aided Engineering Journal,* Vol. 8, 233–240, December 1991.
37. Shirinzadeh, B., A CAD-based hierarchical approach to interference detection among fixture modules in a reconfigurable fixturing system. *Journal of Robotics & Computer-Integrated Manufacturing,* Vol. 12, No. 1, 41–53, 1996.
38. Brost, R. C., and Goldberg, K. Y., A complete algorithm for designing planar fixtures using modular components. *IEEE Trans. on Robotics and Automation,* Vol. 12, No. 1, 31–46, 1996.

4

Integrated Precision Inspection System for Manufacturing Based on CAD/CAM/CAI Environment

Heui Jae Pahk
Seoul National University

4.1 Introduction

The development of CAD/CAM (computer-aided design/computer-aided manufacturing) technologies has had a revolutionary effect on the manufacturing industry in that it has pursued optimization in the design and manufacturing of products. The computer-aided inspection (CAI) technique has emerged after the computer controlled coordinate measuring machines (CMMs) have been widely introduced.

In this chapter, an integrated precision inspection system for manufacturing parts having CAD-defined features is demonstrated. The techniques of precision measurement are demonstrated around CAD/CAI

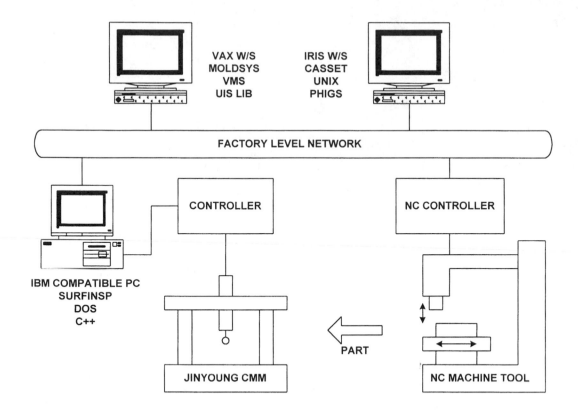

FIGURE 4.1(a)   A typical configuration for computer-aided inspection system of mold manufacturing.

integration for parts having sculptured surfaces with some basic features, such as holes, slots, and bosses. Features to be inspected are chosen in the CAD environment, and inspection planning is performed for each feature. Sampling-point strategies are (1) uniform distribution; (2) curvature dependent distribution; or (3) hybrid distribution of the two, depending on the complexity of the sculptured surface. Line and plane features are divided into subintervals, and measurement points are distributed at random positions in the subinterval. Prime numbers of subintervals are considered for a circle feature, in order to avoid possible periodic distortion of the measurement features.

Prior to measurement path planning, alignment procedures which will relate the workpiece coordinate system (CAD coordinate system) to the CMM coordinate system are generally required. In this chapter, the techniques of two-stage alignment procedures are introduced, enabling alignment even for the products of very thin and sharp geometry, such as turbine blades. The two-stage alignment procedures comprise the rough phase alignment, based on six points probing on clear cut surfaces, and the fine phase alignment, which uses the least squares method based on measurement feedback. For accurate tolerance evaluation, an improved evaluation method is introduced: the actual measurement points are obtained as the closest points on the CAD geometry by the subdivision technique, and the Chebyshev norm is applied iteratively to get optimum alignment of the measurement datum, giving an accurate profile tolerance. Figure 4.1a shows a typical CAD/CAI integration system around networks of computers and machine tools. Figure 4.1b shows the conceptual framework of the inspection system demonstrated in this chapter.[1,5,10]

4.2   Geometric Features and Tolerances for Inspection

Features and Tolerances Defined by ISO

In order to achieve a computer integrated inspection system, a feature-based inspection technique is applied, in which the features to be inspected are chosen in the CAD environment, and several key

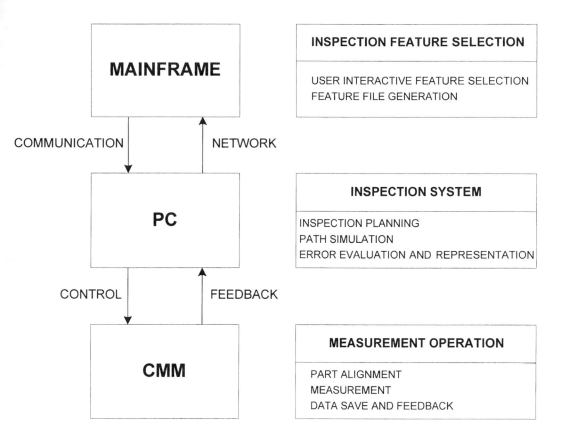

FIGURE 4.1(b) Conceptual framework of the demonstrated system.

TABLE 4.1 Types of Tolerances in ISO

Features and Tolerances		Toleranced Characteristics
Single features	Form tolerances	Straightness
		Flatness
		Circularity
		Cylindricity
Single or related features		Profile of any line
		Profile of any surface
Related features	Orientation tolerance	Parallelism
		Perpendicularity
		Angularity
	Location tolerance	Position
		Concentricity
		Coaxiality
		Symmetry
	Runout tolerance	Circular runout
		Total runout

features are considered for the products. There are several types of geometrical tolerances in ISO,[9] and the types of ISO tolerance are shown in Table 4.1. A geometrical tolerance applied to a feature defines the tolerance zone within which the feature is to be contained. There are four basic types of geometrical tolerances: form, orientation, location, and runout. These four basic types of tolerances can also be classified into two major groups by their relationships to datum: single features and related features. The single features are defined by the single datum; related features are defined by the relationship between the features. Form tolerances state how far the actual features are permitted to vary from the designed nominal form and consist of standard form features and non-standard form features. Standard form features include lines, planes, circles, and cylinders, while non-standard form features include curves and surfaces. Thus the corresponding form tolerances are defined as straightness, flatness, circularity, cylindricity, and profile of curves or surfaces. The form tolerances are described by tolerance zones, which set the limits of the extreme boundaries of features. The orientation tolerances specify the geometrical relationships to datum. Three types of orientation tolerances are (1) parallelism, (2) perpendicularity, and (3) angularity. Location tolerances state the permissible variation in the specified position of a feature in relation to some other features or data: true position, concentricity (coaxiality), and symmetry. Run-out tolerance is the deviation from the perfect form of a part surface of revolution directed by full rotation of the part on a datum axis.[7]

Sculptured Surfaces: Mathematical Description

There are several ways of describing sculptured surfaces mathematically. It is usual to consider a sculptured surface as consisting of several patches of cubic polynomial surfaces. Parametric polynomial surfaces and B-spline surfaces are considered in this section.[2,4]

Parametric Polynomial Surface

When u,v are the principal parameters defining the principal direction in a surface patch, the points data, $P(u,v)$, on the surface can be expressed as the sum of parametric polynomials of the third order. That is,

$$P(u,v) = \sum_{i=0}^{3} \sum_{j=0}^{3} A_{ij} u^i v^j \tag{4.1}$$

where A_{ij} is the coefficient matrix of the column vector defining the polynomial.

B-Spline Surface

The points data, $P(u,v)$, on the surface can be described in the B-spline form as follows:

$$P(u,v) = \sum_{i=0}^{n} \sum_{j=0}^{m} Q_{ij} N_{i,3}(u) N_{j,3}(v) \tag{4.2}$$

where $Q_{ij}(i = 1, 2,\ldots, n; j = 1, 2,\ldots, m)$ are the control points data, $(m + 1)$ and $(n + 1)$ are the number of control points, and $N_{i,3}(u), N_{j,3}(v)$ are the cubic blending functions for the u, v directions, respectively.[6]

Normal Vector to the Sculptured Surface

The normal vector, $\mathbf{N}(u,v)$, at (u,v) location on the surface can be evaluated from the partial derivatives of the surface along the u, v directions. Thus,

$$\mathbf{N}(u,v) = (\partial P(u,v)/\partial u \times \partial P(u,v)/\partial v)/|\partial P(u,v)/\partial u \times \partial P(u,v)/\partial v| \tag{4.3}$$

where the normal vector is divided by the absolute value to make the unit normal vector.

4.3 Measurement Points Sampling and Path Planning

For the selected inspection features, the following inspection planning procedures are performed, i.e., measurement points sampling for sculptured surfaces and basic features.

Measurement Points Sampling and Path Plan for Sculptured Surfaces

The measurement points distribution, i.e., measurement points sampling, is the key process for inspection planning. In this section, three methods are presented for sculptured surfaces: uniform distribution, curvature dependent distribution, and a hybrid distribution of the two.

Uniform Distribution Method

The basic idea of the uniform distribution method is that the grids formed by the measurement points have a nearly square distribution, and this aims uniform coverage for the sculptured surfaces. Suppose N points are to be distributed over a sculptured surface consisting of M surface patches. Let α be the dimensional ratio of the u direction to the v direction of the sculptured surface. Then the number of the measurement points, N_u, N_v, along the u, v directions in a patch can be determined as follows:

$$N_u = \text{round}\,(\sqrt{N/M\alpha}) \tag{4.4}$$

$$N_v = \text{round}\,(\alpha N_u) \tag{4.5}$$

where the round( ) is the function for rounding off the values in the bracket.

The parameters, u_i, v_j along a specified surface patch are determined based on the uniform allocation. That is

$$u_i = (i - 0.5)/N_u (i = 1, 2, ..., N_u) \tag{4.6}$$

$$V_j = (j - 0.5)/N_v (j = 1, 2, ..., N_v) \tag{4.7}$$

where 0.5 is used so that the sampled points are located on the middle of the surface grids.

Curvature Dependent Distribution Method

For the high-curvature surface measurement, it is desirable to have more sampling points than for small-curvature surface measurement. The measurement points can be distributed based on the normal curvature calculation along the sculptured surface. The normal curvature can be defined as:

$$\text{Normal Curvature} = |Curv_u \times \mathbf{N}| + |Curv_v \times \mathbf{N}| \tag{4.8}$$

where the $Curv_u$, $Curv_v$ are the curvature vectors along the u, v direction, respectively, and $\mathbf{N}$ is the unit normal vector of the surface at the measurement points. The $Curv_u$, $Curv_v$ are

$$Curv_u = |P_u|^3/|P_u \times P_{uu}| \tag{4.9}$$

$$Curv_v = |P_v|^3/|P_v \times P_{vv}| \tag{4.10}$$

where P_u, P_v, P_{uu}, P_{vv} indicate the first and second derivative of the sculptured surface along the u, v directions, respectively. The measurement points are distributed so that relatively more points are sampled on the region of higher curvature.

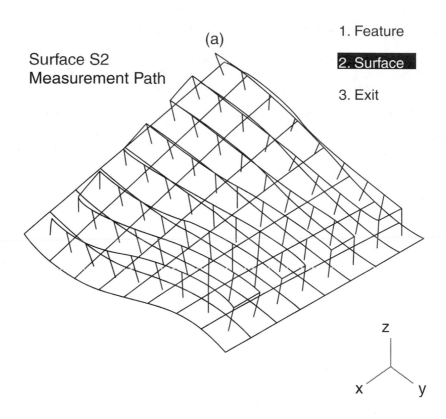

FIGURE 4.2(a) Uniform distribution method.

Hybrid Distribution Method

The hybrid method of measurement points distribution is a combination of the uniform distribution ($N1$ points) and the curvature dependent ($N2$ points) such that $N1$ plus $N2$ makes the total number of measurement points, N. This is to avoid a situation in which too many measurement points are concentrated on the region of high curvature. The assignment of $N1$ and $N2$ of the N points distribution is done heuristically.

Path planning for measurement is based on the normal direction approach for improvement of accuracy on the sculptured surface. Let XS_i ($i = 1, 2,\dots, n$) be the target position for measurement on the surface, and $\mathbf{N}$ be the normal vector at the position. Collision avoidance can be satisfied if the probe approaches and draws back at a sufficiently large distanced from the surface in the normal direction. Thus the approach point XP_i can be calculated as:

$$XP_i = XS_i + d\,\mathbf{N} \qquad (4.11)$$

Figures 4.2a, b, and c show a typical example of the three methods of measurement points distribution for a sculptured surface: uniform distribution, curvature dependent distribution, and the hybrid of the two.

Measurement Points Sampling and Path Plan for Basic Features

The measurement points sampling for basic features such as planes, lines, and circles aims at uniform coverage for the workpiece.[8] The measurement points are chosen to give a true representation of the geometric features. However, the distribution should not be so regular as to follow possible systematic or periodic deformations of the feature.

For *line features*, the line segment is divided into N subintervals of equal length. The measurement point is chosen randomly in each subinterval in order to achieve a nearly uniform distribution of N points on a line segment.

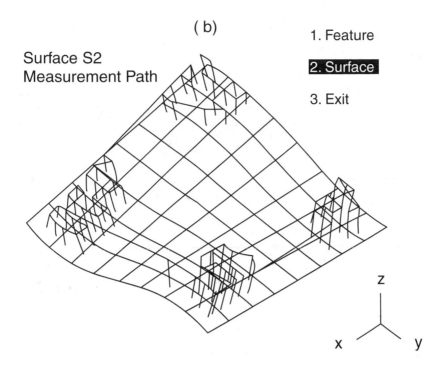

FIGURE 4.2(b) Curvature dependent distribution method.

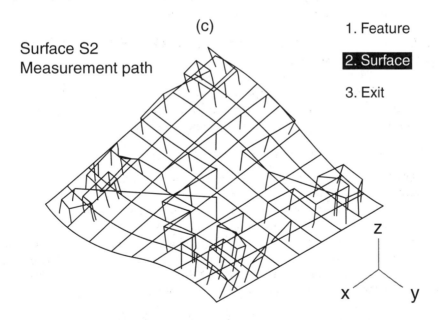

FIGURE 4.2(c) Hybrid distribution method.

For *plane features*, the rectangle is divided into $N1$ by $N2$ subrectangles, where $N1 \times N2$ is approximately equal to the total number of measurement points N in order to achieve a nearly uniform distribution of N points on the rectangular segments of a plane. Measurement points are chosen randomly in every subrectangle so as to avoid possible measurement errors due to systematic or periodic deformation on the feature. A uniform pseudo random generator can be used for this purpose.

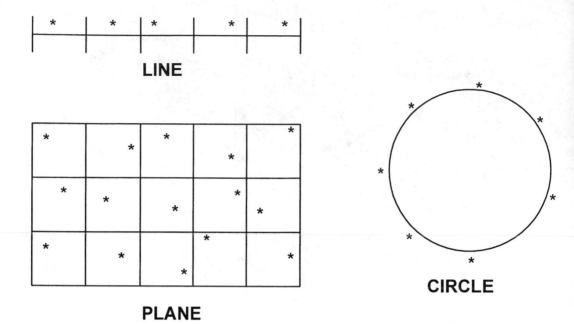

FIGURE 4.3   Measurement point distribution for basic features.

For *holes* and *circles*, the prime number distribution is performed for each circle, in order to detect the possible lobes on the circle. A pseudo random number generator can be used. Figure 4.3 shows a typical measurement points distribution on the line, plane, and circle, respectively.

4.4   Alignment of Coordinate Systems, DMIS Code Generation, and Measurement Operation

In order to perform the CAD-directed inspection successfully, the coordinate alignment has to be performed between the CAD coordinate system and the CMM coordinate system. Two phases of coordinate alignment procedures, the rough phase and the fine phase, are explained below.

Rough Phase Alignment Procedure

Rough phase alignment uses the conventional six points method, which is based on six points probing around the clear cut surfaces of a reference square block. Consider (X, Y, Z) as the CMM coordinate system, and (X', Y', Z') as the CAD coordinate system as shown in Figure 4.4. In this figure, let $A_1(x_1, y_1, z_1)$, $A_2(x_2, y_2, z_2)$, $A_3(x_3, y_3, z_3)$ be the probing points data on the plane 1; $A_4(x_4, y_4, z_4)$, $A_5(x_5, y_5, z_5)$ the probing points data on the plane 2, and $A_6(x_6, y_6, z_6)$ the probing point data on the plane 3. Thus the probing points data, A_1 to A_6, are obtained as the location of probe ball center in the CMM coordinate system when the probe ball contacts the respective planes. When $X'(a_x, b_x, c_x)$, $Y'(a_y, b_y, c_y)$, and $Z'(a_z, b_z, c_z)$ are the base vectors along the CMM coordinate axis, the base vectors can be calculated as

$$Z' = (A_2 - A_1) \times (A_3 - A_1)/|(A_2 - A_1) \times (A_3 - A_1)|$$
$$= (a_z, b_z, c_z) \tag{4.12}$$

$$Y' = (A_5 - A_4) \times Z'/|(A_5 - A_4) \times Z'|$$
$$= (a_y, b_y, c_y) \tag{4.13}$$

$$X' = (Y' \times Z')/|Y' \times Z'|$$
$$= (a_x, b_x, c_x) \tag{4.14}$$

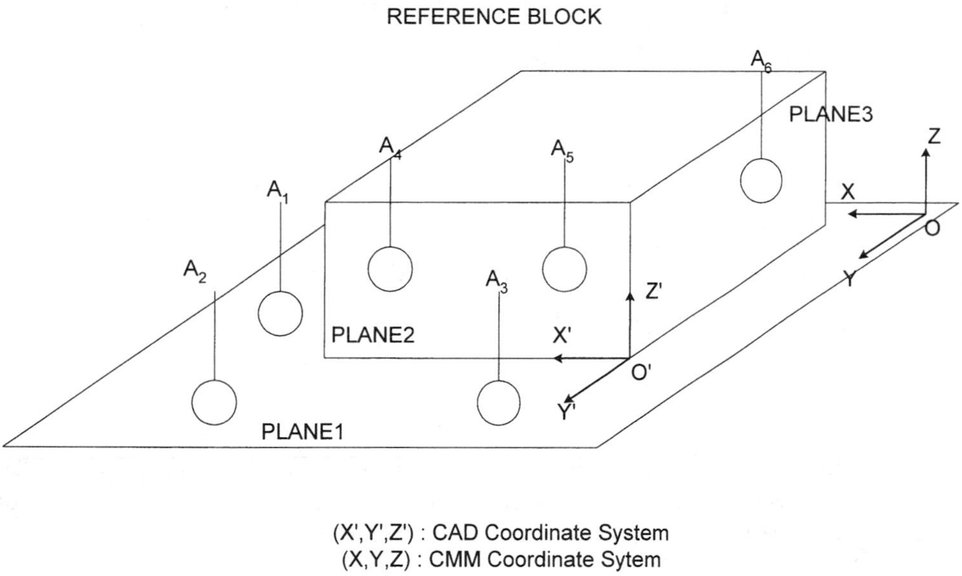

REFERENCE BLOCK

(X',Y',Z') : CAD Coordinate System
(X,Y,Z) : CMM Coordinate Sytem

FIGURE 4.4   Rough phase alignment based on 6 points probing.

When the base vectors of the CMM coordinate axis are known, the probing point data, A_1 to A_6, can be transformed to the contact point data, A_1' to A_6', considering the probe ball radius as

$$
\begin{aligned}
A_1' &= A_1 - r(a_z, b_z, c_z) \\
A_2' &= A_2 - r(a_z, b_z, c_z) \\
A_3' &= A_3 - r(a_z, b_z, c_z) \\
A_4' &= A_4 - r(a_y, b_y, c_y) \\
A_5' &= A_5 - r(a_y, b_y, c_y) \\
A_6' &= A_6 + r(a_x, b_x, c_x)
\end{aligned}
\tag{4.15}
$$

where r is the effective radius of the probe ball.

In order to complete the rough phase coordinate alignment, the origin of the CAD coordinate system with respect to the CMM coordinate system is also required. The origin, $O'(O_x', O_y', O_z')$ of the CAD coordinate system can be evaluated by observing that $O'A_1'$, $O'A_4'$, and $O'A_6'$ vectors are perpendicular to the Z', Y', and X' axes, respectively. Thus,

$$
\begin{aligned}
O'A_1' \cdot Z' &= (A_1' - O') \cdot (a_z, b_z, c_z) = 0 \\
O'A_4' \cdot Y' &= (A_4' - O') \cdot (a_y, b_y, c_y) = 0 \\
O'A_6' \cdot X' &= (A_6' - O') \cdot (a_x, b_x, c_x) = 0
\end{aligned}
\tag{4.16}
$$

where A_1' to A_6' are the coordinates of the contact points around the reference block. Thus the origin point $O'(O_x', O_y', O_z')$ of the CAD coordinate system can be obtained from Equation 4.16.

The relationship is therefore established between the CMM coordinate and the CAD coordinate and can be expressed using the transformation matrix T_1 such that $M = T_1 D$ is satisfied. In the matrix expression, **M**, **D** indicate CMM coordinate and CAD coordinate, respectively, and they are

$$\mathbf{M} = \begin{bmatrix} M_x \\ M_y \\ M_z \\ 1 \end{bmatrix} \qquad \mathbf{D} = \begin{bmatrix} D_x \\ D_y \\ D_z \\ 1 \end{bmatrix}$$

$$T_1 = \begin{bmatrix} a_x & a_y & a_z & O'_x \\ b_x & b_y & b_z & O'_y \\ c_x & c_y & c_z & O'_z \\ 0 & 0 & 0 & 1 \end{bmatrix}$$

(4.17)

where the T_1 matrix of 4×4 is considered for the rotation/translation movement between the two coordinate systems. For the rotation movement between the two coordinate systems, the reduced transformation matrix, RT_1 of 3×3, is considered. The components are

$$RT_1 = \begin{bmatrix} a_x & a_y & a_z \\ b_x & b_y & b_z \\ c_x & c_y & c_y \end{bmatrix}$$

(4.18)

Once the relationship between the two coordinate systems based on the rough phase alignment has been set up, the CNC code for the inspection can be obtained by multiplication of the T_1 matrix and the CAD coordinate data of measurement points. That is

$$\mathbf{M} = T_1(\mathbf{D} + r\mathbf{N})$$

(4.19)

where **D** is the CAD coordinate, r is the probe radius of the CMM, and **N** is the unit vector at the target points defined by Equation 4.3. $\mathbf{D} + r\mathbf{N}$ is used for considering the outward offset points from the measurement target points. The approaching direction of CMM, **DM**, is also calculated from multiplication of the RT_1 matrix and the normal vector components on the measurement target points. Thus

$$\mathbf{DM} = RT_1\mathbf{N}$$

(4.20)

Therefore the rough phase of alignment is performed and the CNC code is generated in a widely accepted machine code format such as DMIS (Dimensional Measuring Interface Specification),[13] and downloaded into CMM. The initial measurement procedure is followed on the curved parts.

Fine Alignment Procedure Based on the Iterative Least Squares Technique

Although the rough phase alignment is performed using the six points probing around the reference block, the CMM often fails to measure the parts of thin geometry such as edges because of residual misalignment error due to the form error of the reference block, etc. The residual misalignment error can be calculated and compensated for by introducing a second transformation matrix, T_2, based on initial measurement data for the curved surfaces. When the measured data **MM** of the curved surfaces are converted to the CAD coordinate system, they can be compared with the nominal CAD data, resulting possibly in a slight deviation between them. Thus, the second transformation matrix, T_2, can be introduced to link the deviation between the nominal CAD data and the converted measurement data. The measured data can be converted to the nominal CAD coordinate system, by multiplying T_1^{-1} to the measured data, **MM**. Thus, the relationship can be formed as,

$$T_1^{-1}MM = T_2(D + rN) \qquad (4.21)$$

The T_2 matrix of 4×4 can be determined by the least squares technique, minimizing the sum of squares of distance between the converted measurement data $(T_1^{-1}MM)$ and the nominal CAD data $(D + rN)$. The sum, E, is

$$E = S|T_1^{-1}MM - T_2(D + rN)|2 \quad \text{for all measurement points} \qquad (4.22)$$

The variational principle can be applied to solve the components of T_2 matrix. A convenient method for the evaluation of T_2 matrix is to use the pseudo-inverse technique for computational efficiency. Therefore, T_2 matrix can be calculated as follows

$$T_2 = T_1^{-1}MM(D + rN)T[(D + rN)(D + rN)T] - 1 \qquad (4.23)$$

where MM is the CMM measurement data for the surface, and $(\ )T$ indicates the transposed matrix. The calculated T_2 matrix is now used to generate the updated CNC codes downloaded into the CMM, and the second measurement data, MM(2), are obtained. When the generated measurement path still fails to measure the thin section of parts, further iterative procedures can be introduced.

Based on the second measurement data $MM(2)$, the second iterative $T2(2)$ matrix can be evaluated in that case. Prior to going to further iterations, ERR, the sum of squares of distance can be used as a criterion. That is,

$$ERR \pounds TOL \qquad (4.24)$$

where ERR = sum of squares of distance

$$= \sum |T_1^{-1}\mathbf{MM}^{(k)} - T_2^{(k)}(\mathbf{D} + r\mathbf{N})|^2 \qquad (4.25)$$

and $\mathbf{MM}^{(k)}$ is the *kth* iterative measurement data, and $T_2^{(k)}$ is the *kth* iterative T_2 matrix. When the criterion is not met, further iterations can be performed, and finer alignment procedures can be followed. Figure 4.5 shows the flowchart for a typical alignment system using the rough- and fine-phase alignments. The illustrated alignment technique can be applied to practical manufacturing of turbine blades, for example. A turbine blade is modeled into four features such as two chord-length-based cubic spline curves and two very small edge circles of around 0.15 mm radius. After locating on a CMM, the turbine blade is probed around the reference block by the six points based on the rough alignment procedure. The transformation matrix T_1 of the rough phase alignment is calculated, and initial measurement procedures are performed on the pressure/suction surfaces. Based on the measurement data, the T_2 transformation matrix is calculated using the least squares technique. The sum of squares of distance are calculated as a criterion for further iterative measurement procedures. Figure 4.6 shows a typical distance deviation between the nominal CAD data points and the measured points. It is noteworthy that, in this figure, the sum of squares of distance is greatly reduced at the first iteration, and further iterations are not needed. Figure 4.7 shows a typical measurement path before and after the fine phase alignment; the measurement path is slightly changed so as to measure the thin edges after the fine alignment procedure.[12]

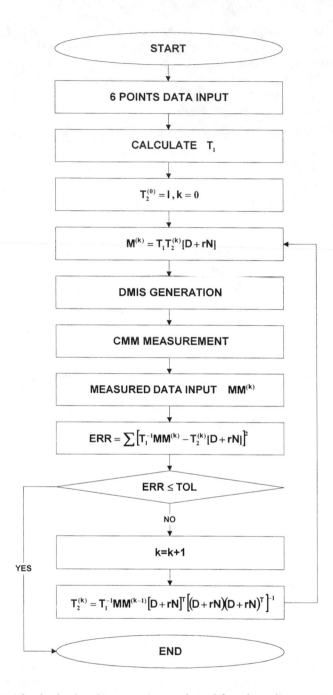

FIGURE 4.5 Flowchart for the developed system using rough- and fine-phase alignment.

CNC Code in DMIS Format

The calculated measurement path plan is generally prepared in a proper CNC code format such as DMIS. DMIS (Dimensional Measuring Interface Specification) code[13] is a widely accepted machine code for bidirectional communication of inspection data between computer systems and inspection equipment. The specification is a vocabularary of terms establishing a neutral format for inspection programs and inspection results data. The DMIS vocabularary is similar in syntax to the APT NC programming language, with major and minor words separated by slashes. There are two basic types of DMIS

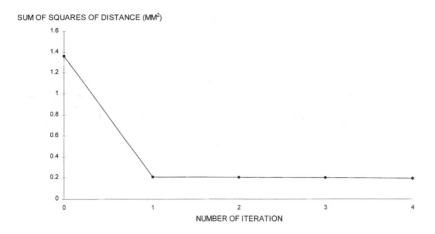

FIGURE 4.6(a)   Sum of squares distance for fine-phase alignment vs. iteration number.

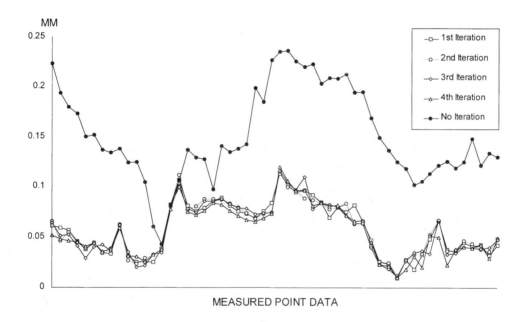

FIGURE 4.6(b)   Distance deviation between the nominal and actual points.

statements: process-oriented commands and geometry-oriented definitions. The process oriented commands consist of motion and machine parameter statements, as well as other statements unique to the inspection process itself. DMIS vocabulary consists of ASCII characters which are combined to form words, labels, parameters, and variables. Table 4.2 shows a typical CNC code for measuring a turbine blade in DMIS format.

4.5   Error Evaluation

In this section, the measurement results are fed back to the computer and error analysis is performed for the features inspected. The error evaluation algorithms are explained for the sculptured surfaces and the basic features.

(a)

(b)

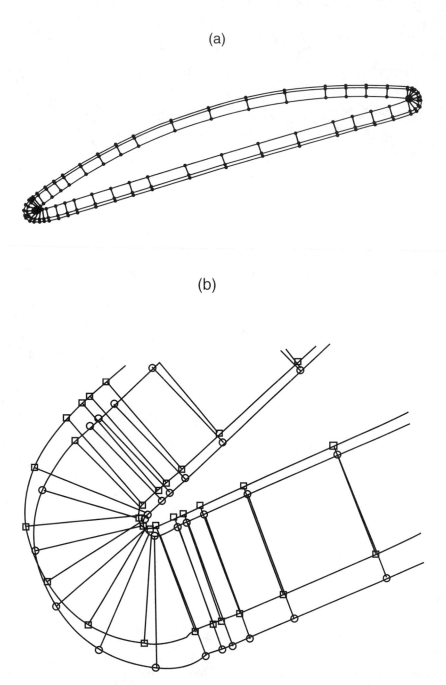

FIGURE 4.7 Typical measurement paths for a turbine blade before and after the fine-phase alignment ($\diamond$: before fine phase alignment, O : after fine phase alignment). (a) Turbine blade airfoil, (b) leading edge.

ISO Profile Tolerance

The error calculation is based on comparison of the measured data with the CAD data. The profile tolerance of the surface is defined in ISO as "the tolerance zone which is limited by two surfaces enveloping spheres of diameter, t, the centers of which are situated on a surface having the true geometrical form."[9] Figure 4.8a shows the ISO-defined profile tolerance for a surface. It is of metrological interest to note

(c)

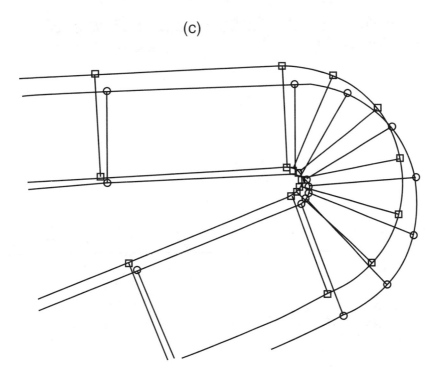

FIGURE 4.7(c) Trailing edge.

that the profile tolerance is the absolute maximum distance between the true geometrical form and the real surface measured, while the flatness tolerance is defined as the deviation between the maximum and the minimum distances from a mathematically defined reference plane. The profile tolerance definition can be applied to the case of CAD-integrated inspection, when the surface of true geometrical form can be replaced by the CAD surface.

Determination of Actual Measurement Points Considering Probe Radius

CMM measurement points are not usually identical to the planned points at the planning stage; thus measurement points on the surface have to be determined for the profile tolerance calculation. There are some reasons why CMM measurement points are not identical to the planned points, e.g., the discrepancy in the normal direction approach and the unavoidable error in the positional control of the CMM. In practice, the control accuracy of CMMs of the CNC type is much lower than the measurement accuracy. Therefore, the actual measurement points corresponding to the measured data have to be found on the sculptured surface, and a method for determination of the actual measurement points is desirable. A feasible method is introduced below.

As illustrated in Figure 4.8b, when the measurement probe touches the surface, the measured data represents the center point of the probe, which is located on the offset surface to the sculptured surface, where the offset length is equal to the effective radius of the probe. The actual measurement points of the probe center can be determined on the offset surface as the minimum distance point from the measured data.

In this figure, let $Q_i(i = 1, 2, ..., N)$ be a point on the CAD-defined surface, and R_i be the normal vector at the Q_i point, whose magnitude is the effective radius of probe R. Then the point on the offset surface can be represented as

$$\text{Points on the offset surfaces} = Q_i + R_i \ (i = 1, 2, ..., n) \tag{4.26}$$

where R_i equals to $r\,\mathbf{N}$, and r is the effective radius of the measurement probe and $\mathbf{N}$ is the unit normal vector.

TABLE 4.2   A Typical DMIS Format for Turbine Blade Measurement

```
DMISMN/'s111250.dmh'
V(1) = VFORM/ALL
DISPLAY/TERM,V(1),PRINT,V(1),STOR,DMIS,V(1)
FILNAM/'c0'
UNITS/MM,ANGDEC
PRCOMP/OFF
SNSET/APPRCH,2.000000
SNSET/RETRCT,3.000000
SNSET/SEARCH,1.000000
FEDRAT/MESVEL,MPM,0.100000
FEDRAT/POSVEL,MPM,0.500000
FINPOS/ON
F(0) = FEAT/GSURF
MEAS/GSURF,F(0),68
GOTO/−3.914192,−299.915039,381.086334
SNSLCT/S(3)
GOTO/46.363110,−223.356903,296.360962
PTMEAS/CART,49.0078,−220.3745,296.0281,0.6612,0.7456,  0.0832
PTMEAS/CART,49.0859,−220.4437,296.0279,0.6613,0.7455,−0.0830
PTMEAS/CART,49.2420,−220.5823,296.0275,0.6614,0.7455,−0.0825
PTMEAS/CART,49.6338,−220.9296,296.0265,0.6603,0.7466,−0.0812
PTMEAS/CART,50.4225,−221.6233,296.0242,0.6555,0.7511,−0.0784
PTMEAS/CART,51.2175,−222.3112,296.0216,0.6490,0.7571,−0.0754
PTMEAS/CART,52.0172,−222.9911,296.0188,0.6429,0.7626,−0.0721
PTMEAS/CART,53.6289,−224.3298,296.0132,0.6317,0.7725,−0.0654
PTMEAS/CART,55.2584,−225.6423,296.0072,0.6200,0.7824,−0.0584
PTMEAS/CART,56.9065,−226.9278,296.0009,0.6079,0.7924,−0.0509
PTMEAS/CART,58.5716,−228.1857,295.9941,0.5964,0.8015,−0.0431
PTMEAS/CART,68.9035,−235.2125,295.9449,0.5278,0.8493,0.0140
PTMEAS/CART,72.4716,−237.3643,295.9250,0.5053,0.8622,0.0368
PTMEAS/CART,76.0937,−239.4295,295.9036,04847,0.8725,0.0612
.........
.........
PTMEAS/CART,77.9260,−240.4320,295.8924,0.4732,0.8778,0.0739
PTMEAS/CART,78.3022,−240.6339,295.8901,0.4708,0.8789,0.0766
PTMEAS/CART,78.6787,−240.8349,295.8878,0.4688,0.8797,0.0792
PTMEAS/CART,79.0515,−241.0328,295.8854,0.4670,0.8805,0.0819
PTMEAS/CART,79.4166,−241.2259,295.8832,0.4654,0.8810,0.0845
PTMEAS/CART,50.4142,−218.5189,295.8688,−0.5410,−0.8375,0.0766
PTMEAS/CART,50.2405,−218.4068,295.8677,−0.5410,−0.8375,0.0775
PTMEAS/CART,49.7022,−218.2227,295.9388,−0.0751,−0.9972,0.0063
PTMEAS/CART,49.1415,−218.3221,295.9846,0.4104,−0.9110,−0.0396
PTMEAS/CART,48.6978,−218.6780,296.0216,0.7942,−0.6027,−0.0766
GOTO/41.549633,−213.253281,296.711426
GOTO/−3.914192,−159.915024,381.086334
SNSLCT/S(3)
GOTO/−3.914192,−299.915024,381.086334
GOTO/42.843388,−226.820557,296.575195
PTMEAS/CART,48.47813,−219.201523,296.039856,0.984173,−0.149599,−0.094996
PTMEAS/CART,48.533371,−219.766235,296.034912,0.93643,0.339085,−0.090116
PTMEAS/CART,48.850739,−220.237427,296.005402,0.661991,0.747058,−0.060607
GOTO/42.892818,−226.960953,296.550873
GOTO/−3.914192,−299.915024,381.086334
ENDMES
OUTPUT/F(0)
ENDFIL
```

(a)

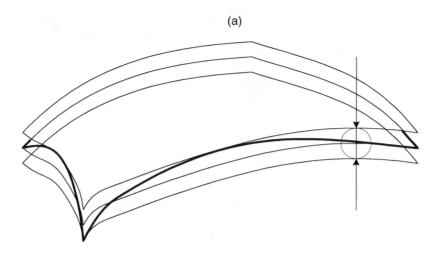

FIGURE 4.8(a) Profile tolerance of surface defined by ISO.

(b)

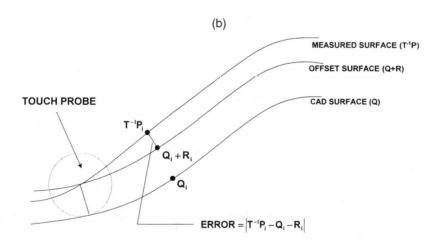

FIGURE 4.8(b) Deviation of sculptured surface.

Assume that P_i is the measured data which correspond to the point $Q_i + R_i$, on the offset surface, and there exists a transformation matrix T between the coordinates of the measured data (CMM coordinates) and the CAD coordinates, satisfying $P_i = T(Q_i + R_i)$. Thus P_i can be converted to the CAD coordinate data by $T^{-1}P_i$, and the Euclidean distance between P_i and $(Q_i + R_i)$ forms the deviation. Therefore, the Euclidean distance between the two points can be calculated as

$$\text{Euclidean distance} = \left| T^{-1}P_i - (Q_i + R_i) \right| \qquad (4.27)$$

In a practical measurement case, the CMM measured data, P_i, is obtained from the machine reading display and the corresponding point, $(Q_i + R_i)$, on the offset surface is found as the minimum distance point from the measured data. The subdivision technique can be used for finding the corresponding point, $(Q_i + R_i)$, as follows:

1. Divide the surface into 16 subsections at 1/4, 1/2, 3/4 and 1 location in terms of (u,v) parameters defining the surface.

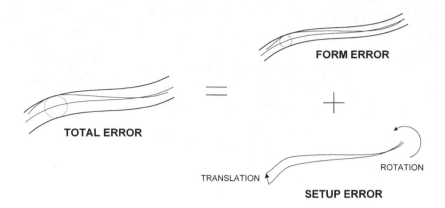

FIGURE 4.9   Form error and setup error for sculptured surface.

2. Calculate the Euclidean distance from the measured point to the points of 1/4 and 3/4 location of (u, v), respectively.
3. Assign the shortest distance point as the nearest point $(Q_i + R_i)$.
4. Choose the subsection having the shortest distance point for further subdivision.
5. Calculate $\delta = |(Q_i + R_i)^{(n)} - (Q_i + R_i)^{(n-1)}|$, where $(Q_i + R_i)^{(n)}$ is the currently determined nearest point and $(Q_i + R_i)^{(n-1)}$ is the previously determined nearest point.
6. Repeat steps 1 to 5 until the calculated δ is not greater than the assigned tolerance.

Profile Tolerance (Form Error) Evaluation

When the actual measurement points are found, the Euclidean distance can be calculated using Equation (4.27), and the dimensional errors for the parts can be obtained from the Euclidean distance data. There are two types of error in the sculpture surface affecting dimensional accuracy of the products: (1) The form error (profile tolerance), which is one of the characteristics unique to the sculptured surface, and (2) The setup error involved in tilting or translation of the sculptured surface. Generally, the form error and the setup error are combined to give the dimensional errors for the sculptured surface. The relationship between the profile tolerance (form error) and the setup error is illustrated in Figure 4.9. In the case of mold cavity manufacturing, the profile tolerance is unique to the machined cavity of the sculptured surface, and the setup error is associated with the adjustment of the cavity in the mold.

As an example of the metrological meaning of the setup error, consider the case of flatness analysis of a plane. A mathematically defined reference plane is derived so that the maximum deviation is minimum for all possible transformations, and thus the evaluated maximum deviation is unique to the plane. Similarly, a mathematically defined sculptured surface having a setup error from the CAD-defined sculptured surface can be determined so that the maximum Euclidean distance is minimum for all possible transformations.

For the given total error of sculptured surface, the form error and the setup error are dependent on the transformation matrix T because the Euclidean distance between the measurement points and the CAD-defined points is dependent on the T matrix as in Equation (4.27). The transformation matrix T is also dependent on the actual measurement points, $(Q_i + R_i)$. Thus recursive iteration can be used for determining the transformation matrix and the Euclidean distance. For a trial transformation matrix T the trial form error D can be calculated as the maximum distance between the measurement points and the points on the offset curve:

$$D = \max|T^{-1}P_i - (Q_i + R_i)| \qquad \text{for} \quad i = 1, 2, \ldots, n \qquad (4.28)$$

An optimal transformation matrix T can be found such that the trial form error D is minimum, which is the profile tolerance. Thus, the profile tolerance is

$$\text{Profile tolerance} = \min D \text{ for all possible transformation} \qquad (4.29)$$

The calculated minimum form error is the profile tolerance which is unique to the sculptured surface. The optimal transformation matrix giving the minimum form error then becomes the optimal setup.

In this section, an iterative algorithm, MINIMAXSURF[11], is explained using the Chebyshev norm between the measured points and the corresponding closest points. Again, let $\mathbf{P}_i$ be the measured points data, $\mathbf{Q}_i + \mathbf{R}_i$ be the calculated closest points on the offset surface defined by the CAD geometry, and T be the transformation matrix between the two data. The Chebyshev norm of power p, L_p, can be defined as

$$L_p = \left[\sum |T^{-1}\mathbf{P}_i - (\mathbf{Q}_i + \mathbf{R}_i)|^p \right]^{1/p} \tag{4.30}$$

The L_p can be minimized with respect to the transformation matrix T and the optimum transformation matrix T_0 can be found. The profile tolerance E is then

$$E = \max |T_0^{-1}\mathbf{P}_i - (\mathbf{Q}_i + \mathbf{R}_i)| \quad \text{for} \quad i = 1, 2, ..., N \tag{4.31}$$

The algorithm for the profile tolerance evaluation is implemented as

1. Set the iteration index k as 0, and the initial transformation matrix $T^{(0)}$ as the multiplication of T_1 and T_2 in case of the rough-phase and fine-phase alignment.
2. Input the measured point data $\mathbf{P}_i$ for $i = 1, 2, ..., n$, where n is the total number of measured points on the surface.
3. Assign the initial closest point $(\mathbf{Q}_i + \mathbf{R}_i)^{(0)}$ corresponding to the $\mathbf{P}_i$ as the measurement target point.
4. Calculate the initial profile error $E_o = \max |[T^{(0)}]^{-1}\mathbf{P}_i - (\mathbf{Q}_i + \mathbf{R}_i)^{(0)}|$ for $i = 1, ..., n$.
5. Increase the iteration index k by 1.
6. Find the closest point $(\mathbf{Q}_i + \mathbf{R}_i)^{(k)}$ corresponding to the transformed $[T^{(k-1)}]^{-1}\mathbf{P}_i$
7. Calculate the kth iterative transformation matrix, $T^{(k)}$, such that the Chebyshev norm, $[\sum |[T^{(k)}]^{-1} \mathbf{P}_i - (\mathbf{Q}_i + \mathbf{R}_i)^{(k)}|^p]^{1/p}$ be minimum using the universal alignment algorithm.[11]
8. Calculate the kth iterative profile error, $E^{(k)} = \max |[T^{(k)}]^{-1}\mathbf{P}_i - (\mathbf{Q}_i + \mathbf{R}_i)^{(k)}|$ for $i = 1, 2, ..., n$.
9. Evaluate the incremental maximum deviation, $DE_k = |E^{(k)} - E^{(k-1)}|$.
10. If the incremental maximum deviation is less than the assigned tolerance TOL then proceed to the end; otherwise increase k by 1.
11. Repeat steps 5 through 10 until the criterion is met.

The flowchart of the implemented algorithm is shown in Figure 4.10.

Error Evaluation for Basic Features

Flatness is the deviation of measured points with respect to the mathematically determined reference plane; the flatness error is evaluated as the distance between the maximum and minimum deviation. Squareness tolerance is the tolerance zone limited by two perpendicular planes to the given reference plane, where the two perpendicular planes contain the whole measured feature. The reference plane can be determined as the best fit plane by using the least squares technique. Parallelism tolerance is defined as the tolerance zone limited by two parallel planes which are parallel to the reference plane; the reference plane can be constructed as the best fit plane by using the least squares technique. The straightness tolerance is defined by the minimum distance between two parallel lines containing the measured datum. The slope of two parallel lines can be mathematically determined as the best fit line by using the least squares technique. The roundness tolerance is defined as the tolerance zone limited by two concentric circles; the circles can be calculated by the least squares circle (LSC), the minimum zone circle (MZC), the minimum circumscribed circle (MCC), and the maximum inscribed circle (MIC).

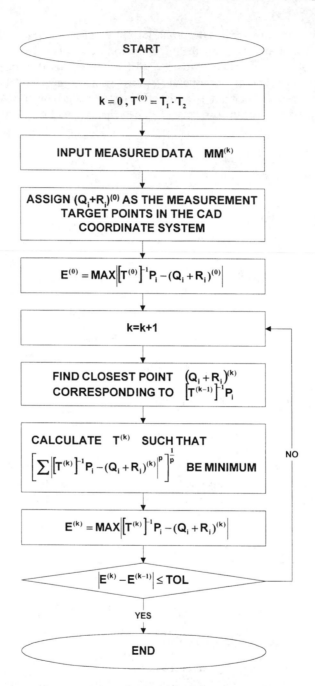

$$k = 0, T^{(0)} = T_1 \cdot T_2$$

INPUT MEASURED DATA $MM^{(k)}$

ASSIGN $(Q_i + R_i)^{(0)}$ AS THE MEASUREMENT TARGET POINTS IN THE CAD COORDINATE SYSTEM

$$E^{(0)} = MAX\left|\left[T^{(0)}\right]^{-1}P_i - (Q_i + R_i)^{(0)}\right|$$

$$k = k + 1$$

FIND CLOSEST POINT $(Q_i + R_i)^{(k)}$ CORRESPONDING TO $\left[T^{(k-1)}\right]^{-1}P_i$

CALCULATE $T^{(k)}$ SUCH THAT

$$\left[\sum_i \left|\left[T^{(k)}\right]^{-1}P_i - (Q_i + R_i)^{(k)}\right|^p\right]^{\frac{1}{p}}$$ BE MINIMUM

$$E^{(k)} = MAX\left|\left[T^{(k)}\right]^{-1}P_i - (Q_i + R_i)^{(k)}\right|$$

$$\left|E^{(k)} - E^{(k-1)}\right| \le TOL$$

NO

YES

END

FIGURE 4.10   Flowchart of the profile tolerance algorithm based on the iterative Chebyshev norm.

4.6   Practical Application to Real Products and Discussion

The explained system has been applied to real products, e.g., a mold having a bicubic B-spline surface with some basic features, a turbine blade having very thin features such as sharp edges.

A Mold Having a Bicubic B-Spline Surface with Some Basic Features[3]

A mold base having the dimension of 250 mm $\times$ 250 mm $\times$ 50 mm was selected as a test mold as shown in Figure 4.11a. A cavity block of a bicubic B-spline surface is inserted and 12 holes are drilled for guide

(a)

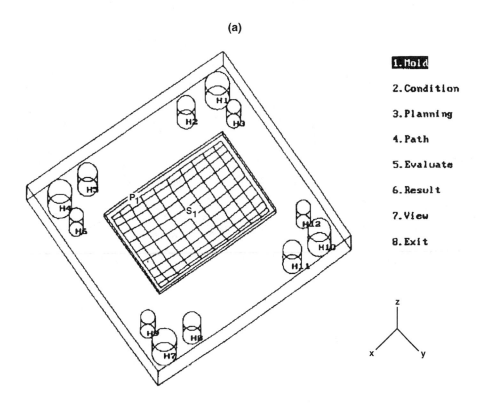

FIGURE 4.11(a)   A typical mold having a sculptured surface, block, and holes.

(b)

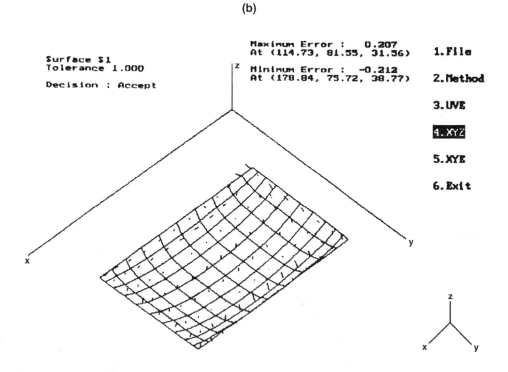

FIGURE 4.11(b)   Error analysis of a bicubic B-spline surface.

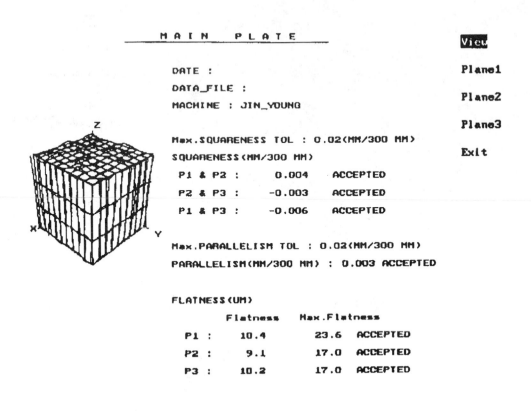

(c)

FIGURE 4.11(c)   Inspection results for a mold plate.

pins on the mold base. Following the inspection planning procedures, the measurement operations are performed and the errors are calculated. Figure 4.11b shows the error analysis of the sculptured surface: the form error is found as 0.423 mm, with the maximum error of 0.207 mm (under cut) and the minimum error of −0.212 mm (over cut). Figure 4.11c depicts the inspection results for the mold base feature, giving squareness, parallelism, and the flatness errors. Figure 4.11d shows the inspection results of a pocket for the cavity block, giving the straightness, parallelism, squareness error, and dimensional accuracy of the pocket.

A Turbine Blade Having Very Thin Features

The CAD data base of a turbine blade is assessed and the geometric features of the blade are reconstructed. Generally, a turbine blade consists of an airfoil fairing in the direction of air flow, a spanwise fairing in the perpendicular direction, and a base block as the reference. A 3-D representation of a turbine blade is shown in Figure 4.12a. Each airfoil section is then stacked in the spanwise direction with respect to the stacking point, where the stacking point is the reference point for generating the momentum of air flow. As shown in Figure 4.12b, the airfoil section is usually composed of several features: pressure curve, suction curve, leading edge, trailing edge, etc. The pressure/suction curves are generally modeled as the smooth spline curves, and the leading/trailing edges are modeled as circles, where the edges have very small radius compared to the pressure/suction curves. After the

(d)

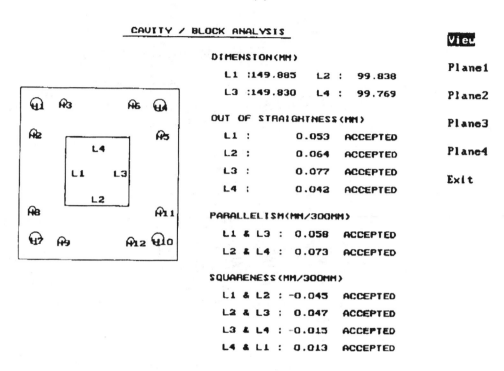

CAVITY / BLOCK ANALYSIS

DIMENSION(MM)

L1 :149.885   L2 : 99.838

L3 :149.830   L4 : 99.769

OUT OF STRAIGHTNESS(MM)

L1 :   0.053   ACCEPTED

L2 :   0.064   ACCEPTED

L3 :   0.077   ACCEPTED

L4 :   0.042   ACCEPTED

PARALLELISM(MM/300MM)

L1 & L3 : 0.058   ACCEPTED

L2 & L4 : 0.073   ACCEPTED

SQUARENESS(MM/300MM)

L1 & L2 : -0.045   ACCEPTED

L2 & L3 : 0.047   ACCEPTED

L3 & L4 : -0.015   ACCEPTED

L4 & L1 : 0.013   ACCEPTED

View

Plane1

Plane2

Plane3

Plane4

Exit

Number of measurement cycles

FIGURE 4.11(d) Inspection results for a pocket.

rough- and fine-phase alignment procedures, the turbine blade is measured by the CMM having DMIS features and analyzed in terms of the profile tolerance. Figure 4.12c shows the profile tolerance, i.e., maximum deviation, where the iteration number indicates the number of iterations for the transformation matrix calculation. When the power index of the Chebyshev norm increases, the rate of convergence decreases, and smaller values of profile tolerance are obtained as shown in Figure 4.12c. For a typical section of blade, the profile tolerance values for the pressure and suction surfaces are found to be 0.168 mm and 0.171 mm, respectively. The nominal profile tolerance is 0.200 mm for both sides. For the radius of the leading and trailing edges, the mean radius is calculated for both edges based on the measured data points, giving 0.148 mm and 0.154 mm, respectively, for the leading and trailing edges, while the nominal radii are 0.150 ± 0.05 mm for both edges.

4.7   Concluding Remarks

An integrated precision inspection system has been demonstrated for the manufacturing parts having CAD-defined features. For the inspection of a sculptured surface, three types of sampling point strategies have been shown: uniform distribution, curvature dependent distribution, and hybrid distribution. The line and plane features can be divided into subintervals, and the measurement points are distributed at random positions in the subinterval. Prime numbers of subintervals can be considered for the circle/hole features, in order to avoid possible periodic distortion of the

(a)

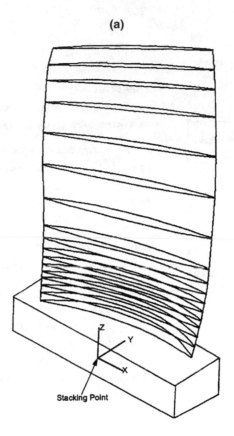

FIGURE 4.12(a)   Three-dimensional representation of a turbine blade.

(b)

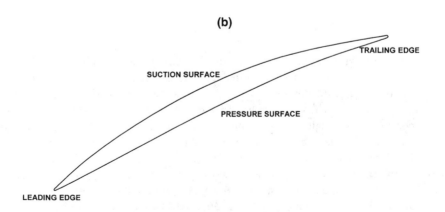

FIGURE 4.12(b)   A turbine blade's airfoil composed of spline and circle features.

measurement features. Two phases of alignment, rough and fine, have been introduced. The rough-phase alignment is based on the conventional six points probing on the clear cut surfaces; the fine-phase alignment is based on the initial measurement of the curved parts, using the least squares technique based on the initial measurement data feedback. Fine-phase alignment is especially useful for the measurement of parts having very thin geometry. It can also be concluded that the CAD/CAI integration technique is essential for the measurement and inspection of complicated parts having

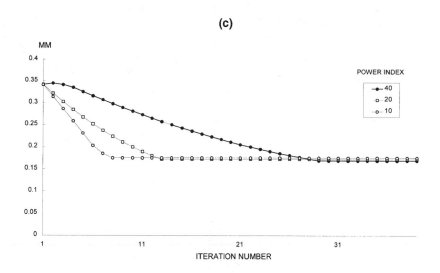

FIGURE 4.12(c)   Maximum deviation (profile tolerance) vs. iteration number.

CAD-defined curved or complicated geometry. For the analysis of profile tolerance of parts, a recursive algorithm has been introduced, based on the Chebyshev norm. As has been demonstrated, the integrated precision inspection method can provide very useful techniques for efficient quality control of parts in CAD/CAM/CAI environment.

References

1. M. Kang, S. G. Kim, and C. W. Lee, CIM for mold factory automation, *Annals of CIRP*, 39, 1985.
2. N. Duffie, J. Bollinger, R. Piper, and M. Kroneberg, CAD directed inspection and error analysis using surface patch databases, *Annals of CIRP*, 33, 1984.
3. N. Duffie, S. Feng, and J. Kann, CAD directed inspection, error analysis and manufacturing process compensation using tricubic solid databases, *Annals of CIRP*, 37, 1988.
4. O. P. Bojanic, D. V. Majstrovic, and R. V. Milacic, CAD/CAI integration with special focus on complex surfaces, *Annals of CIRP*, 41, 1992.
5. C. H. Menq, H. T. Yau, and G. W. Lai, Automated precision measurement of surface profile in CAD directed inspection, *IEEE Transactions on Robotics and Automation*, 8, 2, 1992.
6. M. E. Mortenson, *Geometric Modelling*, John Wiley & Sons, NY, 1985.
7. *Main Plates of Molds for Plastics*, Korean Industrial Standards, KSB 4151–1987, 1987.
8. *British Standard Guide to Assessment of Position, Size, and Departure from Nominal Form of Geometric Features*, British Standards Institution, BS7172:1989, 1989.
9. International Organization for Standardization, *Technical Drawings: Geometrical Tolerancing-Tolerancing of Form, Orientation, Location and Runout-Generalities, Definitions, Symbols, Indications on Drawing*, ISO 1101–1983, 1983.
10. H. J. Pahk, Y. H. Kim, Y. S. Hong, and S. G. Kim, Development of computer aided inspection system with CMM for integrated mold manufacturing, *Annals of CIRP*, 42, 1993.
11. G. Goch and U. Tschudi, A universal algorithm for the alignment of sculptured surfaces, *Annals of CIRP*, 41, 597–600, 1992.
12. H. J. Pahk and W. J. Ahn, Precision inspection system for aircraft parts having very thin features based on CAD/CAI integration, *Int. J. Adv. Manuf. Technol.*, Vol. 12, 442–449, 1996.
13. Dimensional Measuring Interface Specification, Version 2.1, Computer Aided Manufacturing-International, Inc., 1989.

5

Computer-Aided Process Planning (CAPP) in Manufacturing Systems and Their Implementation

P. Nageswara Rao
MARA Institute of Technology

Siva R.K. Jasthi
*Structural Dynamics
Research Corporation*

5.1   Introduction

Developments in the information age have caused the use of computers to spread rapidly throughout the manufacturing arena. With the cost of computing going to ever low levels and increases in its capability going by leaps and bounds, the use of computers has become increasingly important to the manufacturing industries. To this extent, great strides are taking place in the use of computers in design and manufacturing worldwide.

A process planning function establishes the methods and means of converting raw material to a finished part. Thus process planning logically forms a link between the design and manufacturing functions. Computer-aided process planning (CAPP) is the application of computers to assist process planners in the planning function. In this chapter a brief presentation is made of the technologies involved in developing a CAPP system for rotational parts, along with a case study of *GIFTS—a generative, interactive, feature-based, and technology-oriented system*.

5.2   Process Planning

A process is a method by which products can be manufactured from raw materials. Process engineering takes place directly after product engineering has completed the design of a product. From the information received, it creates the plan of manufacture. Process planning is, then, the function of determining exactly how a product will be made to satisfy the requirements specified at the most economical cost.

The importance of a good process plan cannot be over emphasized, particularly for mass production. A few minutes used to correct an error during process planning can save large costs that would be required to alter the tooling or build new tooling. The process plan created must permit the manufacture of a quality product at the lowest possible cost.

The output from the process planning function can be itemized as follows:

1. To determine and select the equipment needed to manufacture the part
2. To determine the order or sequence of operations necessary to manufacture in terms of operation routing, process details and process pictures
3. To specify production tolerances on blanks and on auxiliary surfaces
4. To specify the process parameters for the various manufacturing operations selected
5. Providing the necessary documentation to be used by the shop people.

A typical process plan is shown below with the corresponding part shown in Figure 5.1.

Op. No.	Operation	Tools Used
10	Turning to length and facing the ends	Facing tool, chuck
20	Make the center hole	Chuck, center drill
30	Rough turning of ϕ20 diameter	Rough turning tool
40	Finish turning of ϕ20 diameter	Finish turning tool
50	Forming the radius at one end	Form tool (radius turning)
60	Diamond knurling of the handle	Knurling tool
70	Reverse the part in the chuck and rough turn to ϕ10	Rough turning tool
80	Finish turning of ϕ10 size	Finish turning tool
90	Make the 3 mm radius groove	Form tool
100	Rough turning of ϕ14 diameter	Rough turning tool
110	Finish turning of ϕ14 diameter	Finish turning tool
120	Cutting the external threads M14	Thread cutting tool

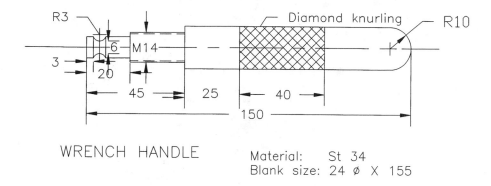

FIGURE 5.1 Typical part drawing whose process plan is given.

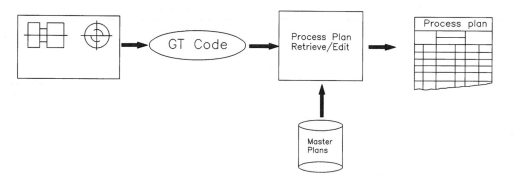

FIGURE 5.2 Variant approach to CAPP.

5.3 Computer-Aided Process Planning

Process planning establishes the methods and means of converting the raw material to a finished part in a manufacturing facility. Computer-aided process planning (CAPP) is the application of computers to assist the process planners in the planning function. There are two basic approaches to CAPP: variant and generative, which are briefly discussed below.

Variant Approach

The variant approach, which is also called retrieval approach, uses a group technology (GT) code to select a generic process plan from the existing master process plans developed for each part family and then edits to suit the requirement of the part (Figure 5.2). Variant approach is commonly implemented with GT coding system. Here, the parts are segmented into groups based on similarity and each group has a master plan.

Several CAPP systems based on the variant approach such as DISAP, TOJICAPP (Zhang and Gao, 1984) etc., are reported in literature. However, this approach is impractical in situations where small batches of widely varying parts are produced. Moreover, this method fails to capture the real knowledge or expertise of process planners, and there is a danger of repeating mistakes from earlier plans that were stored in the data base (Shah, 1991).

Generative Approach

In a generative approach, a process plan for each component is created from scratch without human intervention. These systems are designed to automatically synthesize process information to develop a

Part Descriptive System

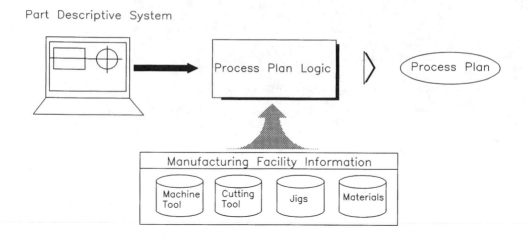

FIGURE 5.3   Generative approach to CAPP.

process plan for a part (Figure 5.3). These systems contain the logic to use manufacturing data bases and suitable part description schemes to generate a process plan for a particular part. Most of the contemporary CAPP systems being developed are generative in nature. Generative approach eliminates disadvantages of variant approach and bridges the gap between the CAD and CAM.

5.4   CAPP: Implementation Techniques

Logical decision is a traditional implementation technique used in CAPP. The simplest approach is to code the process capability in a computer program. A tree-structured classification can be used in the system and each process can be coded as a branch of a decision tree (Chang and Wysk, 1984).

The decision logic should be present in a format that is easy to visualize and check for completeness, contradictions, and redundancy. Generally, each manufacturing process is defined as a separate entity based on the capabilities to generate or modify geometric features or properties. Only values of decision variables change over time or between companies. The techniques for structuring the decision logic are numerous and varied.

The objective of decision logic in a CAPP system is to match the process capabilities with design specifications in an optimal way. Generally, most common decision logic can be classified as one of the three methods: decision tables, decision trees and AI.

Decision Tables

A decision table is partitioned into conditions and actions and is represented in a tabular form. It is a program-structuring tool which provides readable documentation as an automatic by-product. Also, decision tables can be used with preprocessors to eliminate some program coding, and to provide automatic checks for completeness, contradiction, and redundancy.

Decision Trees

A decision tree is a graph with a single root and branches emanating from the root. Decision trees are easier to customize, update, maintain, visualize and develop than decision tables. Decision trees can be represented as computer codes or data. The tree as computer code is converted to a flow chart; the starting node is the root and every branch represents a decision statement which is either false or true.

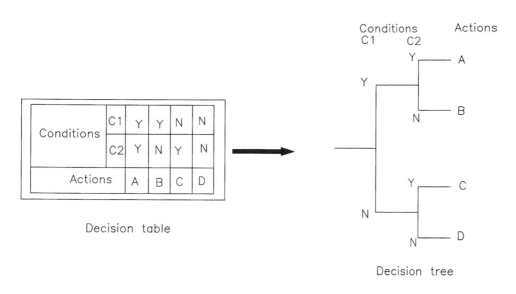

FIGURE 5.4 Decision table and tree as used in CAPP. (From Shah, 1991).

Compared to decision tables, decision trees can be more easily updated and maintained. The flexibility in expanding and contrasting the tree, if necessary, is another advantage. Figure 5.4 illustrates the decision table and tree.

APPAS (Chang and Wysk, 1981) is a good example for a decision tree application. One system that employs decision trees is AUTOPLAN (Vogel and Adard, 1981). TIPPS (Chang and Wysk, 1984) employs decision trees combined with AI techniques.

Expert System Techniques (AI)

The solution to the process planning task depends mainly on the empirical knowledge relevant to the organization, based upon the existing facilities. The popularity of expert systems in CAPP is due to this qualitative, subjective, imprecise and company-specific nature of the process planning knowledge. The expert systems designed to cope with such knowledge characteristics are also much easier to modify and customize than the fixed logic conventional systems, because the knowledge in expert systems is explicitly represented and segregated from the planning (inference) mechanism.

In general, problems in a production system formalism can be represented by an initial state, a goal state, a set of operators, and a control structure. One of the frequently used methods to solve problems in AI is the theorem proving technique. Using this technique, we can proceed from the initial state to the goal state. Proving that the goal state can be reached from the initial state will involve the applications of the operators, thereby providing a solution to the problem.

The process planning problem falls into this category. In the part manufacturing problem, the initial state is the raw material or workpiece from which the part is to be produced and the goal state is the finished part. The set of operators comprises the available machine tools, cutters, etc. Processing of the part involves proceeding from the raw material to the finished part. Application of the operators (machine tools and cutters) moves the problem from one state to another. The various stages represent the workpiece in work-in-progress condition:

$$S_{\text{initial}} \to S_1 \to S_2 \to \ldots \to S_{\text{final}}$$

These stages are non-reversible as once the material is removed, it cannot be added back. However, in process planning, because the actual material removal has not been done, the stages can be reversed,

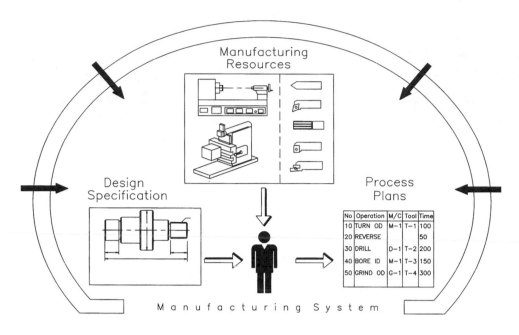

FIGURE 5.5   Conventional process planning in a manufacturing system.

leading to the development of alternative process plans. This strategy is used in the development of expert CAPP systems.

5.5   Modeling for CAPP

In the conventional manufacturing system, a process planner studies the engineering drawing for deciding the plan for its manufacture based on (a) the geometrical and technological constraints present in the part specification and (b) the manufacturing resources available on the shop floor (Figure 5.5). Thus the engineering drawing can be considered the communication link between the design and planning functions, while the process plan forms the link between the planning and manufacturing functions.

The engineering drawings, manufacturing resources (machine and tool specifications), and process plans meant for interpretation by humans using their cognitive skills do not fit in the scheme where computers are used for automating the process planning function. Analogously, the development of a CAPP system thus involves the modeling of the following elements (Figure 5.6):

1. Design specification: The modeling of the design specification involves the representation of the part to be manufactured. The resultant part model establishes the link between design and CAPP functions. This model can be called the *Part Design Internal Representation* (PDIR).
2. Manufacturing resources: Preparation of the process plans is always guided by the status of manufacturing facilities on the shop floor. The information about these resources should be made available to the CAPP system in order to generate realistic plans. The model representing the manufacturing resources can be termed the *Manufacturing Resources Internal Representation* (MRIR).
3. Process plan: Modeling of the process plan involves the representation of the manufacturing instructions in a structured form. This model can be called the *Process Plan Internal Representation* (PPIR).

CAPP can be thought of as the modeling of the above elements and the interaction between these. The following sections describe the various methods of modeling for each of these elements.

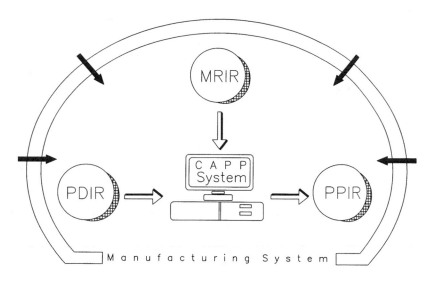

FIGURE 5.6 Interaction of various elements in computer-aided process planning (CAPP). MRIR: manufacturing resources internal representation; PDIR: part design internal representation; PPIR: process plan internal representation.

5.6 Part Modeling for CAPP

One of the problems concerning the automation of the process planning function is that of high input effort needed to describe a part. The reason can be attributed to the fact that the experience and knowledge of humans play a prominent role in interpreting and communicating the part data in a manufacturing system. The translation of the part data from human-interpretable format to computer-interpretable format is complex, but mandatory input for CAPP.

Part modeling, through which a complete and unambiguous definition that captures the design content of the part is to be achieved, has become one of the key research issues since the inception of CAPP. There exist three basic sets of data which completely describe the design content of a part:

- **Geometrical data:** The geometrical data give the basic description of the shape. For example, diameter of a hole, depth of a groove, width of a keyway, etc. constitute this type of data.
- **Technological data:** The information pertaining to tolerances and surface finish can be referred to as technological data, e.g., circularity, diametrical tolerance, runout, etc.
- **General (or global) data:** Certain global characteristics that are applicable to the part as a whole are often added to the design specification. These global attributes include quantity to be produced, work material, design number, part name, and other task-dependent details which normally appear on the drawing and the process plan.

Representation of Geometrical Details

There are many methods followed for part modeling as a part of CAPP system development. These can be categorized as (1) non-CAD models, (2) CAD models, and (3) feature-based models.

Non-CAD Models

These modeling methods are characterized by the absence of CAD systems. These can be classified as

- *Group technology (GT) coding* which is based on GT coding schemes for the retrieval of existing process plans. Detailed process plans for several part families are developed and stored in the system's data base. The major characteristics of the new part are matched with a similar part and the previously stored process plan is retrieved. The retrieved process plan can then easily be modified to suit the new part.

- *Interactive/menu driven models* which pose a series of questions (or menus) and interactively gather the part data from the user. Studying the product range and variation is essential before designing system-user interaction. Although it seems to be simple to answer such questions, the user may not have control over the way in which the system interacts with him, and hence he is compelled to navigate through a series of redundant questions to define even a simple part. This approach is limited to only simple parts and is characterized by the absence of graphic interface.
- *Keywords/description languages* which can provide detailed information for CAPP systems. The output format of the part data can be designed such that the process planning function can easily accomplish its task.

This approach offers limited flexibility to the user in the sense that the user can exercise some control over the system, but, because of its specialized nature, it is limited to simple parts. For complex parts, the translation of the original design to input language can be very tedious. Sometimes the resultant model may not be unique and the sequence with which a part is modeled may affect the planning logic.

CAD Models

Although the modeling methods discussed so far provide alternate solutions for linking design to manufacture, the main problem of generating the part data from a CAD data base is still largely unresolved. The increased use of CAD and the need to integrate it with CAM has led to the use of the internal CAD model as a means of driving the process planning function.

Much of the information needed in the design and manufacturing functions is directly related to the geometric shape of the part. This observation has led to the interest in geometric modeling, with the hope that a solution for the geometrical aspects of product modeling will form the basis for producing effective CAD/CAM systems. Present CAD/CAM Systems can be regarded as the logical outcome of this approach.

Unfortunately, the information stored in the CAD models is in terms of points, lines, arcs, and solids; as such it is not structured to facilitate CAPP. The gap in the abstraction levels of CAD and CAM domains needs to be bridged to obtain integrated CAD/CAM systems (Wright and Hannam, 1989). CAD systems also lack suitable facilities to associate tolerance information and other application-based data (non-geometrical) to the part model. The information in these models is incomplete and low level in nature (Shah and Rogers, 1988b). Hence such models cannot be used directly without further processing for manufacturing applications like CAPP.

Another major problem with these modelers is that they allow a very restricted set of operations on the model. All operations must be expressed as Boolean operations. The primitives customarily supported in solid models do not easily lend themselves to the specification of user-definable features. This is particularly true for compound features such as stepped holes, pockets, etc. The steps involved in describing such features may be quite complex.

Sometimes the model created by the solid models may not be unique because a part constructed by subtraction of operations can also be created using the corresponding union operations (Joshi et al., 1986). Absence of such uniqueness in part representation may severely affect the subsequent process planning logic. Sometimes, a part represented in a CAD system may not be manufacturable. Hence, it is essential to have modeling systems that can check the design for manufacturability (DFM) and geometrical validity of the parts.

Feature-Based Models

In order to overcome the above limitations, the concept of using form features (shape elements) for part modeling received the attention of researchers. It is perhaps fair to state that the concept of features was first introduced by researchers in the process of linking design, and manufacturing.

A feature is a geometric shape specified by a parameter set, which has special meaning to design or manufacturing engineers. Features represent a collection of entities in an intelligent form (like hole, slot, thread, groove, etc.) and hence provide information at a higher conceptual level. The use of such groups of geometry, coupled with the necessary information needed for the process planning function, is seen as a practical means of linking the design and manufacturing (Drake and Sela, 1989; Klein, 1988; Clark and South, 1987; Butterfield et al., 1986).

The attractive prospect in the adoption of form-features is the elimination of exhaustive postprocessing of the part data. Because of the compatibility between the feature-data and the application-data, the development of CAPP systems and other related applications (such as checking the part model for DFM, geometrical validity, etc.) also become relatively easy.

Unfortunately, attempts to define the precise nature of features are fraught with difficulty because of the wide interpretation placed upon the term by different researchers. Features originate in the reasoning processes used in various design, analysis, and manufacturing activities and are strongly associated with application domains (Cunningham and Dixon, 1988; Case and Gao, 1993; Kang and Nnaji, 1993). Researchers are realizing that the feature definition is relative, depending on such factors as application, context, state of the product, structure, and configuration. When representing product knowledge in terms of features, application domains do not stand alone.

Several informal definitions of feature can be found (Wilson and Pratt, 1988; Dixon et al., 1987). However, none of these definitions is entirely satisfactory because the feature definition depends on the application for relevance; therefore it can be surmised that there can be as many definitions as applications.

Though there is no universal definition for a feature, the following views are of interest to CAPP (Shah, 1991):

- From the manufacturing point of view, features represent shapes and technological attributes associated with manufacturing operations and tools.
- From the geometric modeling point of view, features are groupings of geometrical or topological entities that need to be referenced together.
- From the design point of view, features are elements used in generating, analyzing, or evaluating designs.

Three important definitions of features adopted in the modeling systems of CAPP systems are shown in Figure 5.7 and are briefly described below. It can be seen that the amount of information attributed to the feature is increasing with each definition.

Surface Generated after Machining

In this method, the feature represents the final specification of the machined region. For example, consider a through hole (Figure 5.7a). Here, the hole surface and its attributes together define the feature. These are called surface features.

Surface features, however, do not relate explicitly to the extent of the material to be removed. In spite of that, the reason for selecting the surface features is the simplicity they offer during the part modeling (design) stage. The designer can define the feature as given in the part specification and gives little consideration to the machining aspects (e.g., operations, material removal volumes, etc.) while defining a feature, since these are supposed to be the tasks of CAPP. This method is widely followed in CAPP systems dealing with rotational parts. Examples include TECHTURN (Hinduja and Barrow, 1986), CIMROT (Domazet and Lu, 1992), and GIFTS (Jasthi, 1993).

Volume Removed during Machining

In this method, the feature represents the material to be removed to generate the final specification (Figure 5.7b). These can be considered as volumetric features that correspond to machining operations (Henderson and Anderson, 1984; Dong and Wozny, 1990; Kusiak, 1989; Kusiak, 1990). This definition is widely used in CAPP systems dealing with prismatic parts. An Example is MACHINIST (Hayes, 1990).

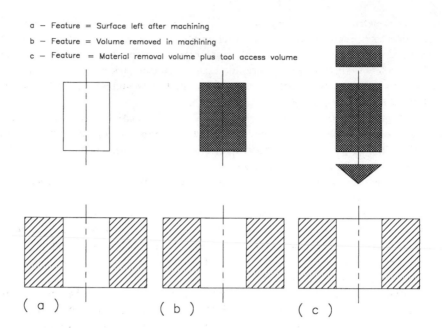

a — Feature = Surface left after machining
b — Feature = Volume removed in machining
c — Feature = Material removal volume plus tool access volume

(a)   (b)   (c)

FIGURE 5.7 Techniques of defining a feature: (a) surface feature, (b) volumetric feature, and (c) material removal plus tool access volumes.

Gindy (1989) proposed a feature taxonomy based on volumetric features and the associated attributes (e.g., external access directions, boundary type, boundary status, etc.). Based on this taxonomy, Gindy et al. (1993) reported a system called GENPLAN which considers the features as component regions that have significance in the context of machining.

Volumetric and surface features can be thought of as complementary to each other if the removal of the former leads to the generation of the latter. In this method, it is possible to provide one-to-one mapping between volumetric features and the machining operations if the features used at the design stage correspond to the features to be used at the manufacturing stage. Volumetric features also help in finding the tool accessibility in a given operation. The constraining factor, however, is that the designer must have some knowledge about the material to be removed in order to get a surface feature.

Material Removal Volume Plus Tool Access Volume

Nau et al. (1993) defined a feature as a pair $f = (r,a)$ where r is the volume of the material removed by the operation, and a is the volume of space needed for access during the machining operation (as shown in Figure 5.7c). In their work, they represented the removal volumes using a library of material removal shape element volumes (MRSEVs) proposed by Kramer (1992a; 1992b). They used this method while generating and evaluating alternative ways to manufacture a proposed design.

When compared to the previous two techniques, this definition is more rigorous, almost bringing the planning domain into the modeling. It amounts to putting the responsibility of planning on the designer.

Realization of Feature-Based Models

Feature-based systems (FBS) can be developed in three basic ways, each method in turn having many variations (Figure 5.8). These can be listed as:

- Feature recognition and extraction systems (FRES)
- Feature-based modeling systems (FBMS)
- Feature-based design systems (FBDS).

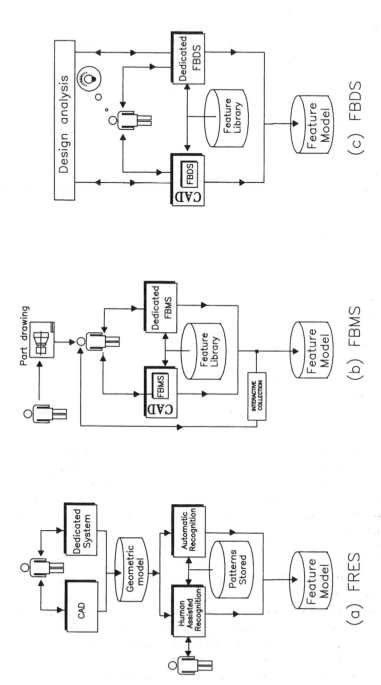

FIGURE 5.8 Realization of form features: (a) feature recognition and extraction system (FRES); (b) feature-based modeling system (FBMS); (c) feature-based design system (FBDS).

Feature Recognition and Extraction Systems (FRES)

In this approach, a feature recognition program examines the CAD data base and makes the deductions about the types of features present. The additional responsibility of interpreting the part representation is shared either by a planning module or by a dedicated postprocessor. Hence, this can be treated as a post-design approach. The method amounts to making explicit what is implicit in a CAD data base.

Feature recognition systems work on a common principle of comparing predefined patterns with portions of the geometric model to retrieve the desired feature. Feature recognition can be based on volumes or surfaces, depending upon the definition adopted. Once the features are recognized, application-oriented information, which was not included in the earlier stages of part representation, can be added to the features for completeness of the model, if necessary. However, the algorithms for recognizing even simple features are fairly complex and are generally modeler-specific. There are very few systems, with exceptions like CADEXCAP (Kalta and Davies, 1992), which can extract the technological (non-geometric) attributes. Although there has been much research and development work in making the CAD data bases more explicit through FRES, this is really a regressive step (Sim and Leong, 1989). Feature recognition and extraction turns out to be a redundant effort which would be unnecessary if a method could be devised for retaining in the modeler all the information available to a designer (Shah and Rogers, 1988).

Because of the complexities associated with FRES, it has become an attractive proposition to make use of features at the time of geometric modeling, which resulted in FBMS and FBDS. Moreover, proponents of other approaches argue that when a designer starts the design process, most of the features are already known; thus, in the process of converting them to a CAD model, the information is lost and has to be recreated by the feature recognizer (Joshi, 1990).

Feature-Based Modeling Systems (FBMS)

FBMS facilitates the interactive conversion of existing designs (usually in the form of drawings) to customized feature-based part models. This is also a post-design modeling method. This method helps in realizing a feature-based model in one step, as it is not necessary to create the geometrical model first. Nevertheless, a geometrical model can be created from the feature model for other purposes. This type of system plays a significant role in situations in which the part design is available only as a blueprint. In such cases, these can provide automated interfaces to CAPP by modeling the part in terms of features, without affecting the design practices followed in the industries. FBMS can be developed in two ways as shown in Figure 5.8.

1. *Using a dedicated system:* Some FBMS embrace system-specific and customized modeling environments. These systems are streamlined to the needs of the CAPP systems and may prove cost-effective as they will not involve the use of any CAD systems. But the development of such systems involves a sizable amount of programming because typical geometry handling facilities (mapping between part geometry and screen coordinates, menu and mouse interface, etc.) are to be built completely from scratch.

2. *Using a front-end feature generator over a CAD system:* Modern CAD systems usually support an external high level language (like C) or an internal programming language (e.g., AutoLISP in AutoCAD, GRIP in UNIGRAPHICS) to enable the users to (a) access the internal CAD model and (b) customize the CAD system for different applications. These capabilities are used in developing FBMS by employing a front-end feature generator as an overlay on an existing CAD system. In this method, FBMS can make use of the typical CAD facilities (like mouse interface, zooming, editing, etc.) offered by the CAD system and thus relieves the programmer or developer from building them from scratch. The constraint on this approach is that the CAD system must be available within the industries if CAD systems cannot support the runtime versions.

Feature-Based Design Systems (FBDS)

Feature-based design facilitates the conceptual designing of mechanical parts in terms of their constituent parameterized features. Here the emphasis is on the conceptualization of a product during the design stage. Feature-based design essentially involves the feedback to the user which can be on the validity of the designs, design for manufacturability (DFM), and other related factors. Iterative loops between FBDS and the user until the user reaches a satisfactory design are crucial here. It can be considered as a predesign method.

Many researchers frequently use the term "feature-based design" to describe post-design modeling methods. (In this chapter, feature-based design denotes the conceptual designing of the part in terms of features). In fact, the characteristic of the feature-based design systems that distinguishes it from other systems is the origin of the model.

Other approaches (FRES and FBMS) start with an existing design (a drawing or CAD data) to build the model, whereas this approach solely depends on the imagination and ability of the designer to use the features for designing the part. Hence, FBDS must provide the designer with the tools necessary to carry out the design analysis. These can also be developed using dedicated systems or general purpose CAD systems.

This offers the total solution because the part modeling and the design representation match with each other (both are in terms of features) and there will be no special need to carry out part modeling. The design representation can serve the purpose of supplying the required data to the CAPP function.

Representation of Technological Details

The information pertaining to tolerances and surface finish can be referred to as technological data. Technological data have considerable influence on the determination of machining sequence, manufacturing procedures, machine selection, and chucking positions (Halevi and Weill, 1985; Weill, 1988). To enable the CAPP system to make realistic decisions, it is necessary to include the technological details in the part model.

Although representation of technological details is as important as geometrical details, it does not seem that this aspect has enjoyed much interest in the initial stages of CAPP system development. The reason for this may be attributed to the adhoc modeling methods followed when CAD systems are not widely used.

In the later stages, CAD systems are used for part modeling. Because of the limitations of CAD systems, FBS has become popular in recent years. In parallel to these developments, issues concerning the representation of technological details in CAD and FBS are addressed by many researchers. Subsequent discussion gives a brief chronological presentation of the work done in this area.

In the early stages, a data base management system (DBMS) for storing technological and material information along with information obtained from the geometric modeler was proposed by Iwata and Arai (1983).

Representing dimensioning and tolerancing in solid models was studied by Requicha (1983). He proposes a tolerancing theory based on the "variational class" concept. This theory treats tolerances as properties or attributes of an object's features (surfaces and edges), the variational information is represented by a graph, called Vgraph. Requicha and Chan (1986) also designed a variational graph (Vgraph) to implement this theory in a CSG-based system. As an independent graph, Vgraph stores all the variational information about an object and is attached to a CSG tree through a set of nominal faces (which are indexed through a suitable method) of the object.

A number of other schemes have also been developed by other researchers such as Kimura et al. (1986), Suzuki et al. (1988), Ranyak and Fridshal (1988), and Gossard et al. (1988). Faux (1988) classified surface features (of ANSI standard) into resolved primitives (size and form can be separately defined from position and orientation) and unresolved primitives. These primitives are then used for the attachment of geometric tolerances to explicit data, called datum reference frame (DRF).

In a recent work, Jasthi et al. (1994) developed a part modeling scheme for rotational parts in which the technological details are divided into two categories: (a) intra-feature data and (b) inter-feature data. In their scheme, feature-specific details are termed intra-feature data (e.g., diametric tolerance, surface finish, circularity, cylindicity, etc.) while inter-feature data denote those tolerances controling a feature in relation to another feature (e.g., runout, concentricity). This scheme is implemented in developing Turbo-Model, a feature-based part modeling system for rotational parts.

Based on the critical study of the published literature in the implementations in several systems, the following conclusions can be drawn:

1. In many representation schemes, the dimensions and tolerances are interpreted as constraints between features. These constraints are also called relationships, links, variations, or lists. It seems that different terms are employed by several researchers to describe the same concept.
2. Tolerance representation schemes are, to a large extent, related to the representation schemes (CSG, B-Rep, Wire frame or hybrid) employed in CAD systems and feature-based systems.
3. The latest works suggest that there is a need to address the issue of representing technological information with reference to the form features in the wake of feature-based systems.
4. To represent dimensions and tolerances, primitive (low-level) features are to be maintained along with the high-level form features. This requirement has led to growing interest in hybrid systems (which can provide multiple abstractions).
5. The primitive features (*cone, cylinder, block*, etc. in CSG models; *point, line, arc* in wire frame models; *surfaces* in B-rep models) are to be managed throughout the part modeling. The study of the relationship between the form-features and primitive features is a prerequisite to designing the data structures for part representation.
6. Proper indexing (linking or referencing) methods are to be established to access the geometrical entities (or features) of the model for attributing dimensions and tolerances.
7. The features classification in these modeling schemes is influenced not only by the application domain (like design analysis, CAPP, assembly planning, interference checking, etc.) but also by the modeling issues of GD&T.

Representation of Global (or General) Details

Along with the geometrical and technological data, the part specification also includes certain attributes such as work material, quantity to be produced, etc. These details affect planning decisions such as the process selection and cutting parameter (speed, feed, depth of cut) selection and need to be represented in the part model. Other details, such as the design number, part name, planner's name, etc. (which normally appear on the drawing and the process plan) can also be included in the part model.

This type of data is purely non-geometrical. It can also be seen that the number and variety of geometrical and technological attributes vary from part to part. Hence, complete control should be given to the user in defining these details in a part modeling system. However, in the case of general attributes, there exists a fixed set of details for all parts in a given manufacturing system. Due to this, these details can be easily modeled through predetermined system-user interaction and can be made available to the subsequent CAPP system.

Part Modeling for CAPP: A Unified Framework

Various part modeling schemes followed in CAPP have been studied in earlier sections and general trends in modeling different types of part details have already been projected. It should be noted, however, that in spite of the numerous part modeling systems reported in literature, a uniform methodology for part modeling is not well established. This is understandable because the selection of a particular modeling scheme and the subsequent system development depends on several factors such as the scope of the CAPP system, part range (rotational, prismatic, etc.), geometrical complexity of the part, the technological details applicable, economics of system development, etc. Each of these factors affects the choice of modeling method in some way. For example, geometrical complexity dictates the feature definition (surface or volume) and representation scheme (wire frame or solid model) while the economics of system development dictates the choice of software (CAD system or programming language) and hardware (PC or workstation) platforms.

Before making an attempt to generalize the methodology for developing feature-based systems, it is necessary to bring out the underlying principles that are applicable to different modeling schemes. The following points are generally valid to all feature-based systems irrespective of the modeling scheme, representation method, part range, or development platforms selected for developing a modeling system:

(a) The complete part specification, comprising geometrical, technological and general details, should be contained in the part model.
(b) Though high-level form features address the issue of CAD-CAPP integration, it is also necessary to maintain the low-level primitive features simultaneously in order to facilitate the visualization of the form features and the representation of the technological attributes.

Based on these common concepts, a framework for containing various types of data in a feature-based model can be given as shown in Figure 5.9, which shows the part model as a set of three types of data. The steps to be followed for the development of a feature-based system within this framework are listed below. Note that the following sequence is not prioritized since the steps involved are inter-dependent. However, it is felt that the given sequence can serve as a guideline in developing a feature-based system.

(a) Decide the feature definition based on the part type. More than one feature definition can be adopted to describe a part.
(b) Catalog the form features (along with the geometrical attributes) identified in the part range.
(c) Decide the representation (wire frame, CSG, B-rep or hybrid) scheme based on factors such as the complexities involved in the part geometry, feature definition or choice of software and hardware.
(d) Decide the set of operators (e.g., Boolean operations) using which the part range under consideration can be modeled as a set of form features.
(e) Study the arrangement of form features in the part range. Some general rules on the arrangement of the features in the part range can be observed. These rules can form the basis to check the validity of the model.
(f) Identify the primitive features that can construct the form features. The relationship between the primitive features and form features must be consistently maintained with suitable indexing.
(g) Decide the set of technological details applicable to the part range. Some general rules on the application of these details to the individual form features and primitive features can be observed. Validation rules can be developed at this stage.
(h) Classify the technological data, based on the number of (primitive or form) features involved in specifying a particular data.
(i) Consider the task-dependent general (or global) data to be added to the feature-based model.
(j) Fix the data structures for representing the geometrical, technological, and global details.
(k) Select the method of realizing the feature-based part model.
(l) Choose the software and hardware platforms for the system development.
(m) Finally, carry out the system development. Flexibility, customization, and user-friendliness are the guidelines in the development.

The development of a part modeling system based on this methodology is presented in later sections.

5.7 Modeling of Manufacturing Resources for CAPP

Process planning is a data-intensive activity. Preparation of the process plans is always guided by the availability and status of manufacturing facilities on the shop floor. The information about these should be made available to the CAPP system to enable it to generate realistic process plans.

As shown in Figure 5.6, MRIR is a collection of information representing manufacturing resources such as machine tools, cutting tools, materials, jigs, fixtures, accessories, inspection gauges, etc. If the scope of the CAPP system is limited to a few manufacturing facilities, these can be directly embedded

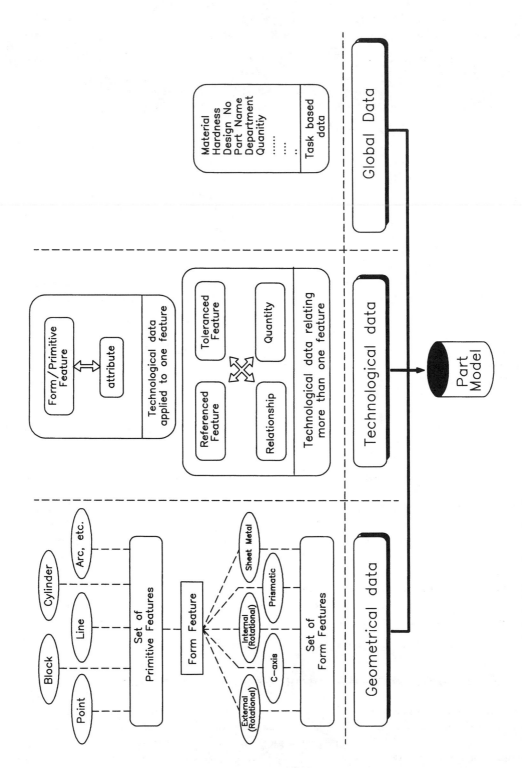

FIGURE 5.9   Framework for building a feature-based system.

within the CAPP system. However, the hard coding of manufacturing data has some serious limitations. For example, if any new machine is added (or removed or breaks down), significant modifications need to be done either in the data files or in the program to reflect the change on the performance of the CAPP system.

It is thus desirable to develop MRIR external to the CAPP system. Other advantages of maintaining MRIR as an independent data base management system (DBMS) are

(a) Flexibility: The same data bases can be used for other manufacturing functions, e.g., for inventory control;
(b) Customization: The data base can be adapted to the unique character of the manufacturing environment at a given instant; and
(c) Enhancements: Without affecting the CAPP systems, these data bases can be modified or changed.

5.8   Modeling of the Process Plan for CAPP

Process planning involves the selection of machine tools, setups, machining operations, cutting tools, process parameters, etc. All experts agree that the decisions about items in this list must be made at some point during the planning process, but whether the order of the decisions in the list has any relevance is a debatable question. Furthermore, the list is highly interdependent. The order in which the planning decisions are made is a factor of a given manufacturing system. Irrespective of this order, there must be some "container" for keeping the partial plans (or the intermediate results of the planning). In this context, process planning internal representation (PPIR) serves the purpose of a container which can be referred to in various stages of process planning.

In addition, the instructions for manufacturing a part can exist in one of several forms:

(a) Textual Process Plans: the plans used by the machinists on a shop floor
(b) Graphical Simulation: the format used by the automated systems for showing the process plan graphically
(c) Pictorial Process Plans: the status of the component (or machining) after each setup for illustrating the part, fixtures, setup, clamping, etc.
(d) NC Programs: the format used by numerical control machines for executing the process plans. The basic set of manufacturing instructions from which these formats can be derived is the same for a given part. To enable a CAPP system to generate the process plans in one or all of these formats, it is necessary to represent the planning details in a structured format. In this context, the role of PPIR in a CAPP system can be appreciated (Jasthi et al., 1995).

To represent the process plan, it is first necessary to gain an insight into the manufacturing instructions. At any instant of manufacturing, the blank (or semi-finished part) is set on a machine tool. In this setup, some portions of the material will be removed so as to move towards the final part specification. These chunks of material can be called machinable volumes (or pockets). The parameters for machining are governed by the part specification and the manufacturing resources. Thus, if (a) the specification of the machine being used, (b) the details of the setup on the selected machine, (c) the number and sequence of pockets being removed in each setup, and (d) the parameters (cutting tools, process parameters, etc.) for machining each pocket are available, then the manufacturing instructions (or partial process plan) at that instant can be specified (Figure 5.10).

To summarize, the set comprising machine, setup, pocket, and parameters can be considered the core of the process planning content in the machining domain. If all such sets applicable to a given part can be obtained, the process plan for that part can be generated. Thus the need to represent these ordered sets in a structured format forms the basis for PPIR (Figure 5.11).

The implementation of this model is explained in later sections.

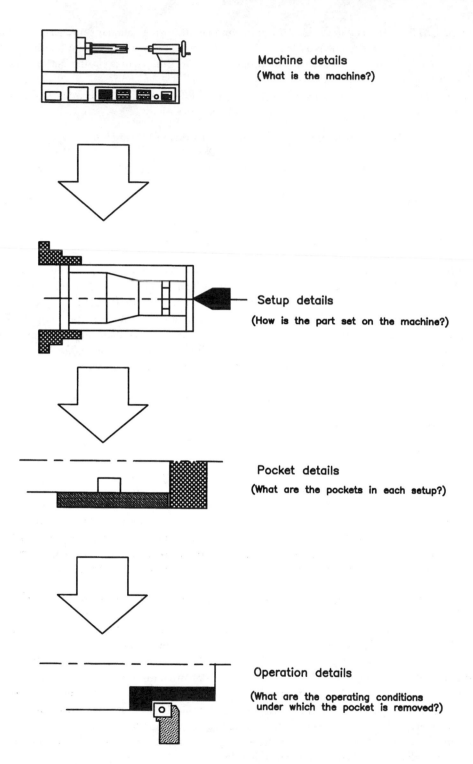

FIGURE 5.10 Process plan content.

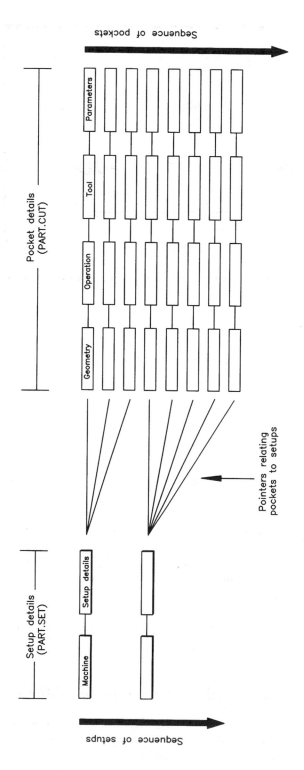

FIGURE 5.11 Schematic diagram of a process plan internal representation (PPIR).

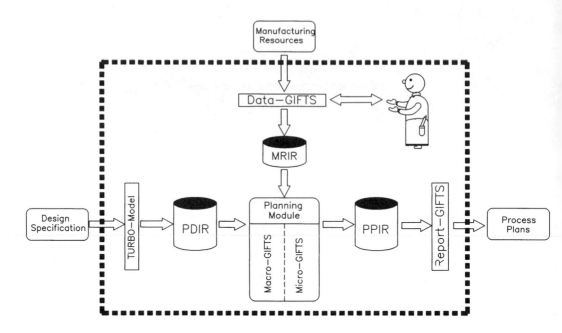

FIGURE 5.12   Schematic sketch of the CAPP system, GIFTS.

5.9   GIFTS: A Generative CAPP System for Rotational Parts

Based on the models described above, this section outlines a PC-based CAPP system for rotational parts conceived as *GIFTS—a Generative, Interactive, Feature-based and Technology-oriented System*. The schematic sketch of GIFTS is shown in Figure 5.12. The objective of GIFTS is to achieve a PPIR for a given PDIR with respect to MRIR. GIFTS is designed in modular fashion with each module responsible for one function of planning as follows.

> **Turbo-Model:** a part modeling module for representing the part data in terms of form features. The resultant part model is PDIR.
> **Data-GIFTS:** a data base management module for modeling the manufacturing resources. These data bases are collectively referred to as MRIR.
> **Macro-GIFTS:** a macro-planning module responsible for the planning functions at the part level. Machine selection, setup planning, and operation sequencing are attempted by this module.
> **Micro-GIFTS:** a module that works at the operation level. Selection of cutting tool, optimization of process parameters, determination of costs and times are attempted by this module.
> **Report-GIFTS:** a module that converts PPIR, which stores planning details generated by Macro- and Micro-GIFTS, to the required external format (e.g., textual process plan, graphical simulations of plans).

5.10   Turbo-Model: Part Modeling System for GIFTS

The proposed CAPP system, GIFTS, is designed to work in a manufacturing system where the part design is available as an engineering drawing. Hence, a post-design approach needs to be employed for part modeling. As shown in Figure 5.13, the second method (FBMS) is employed because it avoids the additional step of creating the geometric model to prepare the input for FRES.

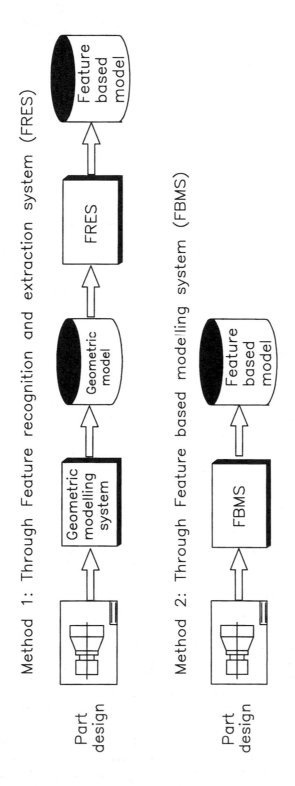

FIGURE 5.13 Methods for post-design part modeling.

Several techniques of defining a feature were presented earlier. Of these, the first definition—"feature is a part surface generated after machining operation(s)"—is adopted in the present work, because (a) it is simple to define surface features; (b) it is necessary to provide a natural way of working with features; and (c) the definition of rotational features in terms of removal volumes is difficult.

Part Representation-Data Structures

The design of data structures for representing the part information (global, geometrical, and technological details) depends on several factors:

- The type and number of form features considered
- The relationship of the form features with the primitive features
- The general arrangement of the features while constructing a part
- The nature and number of technological details considered
- The modeling environment (CSG, B-rep, wire frame etc.)
- The application data.

Global data can be thought of as common data that can be accessed by every part feature. Global data (e.g., component name, material name, material code, drawing code, revision number, process planner's name, department name, date, etc.) need to be represented for the completeness of the model. Some of these data are required to make planning decisions, while other data are used in generating company-specific process plans.

This type of data is purely non-geometrical. This is also non-repetitive in nature as only one set of data exists for a given part. Hence, it is stored as a separate entity in a predetermined order. This can be extended to contain any other relevant data, if the need arises.

With geometrical data, a large number of rotational parts from several industries are analyzed and the constituent geometric features identified. These features can be grouped into two categories based on the geometry. These external features (turn, groove, taper, etc.) and internal features (bore, internal groove, etc.) are shown in Figures 5.14 and 5.15. Since the scope of the present work is limited to rotational parts, 2-D wire frame representation is adopted as it is adequate for visualizing the rotational parts without any ambiguity. In this representation, the point, line, and arc constitute the primitive features upon which the form features are built. Based on these considerations, the parameters required for representing the external and internal rotational features evolve. These are shown in Tables 5.1 and 5.2, respectively. The structures shown in the tables are basically the explicit representation of feature coordinates with a code attached to them.

Technological data—the integral representation of the technological details (tolerances and surface finish) with the geometry—assume significance in the wake of feature-based models. In feature-based systems, a part can be thought of as an ordered compilation of features. In this context, it can be observed that each feature has tolerances associated with its size parameters such as length and diameter. Also, there are form tolerances on features, such as cylindricity, circularity, etc. Apart from these, there will be relational tolerances which involve more than one feature.

In the present work, these technological details are categorized into two groups, (a) intrafeature tolerances and (b) interfeature tolerances, to ensure a simple way of attaching tolerances to the features. This division is based on the number of features involved in defining a particular piece of technological data. The classification of the technological data adopted in the present study is shown in Figure 5.16.

Intrafeature tolerances. It can be observed that many of the tolerances specified are directly related to only one feature at a time. For example, all form tolerances fall into this category. These are feature-specific and are frequently applied to single features or portions of a feature. These tolerances are specified without a common datum reference because the features are not controled in relation to another feature. Intra-feature tolerances can be attached to the feature, along with the geometry, as additional attributes.

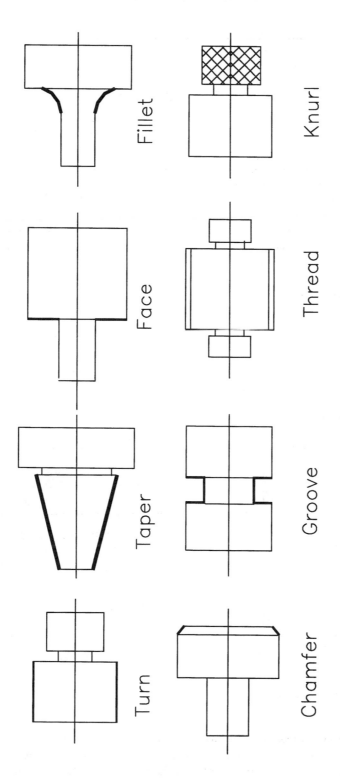

FIGURE 5.14 External rotational features.

TABLE 5.1   Data Structure for Representing External Features

Feature	Code	1	2	3	4	Other Parameters 5	6
Turn	10	Xb	Yb	Xe	Ye		
Taper	11	Xb	Yb	Xe	Ye		
Face	12	Xb	Yb	Xe	Ye		
Arc	13	Xb	Yb	Xe	Ye	R	D
Fillet	14	Xb	Yb	Xe	Ye	R	D
Chamfer	15	Xb	Yb	Xe	Ye		
Groove	16	Xb	Yb	Xe	Ye	Depth	Form
Thread	17	Xb	Yb	Xe	Ye	Pitch	Type
Knurl	18	Xb	Yb	Xe	Ye	Pitch	Type

Xe,Ye: end point; Xb,Yb: start point; R: radius; D: direction (CW/CCW); Type: type of thread/knurl.

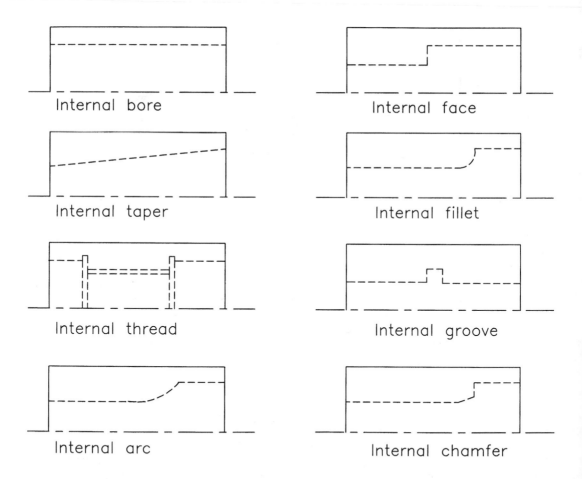

FIGURE 5.15   Internal rotational features.

For rotational components, straightness, circularity (roundness), cylindricity, and angularity are considered. Surface finish and dimensional tolerances which are applied to single features (like diametric tolerance) are also included in this category. Attributing the tolerances to the surface features seems to be rational because the method is in agreement with the definition of the feature adopted (Figure 5.17).

TABLE 5.2 Data Structure for Representing Internal Features

Feature	Code	1	2	3	Other Parameters 4	5	6
Bore	20	Xb	Yb	Xe	Ye		
Taper	21	Xb	Yb	Xe	Ye		
Face	22	Xb	Yb	Xe	Ye		
Arc	23	Xb	Yb	Xe	Ye	R	D
Fillet	24	Xb	Yb	Xe	Ye	R	D
Chamfer	25	Xb	Yb	Xe	Ye		
Groove	26	Xb	Yb	Xe	Ye	Depth	Form
Thread	27	Xb	Yb	Xe	Ye	Pitch	Type

Xe,Ye: end point; Xb,Yb: start point; R: radius; D: direction (CW/CCW); Type: type of thread/knurl.

It should be noted that in rotational parts, many of the tolerances are controled with respect to the axis of the part. There will be no special need to model the axis as it is implied. The resultant data structure for storing the feature information (geometrical + intrafeature data) is shown in Appendix A.

Interfeature tolerances control a feature in relation to another feature. These relational tolerances, as the name specifies, relate two or more features through a tolerance. Of the related features, one feature can be named as the reference feature (REF) and the other as the toleranced feature (TOF). REF and TOF are related by a type of tolerance (relation, REL). The relationship can be denoted by a quantity (QNT). The quantity QNT may be specified by two values (denoting upper and lower bounds) for some tolerances.

Thus, the set of

(TOF) Which is the feature to be toleranced?

(REF) To which feature is TOF related?

(REL) How are these (TOF and REF) related?

(QNT) How much is the relation?

completely describes any type of tolerance which relates more than one feature.

It is important to note here that TOF and REF can be a set of primitive features—an important observation as it forms the basis for evolving the structures to represent the interfeature data. For rotational components, runout and concentricity are considered. Length-wise tolerances are also included in this category. To represent these tolerances, primitive features are to be maintained along with the form features. The data structure for representing the set (TOF, REF, REL, QNT) evolves based on these considerations.

Turbo-Model: Implementation

The objective of the Turbo-Model is to provide a simple, user-friendly environment in which a rotational component can be interactively modeled in terms of features and their technological attributes so as to provide the necessary inputs to the CAPP system (GIFTS). The system-user interaction is designed to facilitate easy navigation of the user through the system. Many facilities such as hierarchical menus, icons, mouse interface, help, prompts, etc. are employed in the system.

Modeling Part Data

The system supports an exhaustive set of commands and external utilities to carry out various activities involved in the modeling process.

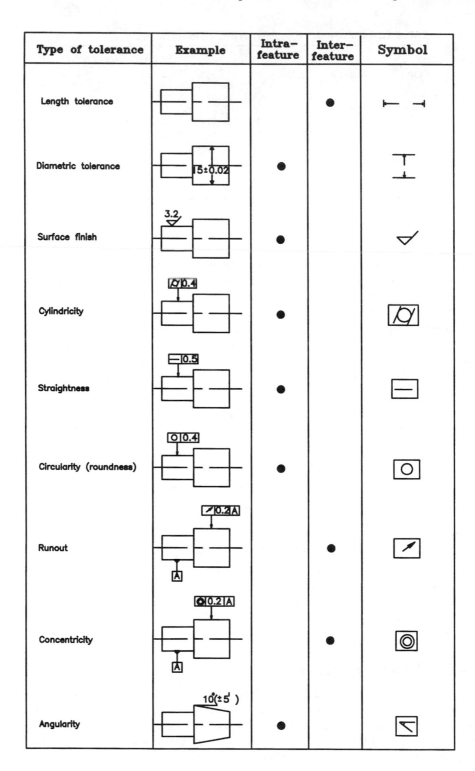

FIGURE 5.16 Classification of the technological data.

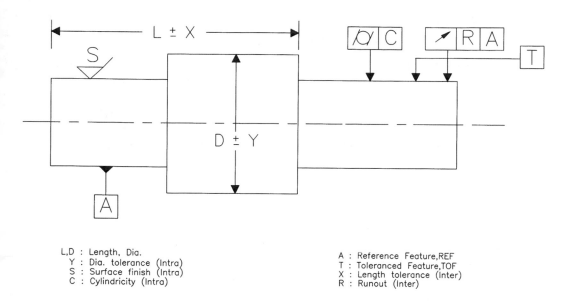

L,D : Length, Dia.
 Y : Dia. tolerance (Intra)
 S : Surface finish (Intra)
 C : Cylindricity (Intra)

A : Reference Feature,REF
T : Toleranced Feature,TOF
X : Length tolerance (Inter)
R : Runout (Inter)

FIGURE 5.17   Interfeature and intrafeature tolerances.

1. *Modeling the global data.* Defining this type of data is simple. On issuing a command GLOBAL, the system poses a series of questions with default values enclosed. A set of commonly applied global data is always kept in a system file. If the global data are not specified during a modeling session, then the default values stored in the system file are attributed to the part. So, it is at the user's discretion whether to define these data or not.

2. *Modeling the geometrical data:* First, the user has to specify the overall length and diameter of the part. Based on these dimensions, the system automatically scans the blank data base for selecting the blank dimensions. Then the external features are defined, starting from left to right, followed by the internal features. Defining the geometrical data of a feature involves the execution of a related command. Simple commands like TURN, GROOVE, etc. are provided to get the geometry of the feature (in terms of length, diameter, width, depth, etc.). A macro, which invokes the system-user interaction to gather the necessary data, is provided for each command. The details given by the user are then stored in the part file.

3. *Modeling the technological data:* Technological attributes like surface finish, form tolerances, and relational tolerances can be defined at any time, wherever necessary, for the completion of the part modeling.

4. *Intrafeature data:* A set of commands (such as DIATOL, FINISH, etc.) for attributing intrafeature data to various features are invoked only on the user's request. The execution of any of these commands involves the selection of a feature and the specification of the attribute. The selected feature is then located in the part file and the attribute is stored in a proper field of the feature record.

5. *Interfeature data:* Several commands like LENTOL, RUNOUT, etc. are provided to define inter-feature data. The execution of these commands essentially involves the selection of two features. The primitive features of the selected form feature will then be indexed based on the proximity of the primitives to the selection point. Once the definition of interfeature data is complete, the relationship will be appended to the file in which information pertaining to relational tolerances is stored.

6. *On-line validation of the modeling process:* The primary requirement of a part modeling system is to provide a valid model to the downstream CAPP system. Some procedures are built into the

system to ensure that the part being modeled is sensible, geometrically feasible and within the scope of the manufacturing system. Some of these are

- *Feature configuration file (FCF):* For each feature, the extreme limits in geometric attributes are recorded in a feature configuration file. This file is referred to by the system during its run time to ensure that the features being defined are within limits.
- *Interference checking:* Care is taken to ensure that only feasible components can be modeled in the environment. For example, at any instant of the modeling, (a) the external envelope of the component must not exceed the blank envelope; (b) the internal envelope must be contained within the external envelope; and (c) the internal-left and the internal-right features must not cross each other (or overlap).
- *Geometrical feasibility:* While modeling a part, the features and their combination must be geometrically feasible. For example, (a) some features cannot be repeated successively (a "turn" followed by a "turn" feature is not allowed); (b) an external groove cannot be the first feature; (c) a groove or face must exist between the features turn and thread; and (d) the external contour must be continuous and closed. These restrictions also lead to the compact process planning algorithms. Each time the user tries to make a feature, the system examines whether previously defined features satisfy the pre-conditions.

To demonstrate the modeling capabilities of Turbo-Model, an example part shown in Figure 5.18 is considered. The graphical modeling of the part is shown in Figure 5.19. The PDIR of the part is given in Table 5.3.

5.11   Data-GIFTS: Modeling Manufacturing Resources in GIFTS

Earlier it was explained why the information pertaining to the manufacturing resources must be maintained as a separate data base external to the main CAPP system. The preparation of the data bases (MRIR) is one of the most time consuming tasks. The data structures for machine tools, cutting tools, inspection gauges, jigs, fixtures, materials, etc. are formulated after several consultations with a few manufacturing industries. The data structure for representing the machine tools is shown in Table 5.4. Along similar lines, representation schema for other manufacturing resources are designed. Data-GIFTS is then developed for managing these resources through a menu-driven interface.

The execution of Data-GIFTS is essentially required when installing the CAPP system for the first time. Afterwards, it is required to run this module only when there is a change in status of the manufacturing resources such as addition/deletion of machine tools, cutting tools, etc. Once Data-GIFTS is installed, all the data bases of MRIR will be ready for subsequent use by the other modules of GIFTS and will be frequently referred to in various stages of planning such as selection of machines, tools, setups, operations and sequences, etc.

5.12   Modeling of Process Plan in GIFTS

In Section 5.8, the model for PPIR is explained. This section discusses how the elements of PPIR set comprising machine, setup, pocket and parameters are represented in GIFTS.

Machine

Once the machine is selected, it can be referred to by its number/code used on the shop floor. Some details of the machine such as power, permissible range of speeds and feeds, etc. are necessary for making subsequent planning decisions. Since the manufacturing resources are separately modeled

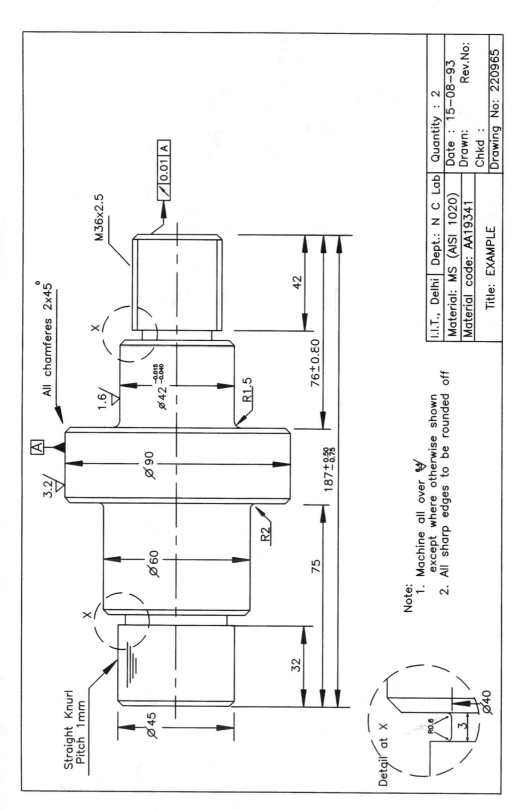

FIGURE 5.18 Component modeled in GIFTS CAPP system.

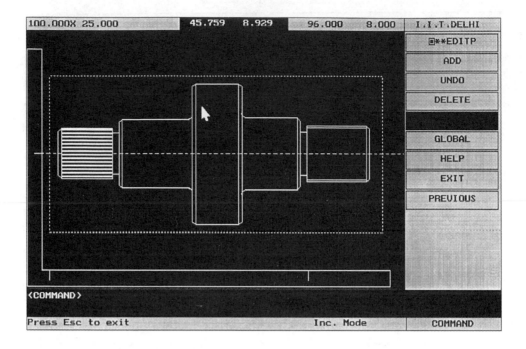

FIGURE 5.19   Graphical model of the part shown in GIFTS.

in Data-GIFTS, it is possible to reference the MRIR based on the machine code/number to get these details.

Setup

Description of a setup will be with respect to the part geometry at a given machining stage. The complexity involved in representing a setup is a factor of part type. For example, the majority of rotational parts can be machined by the between centers method (BC), chuck and center method (CC) and the chuck only method (CO). These setups can be easily represented in PPIR by referencing through codes such as BC, CC and CO.

Pockets

Since the actual machining starts only after the entire planning is done, it can be surmised that the material being removed in each operation (or pockets) must be known in advance. The pockets described here should be understood in the context of the machining. The role of pockets can be appreciated when machining processes are viewed as elements responsible for cutting out the volumes of material from initial stock (or semi-finished part) to produce the final shape of the product.

The definition of a pocket depends upon the machining domain. Three cases of interest are shown in Figure 5.20. The pocket can be defined as the material being removed in a single operation (grooving) involving a single tool (grooving tool). However, this definition might not be convenient in other cases. For example, consider a hole-making operation that involves drilling, boring and reaming operations. Instead of representing the material being removed to produce the hole as three distinct pockets associated with three different cutting tools, it is convenient to represent it as a single pocket. Similarly, in CNC turning operation, defining the pocket as the material being removed in a single setup will facilitate easy representation of a pocket.

TABLE 5.3 Part Data Internal Representation (PDIR) for Part Shown in Figure 5.18

Global (General) Data

1	Test part
2	3
3	220965
4	Mild steel
5	MS-BHEL04
6	200BHN
7	2
8	Jasthi
9	NClab
10	100393

(Geometric + Intrafeature) Information

1	2	3	4	5	6	7	8	9	10	11	12	13	14	15
0	B	00	0.00	0.00	125.00	25.00	0.00	0	0.00	0.00	0.00	0.00	0.00	0.00
1	E	12	3.00	0.00	3.00	20.50	0.00	0	0.00	0.00	0.00	0.00	0.00	0.00
2	E	15	3.00	20.50	5.00	22.50	0.00	0	0.00	0.00	0.00	0.00	0.00	0.00
3	E	10	5.00	22.50	55.00	22.50	0.00	0	0.80	0.50	−0.20	0.01	0.00	0.00
4	E	12	55.00	22.50	55.00	17.50	0.00	0	0.00	0.00	0.00	0.00	0.00	0.00
5	E	16	55.00	17.50	58.00	17.50	3.00	0	0.00	0.54	−0.00	0.00	0.00	0.00
6	E	11	58.00	17.50	93.00	15.00	0.00	0	12.50	0.00	0.00	0.00	0.40	0.00
7	E	12	93.00	15.00	93.00	10.00	0.00	0	0.00	0.00	0.00	0.00	0.00	0.00
8	E	16	93.00	10.00	95.00	10.00	2.00	0	0.00	0.00	0.00	0.00	0.00	0.00
9	E	17	95.00	10.00	120.00	10.00	1.00	0	0.00	0.00	0.00	0.00	0.00	0.00
10	E	15	120.00	10.00	122.00	8.00	0.00	0	0.00	0.00	0.00	0.00	0.00	0.00
11	E	12	122.00	8.00	122.00	0.00	0.00	0	12.50	0.00	0.00	0.00	0.00	0.00
12	L	20	3.00	10.00	13.00	10.00	0.00	0	0.00	0.00	0.00	0.00	0.00	0.00
13	L	22	13.00	10.00	13.00	5.00	0.00	0	0.00	0.00	0.00	0.00	0.00	0.00
14	L	20	13.00	5.00	25.50	5.00	L	0	0.00	0.00	0.00	0.00	0.00	0.00

Interfeature Information

1	2	3	4	5	6	7
1	H	4	1	1	0.2	−0.5
5	H	7	2	1	0.3	−0.0
1	H	13	1	1	0.1	−0.1
1	H	7	1	1	0.1	−0.1
1	T	3	0	0	0.1	

Depending upon the machining domain, the pockets assume different shapes. The scope of GIFTS is limited to rotational parts and the shapes of the pockets generated in rotational machining are shown in Figure 5.21. When the precise representation of the geometry of a pocket is not possible, a symbolic representation (threading, knurling, etc. operations) is employed.

Parameters

Along with pocket geometry, other parameters such as the machining operation, cutting tool, process parameters, etc. need to be included in PPIR for the complete description of the process plan:

(a) Machining operation: The geometric shape of a pocket alone is not sufficient since different machining operations can produce the same shape. It is neccessary to tag the machining operation to a pocket. In GIFTS, all machining operations used in rotational machining are listed and labeled as "TURN," "GROOVE," "BORE," etc.

TABLE 5.4   Data Structure for Representing the Machine Tools

S. No.	Name	Type	Width	Dec	Description
1	MAC_CODE	C	12		The machine code number as used in BHEL.
2	TYPE	C	3		Machine tool type.
					BH—Horizontal boring machine, LC—Center lathe, MH—Horizontal milling machine, PL—Planer, MV—Vertical milling machine, PM—Plano miller, DR—Radial drilling machine, DH—Deep hole drilling, SK—Key way milling machine BV—Vertical borer, F&C—Facing and centering NMV—CNC mill vertical, LCT—Combination turret lathe, SL—Slotting, GCL—Cylindrical grinding NCL—CNC Lathe.
3	MAC_DESC	C	40		Description of the machine tool.
4	STATUS	C	1		Status of the machine tool. A—Available, B—Breakdown, R—Reserved.
5	LENGTH	N	5		Maximum length of the part in mm that can be accommodated in the machine tool. Length between centers for lathes and cylindrical grinders, or maximum table movement in X-direction for mills. Stroke in case of planers and slotters.
6	SWING	N	5		Swing over carriage for lathes and cylindrical grinders, or maximum table movement in Y-direction for mills. For slotters and planers it is the cross travel dimension.
7	HEIGHT	N	5		Maximum height of the part in mm that can be accommodated in the machine tool. For lathes it is swing over carriage. Will not be used for cylindrical grinders, but maximum table movement in Z-direction for mills.
8	POWER	N	6	2	The power available at the spindle in kW.
9	SPIN_HOLE	N	5	1	The size of hole through the spindle in mm.
10	SPIN_TAPER	N	3		The taper in the spindle in terms of Morse taper or ISO.
11	SPEEDS	N	3		The maximum number of speeds available in the machine tool.
12	MAX_SPEED	N	7	1	Maximum speed available in the machine tool, RPM.
13	MIN_SPEED	N	6	1	Minimum speed available in the m/c tool, rpm.
14	FEEDS	N	3		Maximum number of feeds available in the machine tool.
15	MAX_FEED	N	6	3	Maximum feed available in the m/c tool, mm/rev.
16	MIN_FEED	N	6	3	Minimum feed available in the m/c tool, mm/rev.
17	THREAD	N	6	2	Maximum thread making capacity for lathes in mm (lead or pitch).
18	BAR_SIZE	N	5	1	Maximum size of the bar that can go through the spindle.
19	COOLANT	C	1		Coolant capability available in the m/c tool in the form of a code. M—Mist, F—Flooding, A—Air O—Mineral oil, W—Water soluble oil, K—Kerosene
20	ACCURACY	C	4		The accuracy achievable with the m/c tool in terms of tolerance grade.
21	FINISH	N	5	2	The finest surface finish achievable in RMS microns.
22	MIN_SHANK	N	4		Minimum size (shank diameter or size) as representative of tool capacity.
23	MAX_SHANK	N	4		Maximum size (shank diameter or size) as representative of tool capacity.
24	PUR_DATE	N	4		The actual year of purchase of the machine tool.
25	COST	N	10		The purchase cost of the machine tool.
26	MANFACTUR	C	20		Manufacturer of the machine.
27	ACC1	C	4		Accessory available with the machine tool as a code (pointer).
28	ACC2	C	4		Accessory available with the machine tool as a code (pointer).

TABLE 5.4 (continued) Data Structure for Representing the Machine Tools

S. No.	Name	Type	Width	Dec	Description
29	ACC3	C	4		Accessory available with the machine tool as a code (pointer).
30	ACC4	C	4		Accessory available with the machine tool as a code (pointer).
31	REMARKS	C	50		Remarks if any.

(b) Cutting tools: Along with the pocket and machining operation, the cutting tools used in a machining operation can be represented by tool codes. Recall that the cutting tools are modeled in Data-GIFTS. Given a cutting tool code, one can obtain other details from MRIR.

(c) Cutting process parameters: The parameters (speed, feed, and depth of cut) can be represented along with the pocket geometry, machining operation, and cutting tools.

5.13 Macro-GIFTS: Macro-Planning Module

In earlier sections, the implementation of Turbo-Model for achieving PDIR and Data-GIFTS for achieving MRIR is explained. The model of PPIR as employed in GIFTS is also explained. The function of the planning modules of GIFTS is to take PDIR and MRIR as inputs and generate PPIR as output as shown in Figure 5.12. The following sections outline how this is achieved in GIFTS.

The planning functions of CAPP can be broadly divided into: (a) macro planning and (b) micro planning. Macro-planning deals with upstream planning issues and hence is concerned with planning at the part level. Micro planning, also called operation planning, is mainly concerned with the details of specific manufacturing operations. The selection of cutting tools and process parameters falls under the scope of microplanning.

Macro-GIFTS, the macro-planning module of GIFTS, is responsible for operation selection, machine tool selection, setup planning, and operation sequencing.

Operation Selection

One of the most critical issues involved in process planning is the identification of the various operations required for manufacturing the part. This choice is more difficult because there are too many possibilities and the planner has to come up with the optimum choice commensurate with the requirements. Hence it is necessary to combine the product geometry and technology, along with the machining resources capability, to arrive at a possible operation for a given feature or pocket.

There are a number of approaches suggested in the literature. The approaches will also have to be based on the product geometries. We will therefore divide the operation selection into the following categories:

- Cylindrical surfaces (external and internal)
- Plane surfaces
- Hole making processes
- Sculptured surfaces
- Pocket clearance

Each of these categories has its separate methods and algorithms for the purpose of identifying the machinable volumes and the corresponding operation selection and sequencing methods. One major assumption we will be making is that the product data base is organized in the form of features. This helps in making the machinable volume identification an easier task. Even if the original product is in other forms, a feature recognition module can be added in between to make use of the following

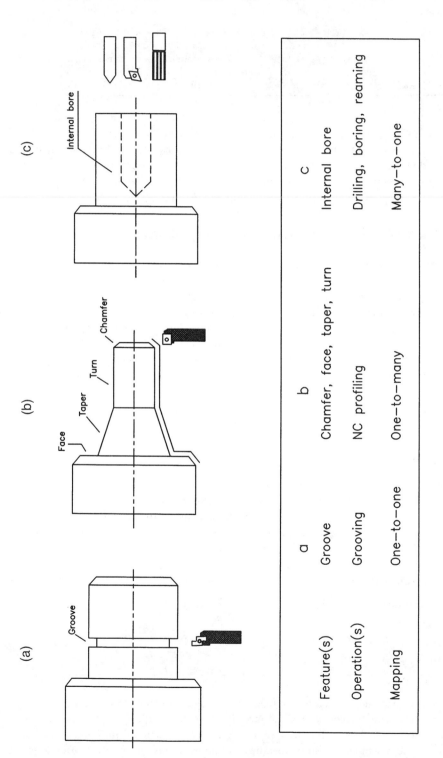

FIGURE 5.20 Mapping between machining operations and features (pockets).

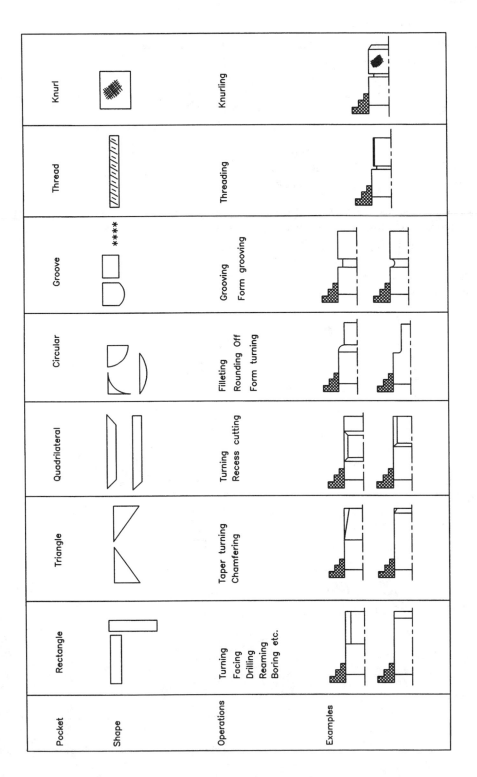

FIGURE 5.21 Pocket shapes in conventional machining.

algorithms. (Some details of the product structures utilized for feature storage are given in Section 5.4: Decision Tables.)

Planning for Cylindrical Surfaces

Cylindrical surfaces are characterized by the operations that are carried out in lathes and turning centers. Some internal cylindrical operations can also be carried out in milling machines or machining centers. The achievable accuracies and surface finishes are given in Figures 5.22 and 5.23 for the external and internal cylindrical surfaces, respectively. While showing the lower limits of the process capability, these figures also show the possible sequence of operations to arrive at the required specifications. For example, to achieve an accuracy of IT grade 7 and surface finish of 0.63 μm, the possible sequences are

Rough turning	Rough turning
Semifinish turning	Semifinish turning
Finish turning	Rough grinding

These could therefore act as a guide to selecting the operations. The given charts can be converted into the form of process capability tables and suitable algorithms could be developed for the process selection method.

A major part of the reasoning that a planner does in the early stages of planning has to do with the selection of processes. The approach in GIFTS consists of three steps: (1) considering the finished part and the raw material, (2) constructing the material to be removed from the raw material to arrive at the

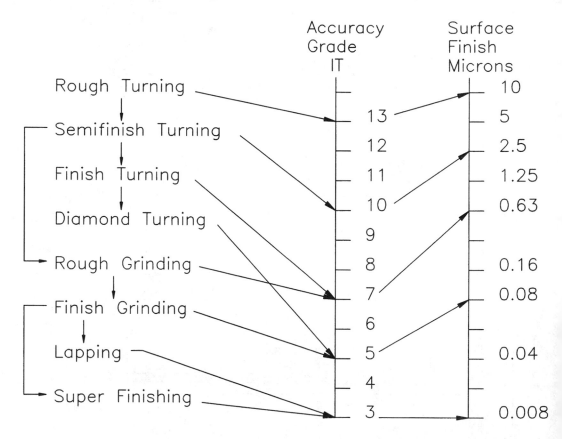

FIGURE 5.22 Achievable lower limits of accuracies and surface finishes based on routing for external cylindrical surfaces.

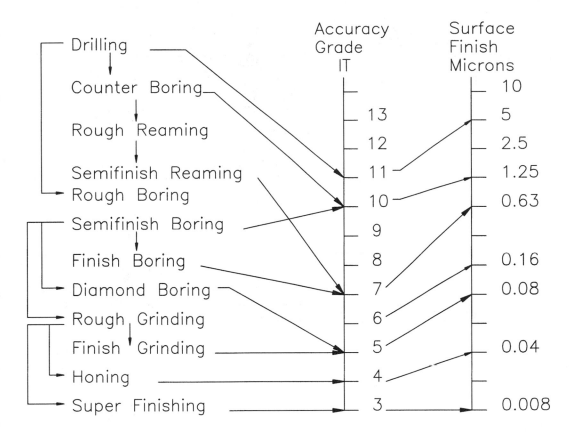

FIGURE 5.23 Achievable lower limits of accuracies and surface finishes based on routing for internal cylindrical surfaces.

finished part, and (3) dividing the total material to be removed into a number of pockets which can be associated with various machining operations.

In GIFTS, backward planning is adopted since it makes planning easier. This is also closely related to the thinking strategy of the process planner. At the end of the operation selection procedure, PPIR is filled with pocket and operation details only. PPIR is partial since other details (machine, tool, parameters, etc.) need to be determined in subsequent modules. For the purpose of process planning the pockets are divided into three categories:

1. *Basic primary pockets*: These are the basic types of pockets, rectangular in nature, which can be machined using the most basic of the lathe operations such as turning, facing and boring. These can be further subdivided into roughing and finishing pockets to take care of the technological specifications given.
2. *Complex primary pockets*: These can be done in the conventional machine tools by using special processes or form tools such as fillet and chamfer. If a CNC turning center is chosen, then the basic and complex primary pockets would be combined.
3. *Secondary pockets*: These require special operations or tools. Examples are grooves, threads, and knurls.

The pocket identification procedure recursively modifies the final part by identifying the pockets until the part reaches the blank shape. As these pockets are identified, the machining operation (e.g., rough turning, threading, etc.) is also attributed to the pocket description. Pocket identification is explained below with the help of an example. The example part containing some representative external features is shown in Figure 5.24. The steps involved in the pocket identification for this part are illustrated in Figure 5.25.

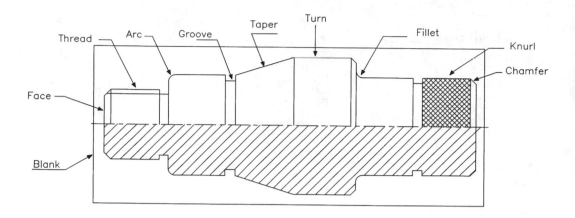

FIGURE 5.24   An example rotational part.

1. *Secondary pockets*: Grooves, threads, and knurls are considered in this stage. There are no roughing or finishing operations associated with these secondary features; hence, one-to-one mapping between these features and the operations can be established. All these features are stored as pockets in the pocket file (Figure 5.25a) and are simultaneously updated to their parent features. For example, a groove is converted to a turn feature. Similarly, the thread is unthreaded and the knurl is unknurled. This step results in an intermediate part (Figure 5.25b) without these secondary features. The material to be removed from the intermediate part (to reach the final part) forms a single pocket in NC machining, which can be subdivided into finishing and roughing profiles. For conventional machining, the pocket identification has to be further carried out as follows.

2. *Complex primary pockets*: It is assumed that machining the chamfers and fillets and rounding off the corners will be carried out as distinct operations on conventional machines using form tools. At this stage, these pockets are stored in the pocket file (Figure 5.25c) and an intermediate component is regenerated (Figure 5.25d).

3. *Finishing pockets*: The finishing operations are not primarily used as material removal operations but are intended to achieve the technological attributes of the surface. Nevertheless, the small amount of the material being removed in a finishing operation is also represented as a pocket to facilitate uniform handling by the system. The stepped component, obtained in the previous step, is then analyzed to offset the elements (usually turn and taper) that require finish machining. Sometimes, the adjoining elements need to be offset along with the turn and taper elements because of the accessibility constraints. These finish pockets (offset volumes) are then appended to the same pocket file (Figure 5.25e). At this stage, what remains is the intermediate part (Figure 5.25f) which needs to be inverse rough-machined.

4. *Roughing pockets*: At this stage, the stepped component contains a series of face, turn and taper features without any technological attributes present. The pocket identification for rough machining is carried out as follows:

   • Face pockets: In this step, the pockets to be removed by a facing operation on either end of the part are marked (Figure 5.25g) leading to the part shown in Figure 5.25h.

   • Taper pockets: Since taper turning is a distinct operation, these are identified as pockets (Figure 5.25i). This step results in an intermediate part shown in Figure 5.25j.

   • Turn pockets: At this stage, the intermediate part contains only face and turn elements. The pattern The patterns these elements make are used to identify the pockets to be machined in the rough-turning operation. Three patterns of interest are shown in Figure 5.26. Identifying such

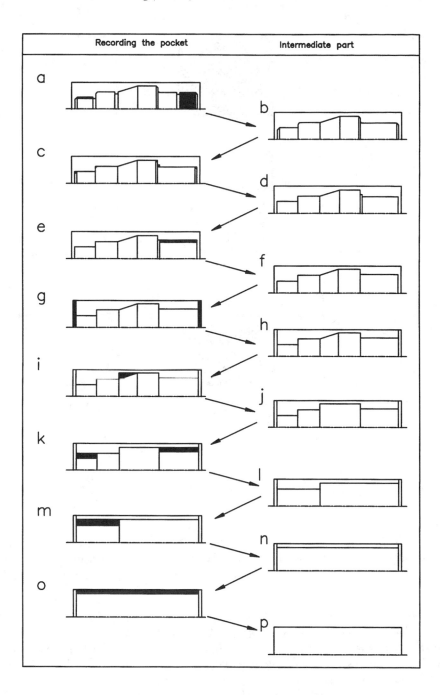

FIGURE 5.25   Steps in pocket identification procedure.

patterns, recording the pockets, and creating the intermediate part are carried out recursively until the stepped component matches the blank dimensions. This recursion is shown from Figure 5.25k to Figure 5.25p. A short version of this recursive procedure is listed below.

(a) To start, construct the volume to be removed in rough turning by a series of face and turn elements.

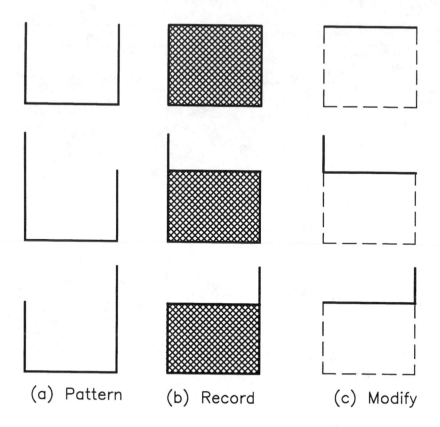

FIGURE 5.26   Down_face-turn-up_face pattern forming a pocket.

  (b) Take the set of first three elements. Check for the "down_face-turn-up_face" pattern as shown in
       Figure 5.26a. If the pattern exists, go to step C; otherwise, go to step E.
  (c) Record the pocket (the cross hatched region in Figure 5.26b) and modify the pattern as shown in
       Figure 5.26c.
  (d) If (pocket_surface = blank_surface) then pocket identification is over. Stop.
  (e) Otherwise, skip these three elements and consider the remaining elements. Go to step B.
  (f) Skip the first element. Consider the remaining elements. Go to step B.

    As soon as the pocket is identified, the corresponding machining operation is also attached to it.
Because of the backward planning strategy employed, the order of pockets is loosely related to the inverse
operation sequence. However, the exact operation sequence can be obtained only after setups are derived.
The above procedure gives rise to a unique set of pockets. No alternate sets of pockets are necessary in
the case of cylindrical components.
    Each of the intermediate shapes generated during the pocket processing are stored in different
files as feature-based models. These models are temporary and do not appear on the design.
Nevertheless, they are essential to carry out subsequent processing, as shown in the above procedure.
The same procedure can be applied to identify the internal pockets also. In that case, the bottom
portion of the contour is to be considered. In backward planning, the blank surface is the limiting
element for external features, while the central axis becomes the limiting element for internal
features.
    At this stage, PPIR is filled with pocket and operation details only. PPIR is partial since other details
(machine, tool, parameters, etc.) need to be determined in subsequent modules.

Machine Selection

The choice of the most suitable machine tool is an important task since the characteristics of the chosen machine tool will reflect on subsequent planning functions, such as setup planning, and the selection of process parameters. The details of the machine tools can be stored in a data base (see sample given in Table 5.4).

Machine tool selection is done by matching the part requirements with the machine tool capabilities (Figure 5.27). A candidate machine tool is selected based on the following four aspects:

Physical aspects: These are concerned with the physical characteristics of the machine tool and the workpiece. For example, the maximum diameter ("swing over carriage" for lathes) and maximum length ("admit between centers" for lathes) that can be accommodated on a machine tool should be greater than those of the part. In case of bar feeding, the machine tool spindle bore must be greater than the diameter of the bar stock.

Technological aspects: As the name suggests, these are concerned with the technological attributes (surface finish and tolerances) present on the part. The achievable accuracy on a machine tool is compared with the required accuracy and only the capable machines are short-listed.

Scheduling aspects: Only available machine tools need to be selected for processing the part. If a machine tool is scheduled to do any other job, it can be made non-available to the CAPP system by simply changing its status in the machine tool data base to either "Reserved" or "Breakdown." In that case, regardless of its capabilities, the machine tool will not be selected. Sometimes, the part may be reserved to a particular machine tool on the shop floor. Recall that such information can be found in the global data of PDIR. If the part is reserved, the machine tool search will no longer be necessary.

Optimization aspects: In spite of the above constraints, it is not unusual to have a large number of qualified machine tools for machining a given part. In such cases, different machine tool selection strategies can be applied, based on the company policy (either minimum production time or minimum production cost), to choose one from the alternatives. The power available, machine tool costs, material removal rates, range of speeds and feeds, are the key parameters used to evaluate the alternatives.

Setup Planning

Once the operations are decided and the machine is selected, the next logical step is to find out how the part is to be set on the machine and what operations are to be carried out in each of these settings. It is also necessary to determine the number of setups, sequencing of setups, and clamping parameters in each setup.

The procedures presented here are based on the works reported by Hinduja and Huang (1989b), Arora (1991) and Jasthi et al. (1994). The execution of these procedures results in some changes in PPIR such as filling up the setup details into the relevant slots and attaching the pockets to different setups.

Deciding the Holding Method

The first task in setup planning is to find a method of holding the job. A particular work-holding configuration will directly affect the operation sequence and must fulfill the technological requirements (like roundness, relational tolerances, etc.) specified on the component. For rotational components, the following methods of holding are possible:

(a) Holding between centers and using face plate and dog as driver (between centers method, BC)
(b) Holding between centers and using chuck as driver (chuck and center method, CC)
(c) Holding in chuck (chuck only method, CO)
(d) Clamping in special fixtures and collets.

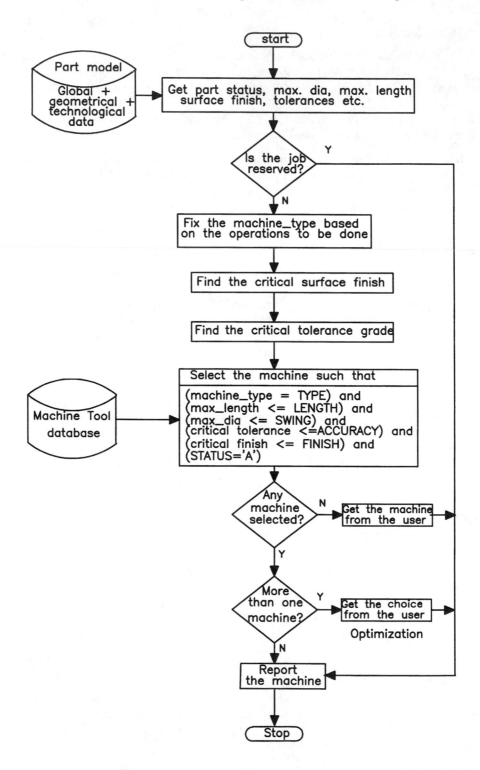

FIGURE 5.27   Procedure for machine tool selection.

The first three possibilities are considered the most common work-holding methods used for one-off or small-batch components. The decision about the holding method is based on a set of rules using length-to-diameter (L/D) ratio, weight, maximum diameter, and minimum diameter of the component. The following rules are employed to classify a component as short or shaft for determining the work-holding method:

(a) If $L/D \le 2$ then the part is a short component
(b) If $(L/D \ge 4)$ and (maximum dia. > 100) then it is a shaft
(c) If $(L/D \ge 4)$ and (maximum dia. ≤ 100) then it is short
(d) If $(2 < L/D < 4)$ and (minimum dia. ≤ 15) then it is short
(e) If $(2 < L/D < 4)$ and (minimum dia. > 15) then stiffness is to be compared. If its stiffness when held between centers is greater than that when held in a chuck, then the part is considered a shaft; otherwise, it is a short component.

After classifying the part, the following guidelines are applied to determine the work-holding method:

- Short components are usually held in chuck (CO method).
- A component classified as a shaft is preferably machined between centers using a dog driver (BC method).
- If the shaft is a heavy component, e.g., more than 350 kg, a chuck is used to drive the shaft (CC method).
- If internal features are present on the shaft, then use of steady rest is necessary.
- The shafts with L/D ratios greater than 12 are considered as non rigid (Kovan, 1959) and two steady rests are employed.
- Steady rest is not used on the side clamped in a chuck.

To calculate L/D ratio, the maximum and minimum diameter of the part can be obtained from the geometrical model of PDIR. Weight of the component can be estimated from these dimensions. Some short-cut methods of finding the holding method are also employed. For example, the part name and the drawing code (which is usually based on the company standards) are available in the global data of PDIR. These attributes can be used as reference for the shop floor practices to determine the holding method. For example, if the name of the part is "rotor shaft," then it is obvious that it has to be held between centers. In such cases, the above calculations need not be carried out.

Deciding the Number of Settings

A setting has been defined as being that part of the machining operation accomplished during a single clamping of the workpiece being processed (Kovan, 1959). In this context, reversing the part on the same machine and shifting the part from one machine to another can also be treated as different settings.

The majority of rotational parts can be machined within three setups. A sample flow chart for determining the number of setups for a "chuck only" component is shown in Figure 5.28. Refer to Hinduja and Huang (1989b) for more details on finding the number of settings.

Attaching the Pockets to Setups

Only in the case of a single setup condition can all of the pockets be machined from the blank to get the final part. If the part needs to be machined in multiple setups, it is necessary to establish the accessibility limits of the pockets in each setup. To relate the pockets to different setups, the concept of demarcation lines used by Hinduja and Huang (1989b) is found to be a suitable approach.

The accessibility of the pockets is determined by placing the demarcation line (DL) at either end of the maximum diameter of the external profile. The pockets lying on either side of DL are placed in different groups. DL thus establishes a limit in such a way that pockets belonging to different groups cannot be accessed in the same setup; however, pockets of the same group can be machined in more than one setup. In the case of internal features, minimum diameter is the criteria to set DL (Figure 5.29).

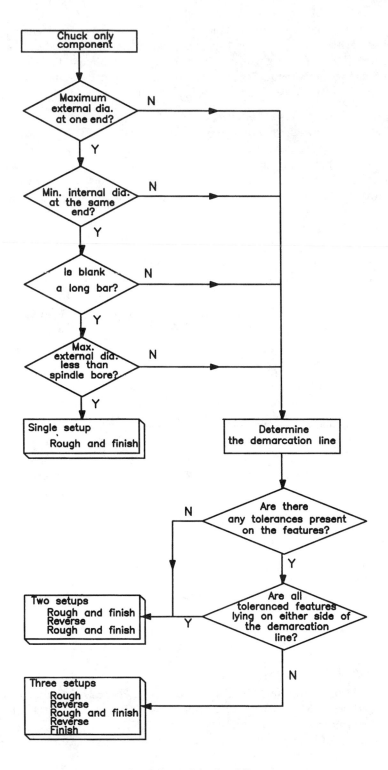

FIGURE 5.28   Determining the number of setups in a "chuck only" component.

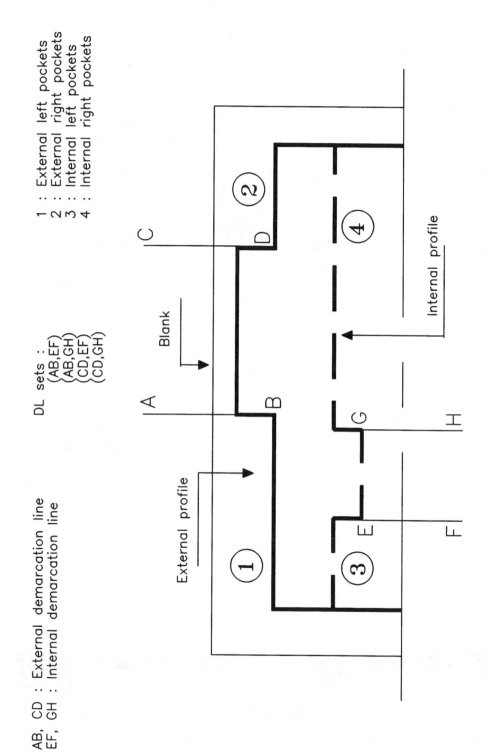

AB, CD : External demarcation line
EF, GH : Internal demarcation line

DL sets :
(AB,EF)
(AB,GH)
(CD,EF)
(CD,GH)

1 : External left pockets
2 : External right pockets
3 : Internal left pockets
4 : Internal right pockets

FIGURE 5.29   Setting the demarcation line (DL).

DL is needed to be set only if the part type is 2 (i.e., part containing through internal features). For other types of parts (3, 4 and 5), the internal pockets can be accessed only from the end to which the internal features are opened.

If the part contains more than one feature with the same maximum external diameter or minimum internal diameter, the demarcation lines give rise to different alternative arrangements. (In Figure 5.29, four different sets of DL are marked.) These alternatives can be evaluated based on certain guidelines:

- The features with relational tolerances (such as concentricity, run out, etc.) should be machined in the same setup.
- The features with high surface finish requirements should be in on one side of the demarcation line as far as possible. This helps in reducing the number of setups since all the finishing operations can be carried out in the last setup, as shown in Figure 5.29.
- The DL should be nearer to the center of the part to balance the component stiffness in different setups.

As shown in Figure 5.29, the number of setups needed to machine a part can be known only at this stage after DL is fixed.

Once the demarcation line is fixed, the pockets are attached to different setups. If any of the pockets spread across the demarcation line, these are split so that they can be placed in different setups (Figure 5.30). While attaching the pockets to the setups (by filling in the slot "pocket.setup"), certain fields in PPIR are referred. These include the fields (a) "pocket.type" which specifies whether the pocket is external or internal; (b) "pocket.operation" and "pocket.nature" which give the operation details; and (c) describing the relative position of the pocket with respect to DL.

Determination of the Clamping Surfaces

Automatic selection of the clamping surfaces depends on the work-holding method and the setup number. Some general guidelines in the form of rules are included in the system for selecting the clamping surface.

Some typical rules of clamping are

- If the surface is conical, it cannot be used for clamping.
- The distance between the cutting point and the clamping point should be minimized.
- The search span, within which the clamping surface must lie, should not extend into the machining span in which the pockets will be machined, etc.

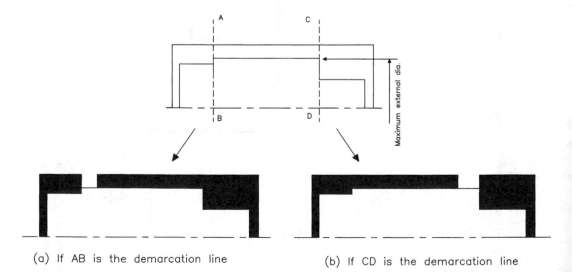

(a) If AB is the demarcation line   (b) If CD is the demarcation line

FIGURE 5.30 Pocket splitting based on demarcation line.

In spite of these rules, more than one clamping surface may qualify. Sometimes, internal surfaces are preferred to external surfaces for the clamping purpose (especially in the case of thin-walled components, in order to avoid the compressive stresses). It is also possible to orient the work-holding devices in different styles, such as reversing the jaws, etc. In the machining domain, the number of valid plans quickly increase with the number of clamping surfaces. In such cases, selection of the best clamping surface will be based on the evaluation of all these alternatives. Some constraints that limit the clamping possibilities, such as cutting and clamping forces and accessing limits of the clamping devices and their availability, will have to be taken into consideration while evaluating these alternatives.

Operation Sequencing

The aim of operation sequencing is to apply sequencing constraints in order to arrive at a feasible plan. In the first instance, it appears that the operation sequence will be the reversed-sequence of the pocket identification since it is based on backward planning strategy. However, it is true only to some extent because the pockets identified earlier are spread over different setups and are placed in different groups to satisfy the relational tolerances and the accessibility constraints. Therefore, these changes will have to be reflected during the operation sequencing.

Operation sequencing is a complex task in CAPP since a great deal of computation is required to evaluate all possible sequences. For example, consider that there are N pockets. In a strict theoretical sense, these can be machined in N! different sequences. In a simple part with 16 pockets (16 in number), it is theoretically possible to have 2.09227×10^{13} (16!) machining sequences. The evaluation of all these alternatives will take years. However, in practice, it is possible to cut this seemingly large search space by applying some constraints. If, for example, a single constraint (that pocket X needs to be machined before pocket Y) can be applied, then the search space is reduced to (N!/2). Similarly, if one operation (e.g., rough turning) needs to be carried out before another operation (e.g., finish turning), the search space will be further reduced.

Multi-level Pocket Sorting

Operation sequencing can be considered as the sequencing (or sorting) of the pockets because the operations are already attached to the pockets. Pocket sorting is attempted at three levels:

1. *Setup level*: At this level, the precedence among the setups is established. For example, the pockets of the second setup cannot be machined until those of the first setup have been machined. Pockets are sequenced (sorted in ascending order) based on the setup to which these are attached (Figure 5.31a)
2. *Operation level*: When identifying the pockets, the threads, grooves, etc. are identified first and are stored at the top of the pocket file. However, these operations are normally the last operations carried out in the sequence. This is resolved by establishing the precedence relationships between pockets belonging to the same setup based on the machining operation (Figure 5.31b). Based on the work shop practices, certain precedence between the operations can be established. A few typical examples are listed below:

   • The machining sequence should prevent destruction of features that have been created by a previous operation. By this yardstick, a thread should be machined only after finishing the adjacent chamfer/groove and parent diameter.

   • Facing operation would normally be the first operation to get a reference for subsequent machining.

   • If internal features are present, as much material as possible is removed by the drilling operation prior to external turning operations.

   • All roughing operations should be performed first, followed by finishing operations.

   • Based on accessibility constraints, grooves will be cut after turning operations.

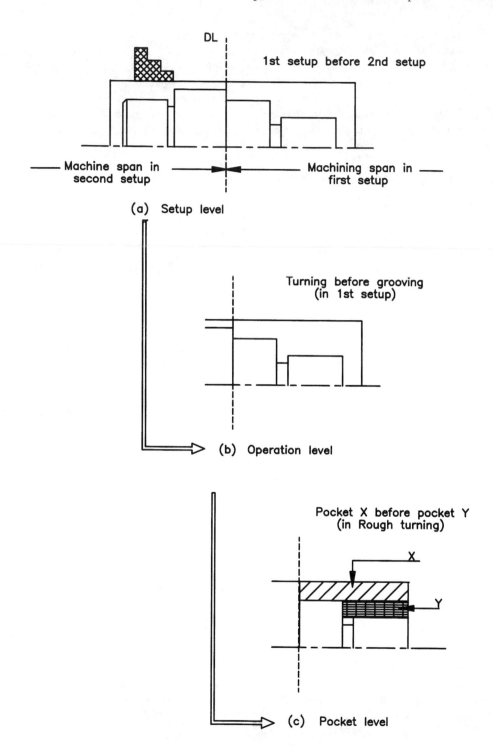

FIGURE 5.31 Operation sequencing constraints for multi-level pocket sorting.

The machining sequence derived from these guidelines is stored in a list, which is referenced while sorting the pockets belonging to the same setup.

3. *Pocket level*: At this level, the pockets belonging to the same setup and same operation are sorted based on their relative position to each other. It may happen that a pocket can be accessed only after machining a pocket lying above (or below, in the case of internal pockets). The relative position of the pockets can be measured with respect to the blank surface or the central axis of the rotational part. In this way, it is ensured that the top pockets are machined before the hidden pockets are made available for the subsequent machining (Figure 5.31c).

Pocket Sorting Procedure

An operation sequencing procedure is developed to sort the pockets as explained above. This procedure, which makes use of the setups, operations_list and pocket details, is presented below:

```
setup_no = 1;
    repeat                                              {setup loop}
        i = 1;
        repeat                                          {operation loop}
            j = 1;
            open the pocket_file
            repeat                                      {pocket loop}
                read pocket;
                if    pocket.setup = setup_no and
                      pocket.operation = operation_sequence[i] then
                begin
                      partial_list[j] = pocket;
                      j = j + 1;
                end;
            until end of the pockets;                   {end pocket}
            sort partial_list on pocket position
            store partial_list in the sort_file
            i = i + 1;
        until operation_sequence[i] = null;             {end operation}
    setup_no = setup_no + 1;
until end of setups                                     {end setup}
pocket_file = sort_file.              {PPIR is sorted}
```

Thus, the pocket sorting at these levels is applied until all setups are covered. The execution of this procedure yields an ordered sequence of pockets in PPIR. At this stage, PPIR consists of the details of machining operation, pockets, machine, and setups.

5.14   Micro-GIFTS: Micro-Planning Module

Micro-GIFTS, the micro-planning module of GIFTS, is responsible for

- Selection of cutting tools
- Selection of optimized process parameters
- Calculation of cutting times and costs.

Cutting Tool Selection

Selection of the best cutting tool for a given operation is one of the complex tasks to be performed by the process planner. When developing a generative process planning system, it is necessary to incorporate mechanisms for automatic tool selection in the system without user intervention.

Many CAPP systems contain a tool selection module with varying degrees of sophistication which gets the required data from other modules of the system. Since a large variety of cutting tools are available, the tool selection problem often becomes complex. It is further complicated by the large number of factors to be considered for tool selection. However, some general guidelines for selecting a cutting tool can be established based on production practices (Chen et al., 1989; Metropoulos and Hinduja, 1991; Yeo et al., 1990; Opitz, 1970; Halevi, G. and Weill, R.D., 1995).

Tool selection essentially involves the specification of tool material, tool holder, and insert (shape, geometry, and grade). The main variables that control the selection process are

- The operation
- The work material and its condition
- The component geometry
- The machine tool
- The production rate
- The manufacturing quality.

Each of the above variables would not only have its independent influence on the selection process but also sometimes be linked with the other variables. Therefore, it is necessary to develop a number of rules, based on the experience on the shop floor, as to how the tool selection is to be carried out.

To make the selection process, it is necessary to store the details of the cutting tools as resources. A typical data structure that could be adopted for this is given in Table 5.5 in the appendix. The tool data base contains information about turning tools, grooving tools, threading tools, boring tools, etc. A separate data base can be made for form tools which can store the features produced by each of these form tools in that data base. The availability of a particular cutting tool on the shop floor can be indicated by changing the field "tool.available."

Selection of Turning Tools

The selection of tools involves the determination of the key parameters and searching the tool data base, based on those key parameters. The guidelines in the tool selection module are based on the practices followed in the participating industry and a tool manufacturer's handbook (Widia, 1989). The inputs coming from the models of the CAPP system such as PDIR, MRIR and PPIR are shown in parentheses below:

1. Based on the operation and nature of cutting (pocket.operation, pocket.nature from PPIR), select the clamping system.
2. Based on the work material (global data from PDIR), nature of cutting, and the clamping system, decide the insert shape and insert clearance angle.
3. Find the tool-access and tool-out directions based on the geometry of the pocket (pocket shape and dimensions from PPIR).
4. Based on the tool-access, tool-out, and feed directions, select the holder style.
5. Based on the pocket dimensions and tool holder style, select the insert size. The steps involved are (a) determine the largest depth of cut from the pocket dimension; (b) based on the approach angle, find the theoretical cutting edge; and (c) find the exact cutting edge based on the insert shape.
6. Find the maximum shank diameter on the machine tool selected. This parameter is obtained by searching the machine tool data base (from MRIR) with the help of machine tool code (from PPIR).

TABLE 5.5 Data Structure for Representing Cutting Tools

No.	Field	Data Type		Description
		Type	Width	
1	Number	Numeric	3	Serial number
2	Toolcode	Character	12	Tool code (it is company specific code, not ISO code), which is stored in PPIR as "pocket.tool".
3	Tooltype	Character	2	Denotes the type of the cutting tool as follows: TT—Turning, TP—Parting, TS—Threading, TG—Grooving, TC—Chamfering, TF—Forming, TR—Radius, TB—Boring, TK—Knurling, DD—Twist drill, DC—Centre drill, DR—Reamer, DB—Counter bore, DS—Counter sink, DP—Spade drill, MF—Face mill, ME—End mill, MS—Slab mill, MD—Slot drill, SF—Side and face mill,
4	Material	Character	1	Specifies the tool material as follows: H—HSS, W—Carbide, M—Ceramic, C—Coated carbide, B—CBN, D—Diamond.
5	Height	Numeric	6,2	These three fields represent the tool holder
6	Width	Numeric	6,2	dimensions. Width and height (or diameter) and
7	Length	Numeric	6,2	length of the tool holder are stored (all in mm).
8	Geo_1	Numeric	6,2	Geometric elements relevant to the
9	Geo_2	Numeric	6,2	process planning, such as approach angle,
10	Geo_3	Numeric	6,2	rake angle, clearance angle, etc.,
11	Geo_4	Numeric	6,2	will be stored in these fields.
12	ISO_Code	Character	13	ISO code for the tool holder.
13	Bit_Code	Character	12	ISO code for the insert.
14	Cost	Numeric	4	Cost of the tool in rupees.
15	Company	Character	20	Name of the tool manufacturers.
16	Available	Character	1	Yes or true : Tool is available. No or false : Tool is not available.
17	Remarks	Character	25	Any remarks about the tool.

7. Scan the tool data base (from MRIR) for the condition

> (tool type = turning
> 1st digit (ISO 1) = clamping system
> 2nd digit (ISO 2) = insert shape
> 3rd digit (ISO 3) = holder style
> 4th digit (ISO 4) = insert clearance angle
> 5th digit (ISO 5) = direction of cutting
> 6th and 7th digits (ISO 6) ≤ Maximum shank
> 8th and 9th digits (ISO 7) ≤ Maximum shank
> 11th and 12th digits (ISO 9) ≤ exact cutting edge)

and report all the tools qualified.

8. Based on the work material, type, and nature of operation, select the insert grade.

Some Comments on Tool Selection

The selection of plain turning tools is shown in the previous section. Depending upon the machining operation to be carried out, different procedures are invoked for selecting the tool. This section highlights some features of the tool selection.

1. A tool is searched for in the data base by three types of search calls:
   • *Searching for the tool by a key parameter.* The search succeeds if the key parameter matches with a field in the data base. This type of search is used for the hole-making tools and form tools (for example, the diameter of drills and the feature code of form grooves).
   • *Searching for the tool which has a key parameter greater than or equal to the specified parameter.* This type of search is used while matching the cutting edge length.
   • *Searching for the tool which has a key parameter less than or equal to the specified parameter.* For example, this search call is applied for grooving tools (whose width should be less than the width of the groove).
2. The system has provision for maintaining an on-line dialogue with the user, thus informing about the status of the tool search. If a tool is not found, the user can specify the tool.
3. General rules for discarding some of the selected tools are employed. Some of these are
   • Selection of a larger insert over a smaller one is preferred because it requires a smaller number of passes, thus reducing the machining time.
   • The tool holder, which is wider and has a shorter length, is preferred because it gives the tool a greater stiffness.
   • In addition to these guidelines, the production practices of the participating industry, such as using different tools for roughing and finishing operations, are included while choosing a tool.
4. There can be a number of pockets attached to each setup. For each pocket, the tooling data base will deliver all tools that could accomplish the operation. The tool selection procedure initially keeps all these tools in a list. In the final stage, the number of tools can be minimized by extracting a common tool set. To summarize, this module tags each machinable pocket with a tool and a machining direction. In the current implementation of the system, the machining direction will be towards the head stock since it is assumed that the pockets lying on either side of the pocket will be machined in different setups. The results are written back to PPIR. These data are used to calculate the process parameters and the time estimates in the next stage of pocket processing.

The functions of the tool selection module in Micro-GIFTS thus include (a) getting the necessary inputs from partial PPIR, PDIR and MRIR; (b) searching the cutting tool data base to decide the tools for machining each pocket; and (c) attaching the tool code to the pocket data in PPIR. The cutting tools available in a given shop floor are stored in the tools data base which is managed by Data-GIFTS. The selection of one of these tools involves the determination of the key parameters and searching the tool data base based on such key parameters.

The execution of this module tags each machinable pocket with a tool and a machining direction. In the current implementation of the system, the machining direction will be towards the head stock since it is assumed that pockets lying on either side of the pocket will be machined in different setups. The results are written back to PPIR. These data are used to calculate the process parameters and the time estimates in the next stage of pocket processing.

Process Parameter Selection

Selection of proper cutting parameters is important as these affect the quality and cost of a machined component. An optimization module in GIFTS is based on mathematical models formulated with an objective function of minimizing the production time (Prasad, 1994). Practical limitations are imposed on speed, feed, and depth-of-cut through constraints. These constraints include the spindle power, workpiece rigidity, tolerances, and surface finish, apart from the permissible ranges of speeds, feeds, and depths of cut. The mathematical models are solved using a geometric programming technique.

Optimization of process parameters is attempted only for primary operations. The parameters for secondary operations (such as grooving, threading, etc.) are specified as a fraction of the turning parameters.

This module gets the necessary inputs from other models of GIFTS and determines the optimum parameters. These parameters are then passed back to PPIR.

With the introduction of sophisticated and highly expensive CNC machines, the need for the optimal utilization of these resources is increasingly felt. This requires consideration of the machining economics in the selection of cutting conditions. It has been realized recently that the optimal process parameter selection becomes one of the important functions of the process planning activity.

Several optimization models are reported in the literature for turning operations (Balakrishnan and DeVries, 1982; Prasad et al., 1993). These models can be broadly divided into two categories: (1) unconstrained optimization models and (2) constrained optimization models.

Unconstrained Optimization Models

Unconstrained optimization has been studied by Brown (1962) and Okushima and Hitomi (1964). The main disadvantage of unconstrained optimization is that it does not represent a realistic machining situation because the practical constraints on the machining variables are not considered.

Constrained Optimization Models

Practical constraints acting on the machining parameters are considered in these models. The constrained models can be classified into single-pass and multipass models. Each of these can further be subdivided into probabilistic and deterministic models. Probabilistic models consider the uncertainties in the empirical relations used in model development.

Several *single-pass models* are reported in the literature. They range from pure graphical solutions to analytical solutions. The graphical solutions include nomograms (Brewer and Rueda, 1963) and the performance envelope concept proposed by Crookall (1969).

The *multipass models* reported by Iwata et al. (1977) and Hati and Rao (1976) come under the category of probabilistic models. The deterministic models have used optimization techniques such as geometric programming (Ermer and Kromodihardjo, 1981), sequential quadratic programming (Chua et al., 1991), dynamic programming (Hayes and Davis, 1979), Powell's unconstrained method (Yang and Seireg, 1992) and partial differentiation (Kals and Hijink, 1978). Hinduja et al. (1985) used tool-specific, depth-of-cut-feed diagrams which represent the possible cutting region to produce easily disposable chips. The selected feed and depth of cut from these diagrams are tested against various velocity-independent constraints, and then the corresponding speed is calculated from the "equivalent chip thickness" form of tool-life equation.

Optimization Approach

Some of the points that should be considered while developing an optimization module as part of a CAPP system are

- *Optimization as part of CAPP systems* The optimization system should form an integrated module with CAPP systems.
- *Flexibility of mathematical models* The mathematical model formulated should be flexible, taking care of only the necessary constraints and omitting the unnecessary ones. For example, during a roughing operation the constraints limiting the minimum values of feed and depth of cut can be ignored, while during a finishing operation the maximum feed, minimum speed, and maximum depth of cut constraints can be eliminated. This reduces the complexity of the problem, thereby yielding faster solutions.
- *Automatic formulation of problem* The optimization system should be capable of automatically formulating the problem, depending on the type and nature of operation being considered, without any human intervention.

Inputs for Optimization Module

The optimization module gets the following inputs from the other modules of the CAPP system, as shown in Figure 5.32.

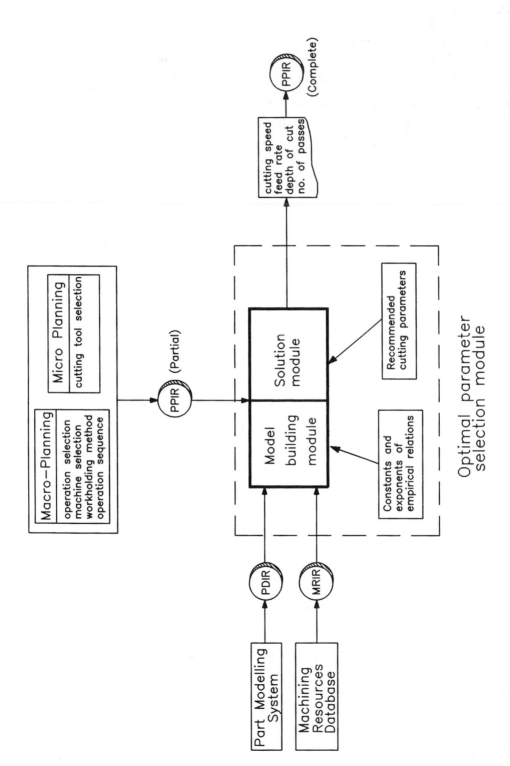

FIGURE 5.32 Inputs to optimization module.

The optimization module will get the details of the part data, both geometrical and technological, from PDIR which is the output of the modeling module. The details pertaining to the material removal volumes (called "pockets"), such as size and shape of the pockets, type of machining operations, and sequence of operations, are obtained from the process planning module. The machine and cutting tool specifications are obtained from the respective modules. The details of work-holding devices and the holding method are taken from the setup planning module. Apart from the part model, the other inputs mentioned above are obtained from process PPIR, which is the internal storage format of the process plan. PPIR gets filled up successively in each of the above mentioned modules and when the control passes through the optimization module, it becomes complete.

In addition to these, the coefficients for the empirical equations used in this module, such as tool-life, cutting power, and cutting force equations, are stored in a machinability data base for different combinations of work material, workpiece hardness, tool material, and tool geometry. The proper coefficients are retrieved from the data base, based on the above mentioned inputs.

Formulation of the Mathematical Model

The mathematical model for the optimization problem involves the formulation of the objective function and the formulation of the constraints.

Formulation of Objective Function

The optimization is usually carried out to achieve any one or a combination of the following objectives: (1) minimization of production cost, (2) minimization of production time or maximization of production rate, and (3) maximization of profit rate. Proper selection of criteria depend on the policy of the company. The minimization of production time is taken as the basis for formulating the objective function below.

The unit production time (T_{pu}) can be expressed as the sum of the loading and unloading time, the machining time, tool changing time, and tool resetting time, i.e.,

$$T_{pu} = T_l + T_m + \frac{T_m}{T} t_{tch} + T_{rs}$$

The machining time for turning operation is given by

$$T_m = \frac{\pi D L}{1000 v f}$$

At low cutting speeds, tools have a higher life but productivity is low, and at higher speeds the reverse is true. This suggests that there is an optimum that balances tool life and cutting speed. Based on experimental work, the following tool life formula is proposed:

$$V T^n = C$$

where T is the tool life in minutes, V is the cutting speed, m/min, and C and n are constants.

Though this is a fairly good formula, it does not take all the affecting parameters into account. As a result, the applicability of the above formula is restricted to very narrow regions of cutting process parameters. This formula was extended to reduce this deficiency:

$$V T^n f^{n_1} d^{n_2} = C$$

This is the most common tool life equation employed by a number of researchers. The constants for the above equation for some common work materials are given in the table below (Widia, 1986). Besides the cutting process parameters, tool life depends on work material as well as tool material. The constants, therefore, are given for each combination of work and tool material.

Tool material ISO Grade	Work material AISI	Exponent for			Constant C
		Tool life, n	Feed, n_1	Depth of cut, n_2	
P01, P10	1020	−0.38	−0.06	−0.10	1150
P20, P30	1020	−0.38	−0.17	−0.11	780
P01, P10	1045	−0.22	−0.21	−0.11	350
P20, P30	1045	−0.22	−0.34	−0.12	226

When these values are substituted, the expression for T_{pu} becomes

$$T_{pu} = T_1 + \frac{\pi DL}{1000vf} + \frac{\pi DLt_{tch}}{1000C} v^{\frac{1}{n}-1} f^{\frac{1}{n_1}-1} d^{\frac{1}{n_2}} + T_{rs}$$

In this equation, T_l is a constant. In a single-pass operation, the term T_{rs} does not have relevance. Also, the depth of cut, d, is fixed and is known in advance. Hence, there are effectively only two variables to be determined in a single-pass operation, i.e., the cutting speed and the feed rate.

The final form of objective function for a single-pass operation will be to minimize

$$T'_{pu} = C_{01}v^{-1}f^{-1} + C_{02}v^{\frac{1}{n}-1}f^{\frac{1}{n_1}-1}$$

where

$$C_{01} = \frac{\pi DL}{1000} \qquad C_{02} = \frac{\pi DLt_{tch}d^{\frac{1}{n_2}}}{1000C}$$

Similarly, the objective function for a multipass turning operation with m passes can be formulated as given below:

$$T_{pu} = T_l + \frac{\pi L}{1000}\sum_{i=1}^{m}\frac{D_i}{v_i f_i} + \frac{\pi Lt_{tch}}{1000C}\sum_{i=1}^{m} D_i v_i^{\frac{1}{n}-1} f_i^{\frac{1}{n_1}-1} d_i^{\frac{1}{n_2}} + m.T_{rs}$$

Formulation of Constraints

Practical limitations are imposed on speed, feed, and depth of cut through constraints. The *power constraint* assumes significance only in case of rough machining and hence can be ignored in the case of finish machining. The power consumed, P, during a turning operation can be expressed as

$$P = \frac{v^{a_p}f^{b_p}d^{c_p}k_c}{60.1000.\eta}$$

This value of cutting power should not exceed the maximum power, P_{max} available on the machine tool. In the above expression, the value of specific cutting force is constant for a particular work material. Hence, the expression for this constraint can be written as

$$C_p v^{a_p} f^{b_p} d^{c_p} \le P_{max}, \quad C_p = \frac{k_c}{60.1000.\eta}$$

The *surface finish constraint* limits the maximum feed that can be used to attain the required surface finish on the machined feature. This constraint becomes active during finish turning. The expression for

CLA value of the geometric surface finish obtained during turning operation with a tool-of-nose radius *r* is given as

$$SF = 1000 \cdot \frac{f^2}{18\sqrt{3}r}$$

Based on the surface finish, SF_{max}, specified on the turned surface, the constraint on feed can be expressed as

$$C_s f^2 \le SF_{max}, \quad C_s = \frac{1000}{18\sqrt{3}r}$$

In *tolerance constraints*, the machining complex, comprising the machine tool, workpiece, work holding device, and cutting tool, will be subjected to deflection under the action of cutting forces. This in turn causes inaccuracies on the diameter of component. The radial component of cutting force is dominant in causing the deflections in the machining complex (Kovan, 1959).

The deflections produced in a plain cylindrical component can be calculated based on the theory of elasticity. However, in the case of stepped components, calculation of deflections becomes difficult. Hence, finite element method (FEM) procedures are used in the present work to compute the radial deflection of the component.

The given component can be converted into a stepped component. In case of taper feature, a step with the mean diameter can be created. Grooves can be considered turn features with diameter as the starting diameter of the groove. The stepped component can then be modeled as an elastic beam with two-noded elements. Unit cutting force is applied on the surface being machined and the corresponding deflections in the component are calculated at different nodes using FEM procedures (Chandrupatla and Belegundu, 1991; Hinduja et al., 1985). The work-holding devices such as the head stock, tail stock, and carriage are assumed to be flexible, since they cannot have an infinite stiffness value (Prasad, 1994).

The maximum deflection, δ_{max}, produced in the component due to the radial component of the cutting force can be expressed in terms of the cutting variables as

$$\delta_{max} = F_r \times \delta_u = C_f v^{a_f} f^{b_f} d^{c_f} \delta_u$$

Twice this radial deflection should not exceed the diametral tolerance specified on the turned surface, D_{tol}. Thus,

$$\delta_{max} \le 0.5 \times D_{tol} \quad - \quad C_f v^{a_f} f^{b_f} d^{c_f} \delta_u \le 0.5 \times D_{tol}$$

The above expression can also be used to check for cylindricity constraint. In this case, maximum deflection, δ_{max}, is equal to the difference between the maximum and minimum deflections produced in the surface being machined.

In rough machining operations, *workpiece rigidity constraint* exists because, due to higher depths of cut and feed rates used, high cutting forces are developed, leading to significant deflections in the component. Hence, the maximum deflection δ_{rm} produced in the machined length of the component during rough machining, due to the radial component of cutting force, should not exceed a preset value, Δ_{max}. This type of check can be made to prevent the incidence of chatter (Hinduja et al., 1985). The final form of the constraint can then be expressed as

$$C_f v^{a_f} f^{b_f} d^{c_f} \delta_{rm} \le \Delta_{max}$$

Usually, *maximum and minimum speed constraints* are imposed by the machine tool. However, in the case of carbide and ceramic tools, certain minimum cutting speeds need to be maintained to avoid the failure of these cutting tools due to burring formation or micro-chipping. Hence, the minimum speed constraint is determined as the larger of the values of minimum cutting speed from machine tool and minimum cutting speed due to cutting tool (V_{min}). Thus, the speed constraint can be expressed as

$$\max\left\{\frac{\pi D N_{min}}{1000}, V_{min}\right\} \le v \le \frac{\pi D N_{max}}{1000}$$

The *maximum feed constraint* for the cutting tool (f_{tmax}) is set per the recommendations given by the cutting tool manufacturer (Widia, 1989). According to these recommendations, f_{tmax} should not exceed 0.4–0.5 times the insert corner radius for triangular inserts and 0.6–0.7 times the corner nose radius for square inserts. The *minimum feed constraint* value is set to the smallest feed rate available on the machine tool. This constraint checks the feed rate during finishing operation from becoming too small. Thus the feed constraints can be expressed as

$$f_{min} \le f \le \min\{f_{min}, f_{t\,max}\}$$

The *maximum depth of cut constraint* has relevance in roughing operations, while the *minimum depth of cut constraint* should be considered in finishing operations. The maximum limit on depth of cut due to cutting tool ($d_{t\,max}$) is set to half the cutting edge length for inserts with included angle of 55° or 60° and for inserts with included angle of 80°–100°, $d_{t\,max}$ is set to two-thirds of the cutting edge length (Widia, 1989). Minimum limit on depth of cut (d_{min}) is set to 0.5 mm for finishing operations, and 1 mm or depth of pocket, whichever is lower, for roughing operations. The depth of cut constraints can thus be expressed as

$$d_{min} \le d \le d_{max}$$

Another important constraint in case of multipass machining is *total depth constraint*. The sum of depths of cut in individual passes, both roughing and finishing, must be equal to the total depth of material to be removed.

$$\sum_{i=1}^{m} d_i = d_t$$

This constraint may not be relevant if equal depths of cut are used for all passes.

Solution Methodology

In the case of multipass optimization, the parameters to be determined are the number of passes, cutting speed, feed, and depth of cut to be used in each pass. As the number of passes is not known *a priori*, the number of variables becomes unknown. This makes a complex problem to solve.

When the total depth of material has to be removed in more than one pass and there is surface finish or diametral tolerance or both specified on the machined surface, the machining of this pocket needs both roughing and finishing passes. As the governing constraints for rough and finish turning operations are different, these two operations are treated separately.

If rigorous optimization has to be carried out for each individual pass, the solution obtained will be computationally expensive. In any case, the cost of employing different depths of cut does not appear to make this approach realistic in conventional machining. While it is possible to accommodate changing depths of cut in the case of NC machines, the usefulness of employing varying depths of cut will depend upon whether there is a significant improvement with respect to objective function. It is therefore safe to assume that equal depths of cut will be used for all roughing passes.

The solution procedure can proceed with finding optimal parameters for the finishing operation first. In case of finishing operations, it can be assumed that the finishing operation is a single-pass operation. This is a reasonable approximation as only very little material will be left to be removed in the finishing operation to achieve the required finish or tolerance on the surface being machined.

The parameters can then be determined for roughing operations, based on the remaining depth of material. The optimization for multipass rough turning operations is carried out in two stages. In the first stage, a set of feasible depths of cut, based on the total depth of material to be removed and the permissible limits of depths of cut (resulting in an integral number of roughing passes), can be calculated. To determine the optimal number of passes and hence the depth of cut to be used, optimization will be carried out for each feasible subset of depth of cut and number of passes. Finally, the subset which results in the least production time is chosen to be the optimal pair. The corresponding cutting speed and feed rate are chosen to be the optimal parameters for the roughing operation. The solution methodology is shown in Figure 5.33.

The individual optimization problem is solved using the geometric programming technique. The dual of the geometric programming (GP) problem is solved by using the techniques of separable programming (SP) and linear programming (LP) (Kochenberger et al., 1973). Simplex algorithm is used for the solution of the LP problem. The primal variables are determined from the dual solution using the method given in Beightler and Phillips (1976). The steps in GP solution procedure are given in Figure 5.34. Details on the solution procedure of GP dual problem can be found in Kochenberger et al. (1973) and Prasad (1994).

Similar formulation can be carried out for facing and boring operations (Prasad et al., 1994a, 1994b). The explicit optimization of cutting parameters for secondary operations such as chamfering, grooving, filleting, knurling, and thread cutting is not necessary. Instead, the parameters for these operations are chosen as a fraction of turning conditions. Formulation for milling operations in a similar approach can be found elsewhere (Manidhar, 1995).

Time and Cost Calculation

The time required for machining a component is equal to the sum of cutting, setup, and handling times. Handling time includes the time involved in tool change, tool resetting, and component loading and unloading. Based on the outputs obtained so far (machining operation, pocket dimensions, and process parameters), the cutting times can be calculated using the machining formulas. However, the calculation of setup times and the handling times is a data-intensive activity since each company has its own set of standard elemental times.

5.15   Report-GIFTS: Report Generation Module

When the control passes through all of the above modules, PPIR contains the necessary information a CAPP system is supposed to generate. The PPIR generated for the example part shown in Figure 5.18 is given in Table 5.6.

As explained earlier, PPIR is basically a set of setups, with each setup pointing to a set of pockets. However, PPIR is an internal model and its format is meant for computer processing. Hence, it is not useful for any practical purpose unless it is converted to the required form, comprehensible by the people involved in using the CAPP system.

A report generation module, called Report-GIFTS, is developed to serve this purpose. This module can be regarded as a postprocessor which transforms the data from one format to another format. When one looks at various formats of the process plans, four possibilities can be visualized:

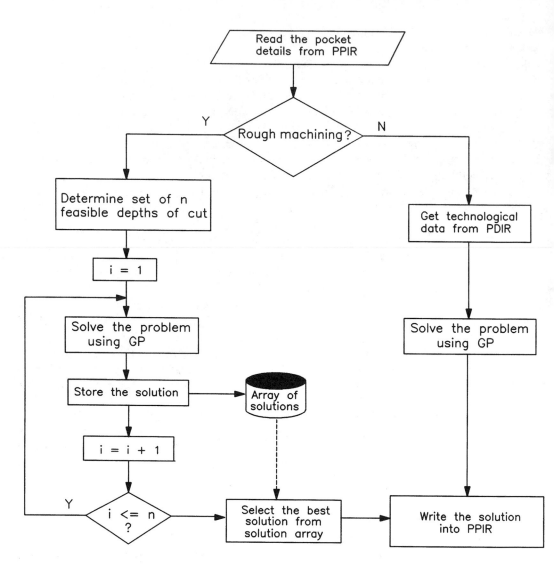

FIGURE 5.33   Solution methodology.

- Textual process plans (in one or more formats)
- Pictorial process plans
- Graphical simulation
- NC programs and/or CL data.

In the present study, these are termed process plan external representations (PPER) and are shown in Figure 5.35.

Textual Process Plans

The ability to generate the process plans in the prescribed formats specified by the company is one feature that is essential in a CAPP system. It is a common practice in industry to use different formats of the process plans depending upon the function of the personnel (or department). For example, the process plan with simple instructions is handed over to the operators on the shop floor while the same

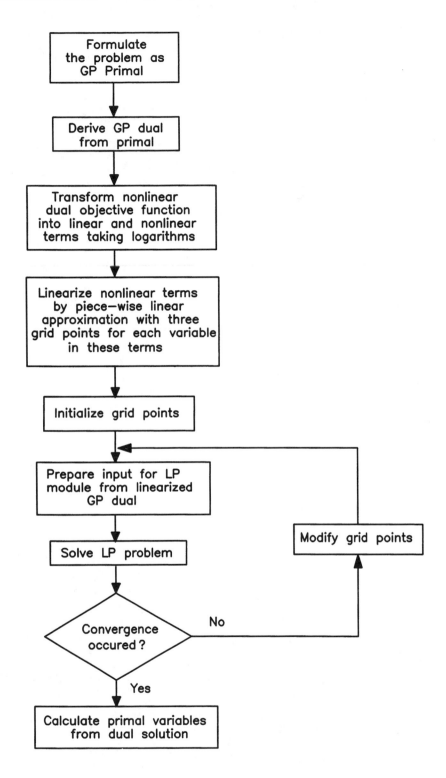

FIGURE 5.34 GP solution procedure.

TABLE 5.6 Process Plan Internal Representation (PPIR) for the Part Shown in Figure 5.18

Machine + Setup Details

1	2	3	4	5	6	7	8	9	10	11	12	13
4-T-15	TRUE	CO	50.00	110	50	0	0	0	0	0	0	0
4-T-15	TRUE	CO	30.00	70	30	0	0	0	0	0	0	0

Pocket + Machining Details

1	2	3	4	5	6	7	8	9	10	11	12	13	14	15	16	17	18	19	20	21	22	23	24	25	26	27	28
15	E	0.00	50.00	6.50	50.00	6.50	0.00	0.00	0.00	—	FA	FACING	ROUGH	U	TT49	244.92	0.50	3.25	2	—	0.00	2.13	0.00	1	1	N	0
24	E	6.50	50.00	81.50	50.00	81.50	45.50	6.50	45.50	—	TU	TURNING	ROUGH	R	TT47	177.65	0.50	4.50	1	—	0.00	1.13	0.00	1	1	N	0
21	E	6.50	45.50	81.50	45.50	81.50	32.00	6.50	32.00	—	TU	TURNING	ROUGH	R	TT47	177.65	0.50	4.50	3	—	0.00	2.86	0.00	1	1	N	0
19	E	6.50	32.00	79.50	32.00	79.50	30.00	6.50	30.00	—	TU	TURNING	ROUGH	R	TT47	261.00	0.50	2.00	1	—	0.00	1.04	0.00	1	1	N	0
17	E	6.50	30.00	41.50	30.00	41.50	22.50	6.50	22.50	—	TU	TURNING	ROUGH	R	TT47	213.16	0.50	3.75	2	—	0.00	2.06	0.00	1	1	N	0
1	E	6.50	22.50	8.50	22.50	6.50	20.50	0.00	0.00	—	CH	CHAMFERING	—		0.00	0.00	0.00	—	0	—	0.00	0.00	0.00	1	1	N	2
4	E	41.50	30.00	43.50	30.00	41.50	28.00	0.00	0.00	—	CH	CHAMFERING	—		0.00	0.00	0.00	—	0	—	0.00	0.00	0.00	1	1	N	6
3	E	38.50	22.50	41.50	22.50	41.50	20.00	38.50	20.00	—	GR	GROOVING	—	B	GR19	0.00	0.00	0.00	0	—	0.00	0.00	0.00	1	1	N	4
5	E	79.50	30.00	81.50	32.00	79.50	32.00	2.00	0.00	—	FI	FILLETING	—		FIR2	0.00	0.00	0.00	0	—	0.00	0.00	0.00	1	1	N	8
2	E	8.50	22.50	38.50	22.50	38.50	0.00	8.50	0.00	—	KN	KNURLING	—		KN	0.00	0.00	0.00	0	—	0.00	0.00	0.00	1	1	N	3
16	E	193.50	50.00	200.00	50.00	200.00	0.00	193.50	0.00	—	FA	FACING	ROUGH	U	TT49	244.92	0.50	3.25	2	—	0.00	2.13	0.00	2	2	N	0
23	E	81.50	50.00	193.50	50.00	193.50	45.50	81.50	45.50	—	TU	TURNING	ROUGH	R	TT47	177.65	0.50	4.50	1	—	0.00	1.20	0.00	2	2	N	0
22	E	117.50	45.50	193.50	45.50	193.50	22.50	117.50	22.50	—	TU	TURNING	ROUGH	R	TT47	173.77	0.50	4.60	5	—	0.00	4.12	0.00	2	2	N	0
20	E	119.00	22.50	193.50	22.50	193.50	21.50	119.00	21.50	—	TU	TURNING	ROUGH	R	TT47	183.82	0.50	1.00	1	—	0.00	1.01	0.00	2	2	N	0
18	E	148.50	21.50	193.50	21.50	193.50	18.00	148.50	18.00	—	TU	TURNING	ROUGH	R	TT47	175.75	0.50	3.50	1	—	0.00	1.03	0.00	2	2	N	0
13	E	81.50	45.50	117.50	45.50	117.50	45.00	81.50	45.00	—	TU	TURNING	FINISH	R	TT47	371.75	0.35	0.50	1	—	0.00	1.08	0.00	2	2	N	11
14	E	119.00	21.50	148.50	21.50	148.50	21.00	119.00	21.00	—	TU	TURNING	FINISH	R	TT47	175.75	0.24	0.50	0	—	0.00	1.09	0.00	2	2	N	15
6	E	81.50	45.00	83.50	45.00	81.50	43.00	0.00	0.00	—	CH	CHAMFERING	—			0.00	0.00	0.00	0	—	0.00	0.00	0.00	2	2	N	10
7	E	115.50	45.00	117.50	45.00	117.50	43.00	0.00	0.00	—	CH	CHAMFERING	—			0.00	0.00	0.00	0	—	0.00	0.00	0.00	2	2	N	12
9	E	146.50	21.00	148.50	21.00	148.50	19.00	0.00	0.00	—	CH	CHAMFERING	—			0.00	0.00	0.00	0	—	0.00	0.00	0.00	2	2	N	16
12	E	191.50	18.00	193.50	18.00	193.50	16.00	0.00	0.00	—	CH	CHAMFERING	—			0.00	0.00	0.00	0	—	0.00	0.00	0.00	2	2	N	20
10	E	148.50	18.00	151.50	18.00	151.50	15.50	148.50	15.50	—	GR	GROOVING	—	B	GR19	0.00	0.00	0.00	0	—	0.00	0.00	0.00	2	2	N	18
8	E	117.50	22.50	119.00	21.00	119.00	22.50	1.50	0.00	—	FI	FILLETING	—		FIR1.5	0.00	0.00	0.00	0	—	0.00	0.00	0.00	2	2	N	14
11	E	151.50	18.00	191.50	18.00	191.50	15.50	151.50	15.50	—	TH	THREADING	—		TH84	0.00	0.00	0.00	0	—	0.00	0.00	0.00	2	2	N	19

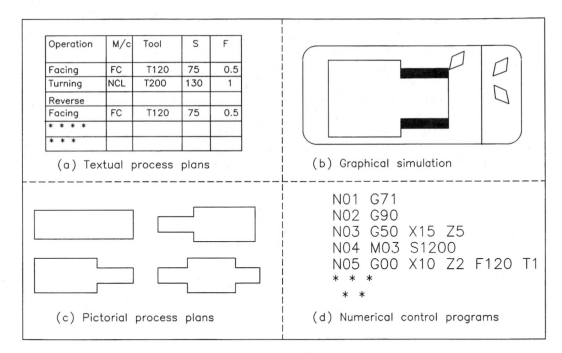

Operation	M/c	Tool	S	F
Facing	FC	T120	75	0.5
Turning	NCL	T200	130	1
Reverse				
Facing	FC	T120	75	0.5
* * * *				
* * *				

(a) Textual process plans

(b) Graphical simulation

```
N01  G71
N02  G90
N03  G50  X15  Z5
N04  M03  S1200
N05  G00  X10  Z2  F120  T1
 *  *  *
    *  *
```

(c) Pictorial process plans

(d) Numerical control programs

FIGURE 5.35   Process plan external representation (PPER).

plan with all details of cutting time and handling time estimates is used by the production planning department.

Graphical Simulation

Graphical support in any application enhances the system-user interaction due to its transparent form of communication. Graphical support during the report generation in CAPP systems is mainly concerned with the animation of the manufacturing processes on the raw material. This gives the process planner a chance to visualize the sequence of machining operations from start to finish. During simulation of the tool movements in machining, the collisions between workpiece, clamping devices, and tools can also be visualized on the screen. This leads to the proving of the machining plan.

Pictorial Process Plans

It is also possible to get pictorial process plans which reveal the status of the component (or machining) after each setup (or even operation). These intermediate process documents combine drawing and text to illustrate the part in various stages of manufacture. This can be considered an extension to graphical simulation. These plans graphically show fixtures, tooling, clamping, setups, etc. and facilitate better understanding of the process plans.

NC Programs

NC code can also be treated as one type of PPER as these are CAM instructions usually written based on the process plan. Generation of CL data, which in turn will be postprocessed to NC code depending upon the machine tool controller, can also be attempted.

In a strict theoretical perspective, a report generation module in a CAPP system should be capable of providing the process plans in one or all of the above formats. This becomes possible only when the basic process plan information is represented in a structured format. In this context, the role of PPIR and Report-GIFTS in the proposed system can be appreciated.

It is to be noted that PPIR alone cannot provide all the details that should appear on the process planning sheets. This is because some fields in PPIR are pointed to other models of GIFTS. These pointers, adopted from the concept of relational data base managements systems (RDBMS), can be thought of as the mechanisms providing meaningful relationships among models. For example, using the machine code and tool code (which are stored in PPIR) as the pointers, the machine tool and cutting tool details can be obtained from the data bases (MRIR). Similarly, the part name, drawing number, batch size, etc. are available as global data in PDIR.

In the current implementation of the system, several macros are developed in Report-GIFTS for automatic translation of PPIR details to predetermined textual formats. Once translated, these are stored in a text file. This file consists of the process planning details (description of setup, machine, and operations, tools to be used, and the process parameters) in a comprehensible form. This file can be retrieved into any word processor for editing the plan if desired and to take the printout of the process plan. Graphical simulation of the plans is also attempted by representing the tool shapes symbolically, the details of which can be found in the next chapter. The process (PPER) generated by Report-GIFTS for the example part shown in Figure 5.18 is given in Table 5.7.

TABLE 5.7   Process Plan for the Part in Figure 5.18—PPER in Textual Format

A. B. C. Ltd.		Process plan for the part EX_GIFTS					QTY:1
No.	Machine/Setup/Operation	Tool	Speed (m/min)	Feed (mmpr)	Doc (mm)	NOP	Time (min)
1	4-A-23 reverse and hold in chuck						
	Face	TT49	198.00	0.16	6.50	1	0.790
	Rough turn dia 100.00 to 64.00 over a length of 75.00	TT49	170.79	0.21	6.00	3	3.700
	Rough turn dia 64.00 to 60.00 over a length of 73.00	TT49	258.52	0.52	2.00	1	0.610
	Rough turn dia 60.00 to 45.00 over a length of 35.00	TT49	161.98	0.17	7.50	1	0.740
	Chamfer dia 45.00		161.98	0.17	2.00	1	0.010
	Chamfer dia 60.00		161.98	0.17	2.00	1	0.010
	Cut groove of width 3 mm	GR19	121.49	0.04	3.00	1	0.070
	Make fillet	FIR2	121.49	0.02	2.00	1	0.190
	Knurl dia 45.00	KNS1	121.49	0.03	1.00	1	1.020
2	4-A-23 reverse and hold in chuck						
	Face to length	TT49	162.00	0.21	6.50	1	0.760
	Rough turn dia 100.00 to 91.00 over a length of 112.00	TT49	161.32	0.33	4.50	1	1.150
	Rough turn dia 91.00 to 45.00 over a length of 76.00	TT49	162.63	0.24	5.75	4	4.300
	Rough turn dia 45.00 to 43.00 over a length of 74.50	TT49	211.86	0.52	1.00	1	0.600
	Rough turn dia 43.00 to 36.00 over a length of 45.00	TT49	187.13	0.38	3.50	1	0.580
	Finish turn dia 91.00 to 90.00 over a length of 36.00	TT47	429.18	0.35	0.50	1	0.580
	Finish turn dia 43.00 to 42.00 over a length of 29.50	TT47	202.43	0.24	0.50	1	0.580
	Chamfer dia 90.00		202.43	0.24	2.00	1	0.010
	Chamfer dia 42.00		202.43	0.24	2.00	1	0.010
	Chamfer dia 36.00		202.43	0.24	2.00	1	0.010
	Cut groove of width 3 mm	GR19	101.21	0.12	3.00	1	0.020
	Make fillet	FIR1.5	151.82	0.05	1.50	1	0.030
	Thread dia 36.00	TH84	40.72	2.50	0.48	10	1.000

TH84: R166N-3232M-310 (Sandvik code); TT49: PDJNR2525M15 (insert: DNMG150608); GR19: Form grooving tool; TT47: PDJNR2525M15 (insert: DNMG150612); KNS1: Straight knurling tool with pitch 1 mm; FIR1.5: Filleting tool with radius 1.5 mm; FIR2: Filleting tool with radius 2 mm.

Acknowledgements

The work reported here has been done at Indian Institute of Technology, New Delhi. The majority of the work reported is done under the sponsorship of the Department of Electronics, Government of India, over a period of five years. A large number of students (A.V.S.R.K. Prasad, G. Manidhar, Vijay Arora, S.K. Gupta, and others) have worked in this period for their masters and doctoral programs in the area of CAPP and are responsible for developing part of the ideas. The assistance received from Bharat Heavy Electricals Limited, Bhopal, in the system development and proving is gratefully acknowledged.

References

1. Arora, V., 1991, Pocket Identification and Setup Planning in Computer Aided Process Planning of Rotational Parts, Unpublished M. Tech. Thesis, NC Laboratory, Department of Mechanical Engineering, Indian Institute of Technology, New Delhi, India.
2. Balakrishnan, P. and DeVries, M. F., 1982, A review of computerized machinability data base systems. *Proceedings of 10th North American Manufacturing Research Conference*, 348–386.
3. Beightler, C. S. and Phillips, D. T., 1976, *Applied Geometric Programming*, John Wiley and Sons, Inc, NY.
4. Brewer, R. C. and Rueda, R., 1963, A simplified approach to the optimum selection of machining parameters. *Engineers' Digest*, 24(9), 133–151.
5. Brown, R. H., 1962, On the selection of economical machining rates. *International Journal Production Research*, 1(1), 1–22.
6. Butterfield, W. R., Green, M. K., Scott, D. C., and Stoker, W. J., 1986, Part features for process planning, CAM-I report: C-85-PPP-03.
7. Case, K. and Gao, J., 1993, Feature technology: an overview, *Int. J. Computer Integrated Manufacturing*, 6(1&2), 2–12.
8. Chandrupatla, T. R. and Belegundu, A. D., 1991, *Introduction to Finite Elements in Engineering*, Prentice-Hall, New Delhi.
9. Chang, T. C. and Wysk, R. A., 1981, An integrated CAD/automated process planning system, *AIIE Transactions*, 13(3), 223–233.
10. Chang, T. C. and Wysk, R. A., 1984, Integrating CAD and CAM through automated process planning, *International Journal of Production Research*, 22(5), 877–894.
11. Chen, S. J., Hinduja, S., and Barrow, G., 1989, Automatic tool selection for rough turning operations, *Int. Journal of Machine Tools and Manufacture*, 29(4), 535–553.
12. Chua, M. S., Loh, H. T., Wong, Y. S., and Rahman, M., 1991, Optimization of cutting conditions for multipass turning operations using sequential quadratic programming. *Journal of Materials Processing Technology*, 28(1/2), 253–262.
13. Clark, A. C. and South, N. E., 1987, Feature based design of mechanical parts, *Proc. AUTOFACT* 87, 169–176.
14. Colding, B. N., 1992, Intelligent selection of machining parameters for metal cutting operations: the least expensive way to increase productivity. *Robotics & Computer Integrated Manufacturing*, 9(4/5), 407–412.
15. Crookall, J. R., 1969, The performance-envelope concept in the economics of machining. *International Journal of Machine Tool Design and Research*, 9, 261–278.
16. Cunningham, J. J. and Dixon, J. R., 1988, Designing with features: the origin of features, *Proc. ASME Computers in Engineering Conf.*, v. 1, San Francisco, USA, 237–243.
17. Dixon, J. R., Cunningham, J. J., and Simmons, M. K., 1987, Research in designing with features, *Intelligent CAD, I*, (Ed.), Yoshikawa, H. and Gossard, D., *Proc. IFIP* TC5/WG 5.2 Workshop on Intelligent CAD, Boston, MA, 137–148.

18. Domazet, D. S. and Lu, S. C. Y, 1992, Concurrent design and process planning of rotational parts, *Annals of the CIRP*, 41(1), 181–184.

19. Dong, X. and Wozny, M. J., 1990, Feature volume creation for computer aided process planning, in: *Geometric Modeling for Product Engineering*, (Ed.) Wozny, M. J., Turner, J. U. and Priess, K., Elsevier Science Publishers B. V., North Holland, 385–403.

20. Drake, S. and Sela, S., 1989, A foundation for features, *Mechanical Engineering*, 111, 66–73.

21. Ermer, D. S. and Kromodihardjo, S., 1981, Optimization of multipass turning with constraints. *Journal of Engineering for Industry, Transactions of ASME*, 103(4), 462–468.

22. Evershiem, W. and Holz, B., 1982, Computer aided programming of NC machine tools by using the system AUTAP-NC, *Annals of the CIRP*, 31(1), 323–327.

23. Faux, I. D., 1988, Modeling of components and assemblies in terms of shape primitives based on standard D & T surface features, in: *Geometric Modeling for Product Engineering*, (Ed.) Wozny, M.J., Turner, J. U., and Preiss, K., Elsevier Science Publishers B. V., North-Holland, 259–275.

24. Gindy, N. N. Z., Huang, X, and Ratchev, T. M., 1993, Feature-based component model for computer-aided process planning systems, *Int. J. Computer Integrated Manufacturing*, 6(1&2), 20–26.

25. Gindy, N. N. Z., 1989, A hierarchical structure for form features, *Int. J. Prod. Res.*, 27, 2089–2103.

26. Gossard, D. C., Zuffante, R. P., and Sakurai, H., 1988, Representing dimensions, tolerances and features in MCAE systems, *IEEE Computer Graphics & Applications*, March, 51–59.

27. Gupta, N. and Kapoor, S., 1986, Computer Aided Process Planning and Part Programming for CNC Turning Centers Using Group Technology, Unpublished B.Tech. Thesis, N C Laboratory, Department of Mechanical Engineering, Indian Institute of Technology, New Delhi, India.

28. Halevi, G. and Weill, R., 1985, Influence of manufacturing tolerances on fixturing of machined parts in process planning systems, *1st CIRP Working Seminar on CAPP*, Paris, 31–34.

29. Halevi, G. and Weill, R. D., 1995, *Principles of Process Planning A Logical Approach*, Chapman & Hall, London.

30. Hati, S. K. and Rao, S. S., 1976, Determination of optimal cutting conditions using deterministic and probabilistic approaches. *Journal of Engineering for Industry, Transactions of ASME*, 98(1), 354–359.

31. Hayes, C., 1990, Machine Planning: a Model of an Expert Level Planning process, Ph.D. thesis, The Robotics Institute, Carnegie Mellon University, Pittsburgh, PA.

32. Hayes, G. M. and Davis, R. P., 1979, A discrete variable approach to machine parameter optimization, *AIIE Transactions*, 11(2), 155–159.

33. Henderson, M. R. and Anderson, D. C., 1984, Computer recognition and extraction of form features: A CAD/CAM link, *Computers in Industry*, 5(4), 329–339.

34. Hinduja, S. and Barrow, G., 1986, TECHTURN: a technologically oriented system for turned components, *Proc. 1st Int. Conf. on CAPE*, Edinburgh, UK, 255–260.

35. Hinduja, S. and Huang, H., 1989a, OP-PLAN: an automated operation planning system for turned components, *Proc. of Inst. of Mech. Engrs.*, 203(B3), 145–158.

36. Hinduja, S. and Huang, H., 1989b, Automatic determination of workholding parameters for turned components, *Proc. of Inst. of Mech. Engrs.*, 203(B2), 101–112.

37. Hinduja, S., Petty, D. J., Tester, M., and Barrow, G., 1985, Calculation of optimum cutting conditions for turning operations. *Journal of Engineering Manufacture, Proceedings of Institution of Mechanical Engineers*, 199(B2), 81–92.

38. Iwata, K. and Arai, E. ,1983, Development of integrated modeling system for CAD/CAM of machine products, in: *Advances in CAD/CAM*, (Ed.) Ellis, T.M.R. and Semenkov, O. I., North Holland Publishing Company.

39. Iwata, K., Murotsu, Y., and Oba, F., 1977, Optimization of cutting conditions for multipass operations considering probabilistic nature in machining processes. *Journal of Engineering for Industry, Transactions of ASME*, 99(1), 210–217.

40. Jasthi, S. R. K., 1993, Some Studies on Form Features in the Development of a CAPP system, Unpublished Ph.D. Thesis, Department of Mechanical Engineering, Indian Institute of Technology, New Delhi, India.

41. Jasthi, S. R. K., Prasad, A. V. S. R. K., Manidhar, G., Rao, P. N., Rao. U. R. K., and Tewari, N. K., 1994, A feature based part description system for Computer Aided Process Planning, *Journal of Design and Manufacturing*, Vol. 4, 67–80.

42. Jasthi, S. R. K., Rao, P. N., and Tewari, N. K., 1995, Studies on Process Plan Representation in CAPP system, *Journal of Computer Integrated Manufacturing Systems*, Vol. 8, No 3, 173–184.

43. Joshi, S., Chang, T. C., and Liu, C. R., 1986, Process planning formalization in an AI framework, *Artificial Intelligence*, 1(1), 45–53.

44. Joshi, S., 1990, Feature recognition and geometric reasoning for some process planning activities, in: *Geometric Modeling for Product Engineering*, (Ed.) Wozny, M. J., Turner, J. M., and Preiss, K., Elsevier Science Publishers B. V., North-Holland, 363–384.

45. Kals, H. J. J. and Hijink, J. A. W., 1978, A computer aid in the optimization of turning conditions in multi-cut operations. *Annals of the CIRP*, 27(1), 465–469.

46. Kalta, M. and Davies, B. J., 1992c, Guidelines to build 2D CAD models of turned components in CAD-CAPP integration, Internal Report, Mechanical Dept., UMIST, UK.

47. Kang, T. S. and Nnaji, B. O., 1993, Feature representation and classification for automatic process planning systems, *Journal of Manufacturing Systems*, 12(2), 133–145.

48. Kimura, F., Suzuki, H., and Wingard, L., 1986, A uniform approach to dimensioning and tolerancing in product modeling, *Proc. 2nd Int. Conf. on Comp. Appl. in Production & Eng.*, Copenhagen, 165.

49. Klein, A., 1988, A Solid Groove: feature based programming of parts, *Mechanical Engineering*, 110, 37–39.

50. Kochenberger, G. A., Woolsey, R. E. D., and McCarl, B. A., 1973, On the solution of geometric programs via separable programming. *Operational Research Quarterly*, 24(2), 285–294.

51. Kovan, V., 1959, *Fundamentals of Process Engineering*, Moscow, Foreign language publishing house.

52. Kramer, T. R., 1992b, A library of material removal shape element volumes (MRSEVs), National Institute of Standards and Technology report (NISTIR 4809).

53. Kramer, T. R., 1992a, Issues concerning material removal shape element volumes (MRSEVs), National Institute of Standards and Technology Report (NISTIR 4804).

54. Kusiak, A., 1989, Process planning: a knowledge based and optimization perspective, *Proc. IFAC on Decisional Structures in Automated Manufacturing*, Genova, Italy, 133–138.

55. Kusiak, A., 1990, Optimal selection of machinable volumes, *IIE Transactions*, 22(2), 151–159.

56. Manidhar, G., 1995, Some Studies in the Design and Development of a CAPP System for Rotational Parts with C-axis Features, Unpublished Ph.D. Thesis, Dept. of Mech. Eng., IIT, New Delhi.

57. Matropoulos, P. G. and Hinduja, S., 1991, Automatic tool selection for rough turning, *International Journal of Production Research*, 29(1), 1185–1120.

58. Nau, D. S., Gupta, S. K., Kramer, T. R., Regli, W. C., and Zhang, G., 1993, Development of machining alternatives based on MRSEVs, *Proc. of ASME Computers in Engineering Conference*.

59. Okushima, K. and Hitomi, K., 1964, Study of economical machining—analysis of maximum profit cutting speed. *International Journal of Production Research*, 3(1), 73–78.

60. Opitz, H., 1970, *A Classification System to Describe Workpieces*, Pergamon Press, Elmsford, New York.

61. Prasad, A. V. S. R. K., 1994, Optimal Selection of Cutting-process Parameters in a Computer Aided Process Planning System, Unpublished Ph.D. Thesis, Dept. of Mech. Engg., IIT, New Delhi.

62. Prasad, A. V. S. R. K., Jasthi, S. R. K., Manidhar, G., Rao, P. N., Rao, U. R. K., and Tewari, N. K., 1993, A survey of approaches to the optimization of process parameters in turning operations. *Proceedings of 2nd International Conference on Computer Integrated Manufacturing*, Singapore, 916–920.

63. Prasad, A. V. S. R. K., Rao, P. N., and Rao, U. R. K., 1994a, Computer Aided selection of Optimal Parameters for Boring process, *Proc. 10th National Convention of Mechanical Engineers, Organized by Institution of Engineers, India at Hyderabad*, 207–214.
64. Prasad, A. V. S. R. K., Rao, P. N., and Rao, U. R. K., 1994b, Optimization of process parameters for facing operations, *Proc. of ISME Conference*, Roorkee, India.
65. Ranyak, P. S. and Fridshal, R., 1988, Features for tolerancing a solid model, *Proc. ASME Computers in Engineering Conference*, 1, 263.
66. Requicha, A. A. G. and Chan, S. C., 1986, Representation of geometric features, tolerances and attributes in solid modellers based on constructive geometry, *IEEE Journal of Robotics and Automation*, RA-2(3), 156.
67. Requicha, A. A. G., 1983, Toward a theory of geometric tolerancing, *Int. J. of Robotics Research*, 2(4), 45–60.
68. Shah, J. J. and Rogers, M. T., 1988b, Functional requirements and conceptual design of the feature based modeling system, *Computer Aided Engineering Journal*, 5, 9–15.
69. Shah, J. J., 1991, Assessment of features technology, *Computer Aided Design*, 23(5), 331–343.
70. Sim, S. K. and Leong, K. F., 1989, Prototyping a feature based modeling system for automated process planning, *Journal of Mechanical Working Technology*, 20, 195–205.
71. Suzuki, H., Inui, M., Kimura, F., and Sata, T., 1988, A product modeling system for constructing intelligent CAD and CAM systems, *Robotics & CIM*, 4(3/4), 483–489.
72. Vogel, S. A. and Adard, E.J., 1981, The AUTOPLAN process planning system, *Proc. of 18th NC Society*, Annual Meeting and Technical Conference, Dallas, TX, 422–429.
73. Weill, R., 1988, Integrating dimensioning and tolerancing in computer aided process planning, *Robotics & CIM*, 4(1/2), 41–48.
74. Widia, 1986, *Recommended Data for Turning Ferrous Materials* (Publication No. W 5.3-10.3 e 486).
75. Widia, 1989, *Widax Tools for External and Internal Machining.* Widia (India) Limited, WMC-042-89.
76. Wilson, P. R. and Pratt, M. J., 1988, A taxonomy of features for solid modeling, in *Geometric Modeling for CAD Applications*, (Ed.) Wozny, M. J., McLaughlin, H. W., and Encarnacao, J. L., Elsevier Science Pub. B.V. (North-Holland), 125–136.
77. Wright, T. L., and Hannam, R. G., 1989, A feature based design for manufacture: CAD/CAM package, *Computer Aided Engineering Journal*, 6, 215–220.
78. Yang, D. Y. and Seireg, A., 1992, Machining parameter optimization for specified surface conditions. *Journal of Engineering for Industry,* Transactions of ASME, 114(2), 254–257.
79. Yeo, S. H., Rahman, M., and Wong, Y. S., 1990, A frame-based approach for the making of holes in turned parts and its further development, *Journal of Materials Processing Technology*, 232, 149–162.
80. Zhang, S., and Gao, W. D., 1984, TOJICAP: A system of computer aided process planning system for rotational parts, *Annals of the CIRP*, 33(1), 299–301.

6

Computer-Aided Design (CAD) Training Programs and Their Application in Electronics and Telecommunication Manufacturing Systems

Christopher McDermott
Rensselaer Polytechnic Institute

6.1 Introduction

The successful adoption of computer-aided design (CAD) technology requires a company to make a number of choices that implicitly determine its implementation strategy. These implementation decisions affect the ease of assimilating CAD into the firm's design and manufacturing processes, the level of satisfaction the firm experiences with the technology, and the ultimate benefits that are achieved by the firm. Decisions regarding training are among the most important in the CAD adoption process. Although it is the consensus of research[1,2,7,21,22,26,35] that the successful adoption of new technology requires a firm to implement new training policies, there are few guidelines that indicate how companies with differing needs and expectations should design training programs in order to meet their own desired objectives.

The purpose of this chapter is to explore the organizational factors affecting the choice of a CAD training program and the associated cost savings. Specifically, we evaluate the variations in the style, format, and techniques in training programs used by companies that have successfully adopted PC-based

CAD systems. The sample consists of companies primarily associated with the electronics and telecommunication industries who represent different sizes and organizational structures. This research proposes a model that depicts how training programs are affected by organizational factors and how the choice of a training program may lead to specific benefits for the company. Empirical data from intensive clinical case studies are included to highlight several of the important and interesting alternatives that managers face in their selection of appropriate training methods. The research provides helpful information regarding the strengths and weaknesses associated with different training methods employed under different working environments. The findings of this study raise a number of issues related to the development of CAD training programs, as well as the costs and benefits of CAD.

6.2   Issues in CAD Training

Given the complexity of many high-tech product designs, it is often impossible to produce and modify the requisite designs without using a CAD system. As CAD use becomes prevalent in a number of industries, particularly the electronics, telecommunications, and automotive industries, the company that uses CAD can no longer assume that it holds a competitive advantage. Instead, it may find that its competitors are also using CAD and also realizing the benefits of both time savings and improved quality in design work. In these industries, CAD use is reduced to a *qualifying* requirement for sustained design competitiveness. However, given the intense cost pressures these industries are experiencing due to global competition, a company can achieve significant cost savings by developing a CAD training plan that enhances the value added by design work. Thus, when CAD becomes a core technology, the issue for management is no longer whether to use CAD, but rather how to strategically leverage the full potential of the CAD system through cost-effective training programs aimed at improving worker skills and abilities.

It has been shown that CAD helps companies increase design productivity and improve the quality of designs. Forslin and Thulestedt[19] found that 90% of CAD users surveyed felt that their CAD system had produced time savings. However, successful use of CAD does not come without proper worker training. Majchrzak and Salzman[30] also stress the importance of developing worker skills and training programs as critical social and organizational dimensions that are crucial to the firm in reaching the full potential of CAD systems.

Training programs focus on improving skills so that the worker will be able to operate the new system efficiently. "Skill relates to the ability to perform a particular task in a particular way and is built up by an individual over time. A technical innovation like CAD requires the acquisition of certain new skills while rendering others largely redundant."[26] Bretz and Thompset state that, although it is estimated that U.S. organizations spend over $44 billion per year on employee training, "training methods are often seen as fads, training program evaluation is rare, and rigorous evaluation is virtually nonexistent."[10] Thus, a company's selection of the *proper* training program or method may lead to significant cost savings relative to other companies that have made suboptimal choices with respect to training. However, companies rarely perform a systematic evaluation of their training processes in order to determine how to improve the selection of training methods so that future training programs are more effective.[38]

A review of recent research relating to CAD systems and their associated training programs is useful in understanding the complexity of the major issues. The review of the CAD acquisition literature is broken down into two categories of issues: pre-adoption and implementation. Pre-adoption training issues are those in which training is considered prior to a firm's final decision regarding purchase of the new technology. These issues are critical because unless a firm is attentive to the important "people" issues, (e.g., the extent to which the entire workforce *needs* to be trained, or whether the present workforce is willing or able to *be* trained for the new technology) prior to the acquisition, there is a significant risk that the adoption will not be successful. Implementation issues are those which occur after the firm has *already committed* to the adoption of the system. A key issue is the design of the implementation process in order to maximize the usefulness of the system, while keeping costs in check. Examples of implementation issues (to be discussed later in this section) include decisions regarding the training method that will be used and the deployment of lower-skilled workers in using the new technology (de-skilling). Table 6.1 displays the issues according to their stage.[37]

TABLE 6.1   Issues of CAD Adoption

Stage	Question	Issues
Pre-Adoption	1. What is the feasibility of training?	What is an estimation of cost?
		When to give training?
		Where to get training?
	2. How to gain expected benefits?	Who needs training?
		Are funds available?
	3. How will it impact work flow?	Worker attitudes
		Worker morale
		Worker skills
Implementation	1. What type of training?	Formal or informal training?
		Off- or on-site?
		Vendor or consultant?
		Customized or generic?
	2. Who is to be trained?	Skilled or semi-skilled workers?
	3. Will management be trained?	Supervisors or higher-level managers?
	4. Is training included in the budget?	What type of training can be afforded?

Pre-Adoption Issues

Pre-adoption training issues often center around the tradeoffs between the anticipated out-of-pocket costs and the associated benefits of CAD. One of the primary managerial challenges is to assess accurately the training needs associated with the proposed CAD system and to convert them into a budget of cash expenditures. Some of the major factors that affect training costs are discussed in this section. They include: (1) determining the types and number of workers to be trained; (2) the technological aptitude or "trainability" of the workers; (3) the training method to be employed; and (4) the loss of productivity during the learning period. The benefits of CAD often include the expected cost savings or additional revenues accrued when the current methods are replaced by the proposed system. However, some anticipated benefits may be difficult to quantify accurately. For instance, dynamic market conditions can present a compelling justification for the need for a CAD system; the benefit may be survival through sustained competitiveness of the firm. On the other hand, problems with worker attitudes and morale can jeopardize the success of the implementation of the CAD system and create an opportunity cost with respect to unrealized benefits from the technology. These issues are discussed below.

One major issue in the adoption decision is whether it is possible to re-train the present workforce to perform the tasks required under the new CAD system. This question involves examining both the "person" and "task" components of the training decision.[21] It is possible that the tasks and procedures used in the CAD system are so high level that some of the potential users may not possess the necessary skills and education to learn the new system within a reasonable time period. Worker receptivity may also be a problem if the existing workers exhibit non-adoptive behavior and decide that they are not interested in learning or using the new technology. Beatty and Gordon[7] found evidence of such non-adoptive behavior in certain instances of CAD implementation. Lack of worker support can jeopardize any hope of success in implementing the new CAD technology. It is critical, therefore, that the company accurately inventory its workers' skills and abilities and assess workers' attitudes toward using the new system.

A company considering the adoption of a CAD system must also include an estimate of the total expense of educating the workforce. For example, the acquisition cost of a CAD system is much more than just the price of the computer hardware and software. A more complete estimate of expenses includes both training program expenses and a "cushion" for the loss of productivity experienced as employees learn to master the new system. An acquisition budget that does not include these "soft" costs can cause financial problems for an adopter firm that hasn't anticipated these expenses. Research has found that the underestimation of training and education expenses is a major reason for failure in CAD systems implementation.[3] The underestimation of CAD costs has created situations in which the firm was overextended to the point it could no longer remain competitive; in some instances, the result was the demise of the business.

A related issue is the extent to which training is offered throughout the firm. It is often critical to train employees besides those who are involved in design in order for the firm to truly understand and use the technology to its full potential. Examples of such employees may include CAD supervisors, secretarial staff, and manufacturing employees. Sometimes a new system may be so radically different from the existing one that any employee who has the slightest contact with CAD should receive training. A recent survey of firms using CAD systems showed that 50% of all companies with CAD/CAM training provided training to a wide range of occupational groups, including all shop-floor staff and managers, materials management, repair and maintenance, manufacturing engineering, and, of course, design engineering.[29] CAD systems were found to be so different from traditional design methods that workers who did not receive any CAD training were at a tremendous disadvantage when they came into contact with the system. Thus, the need for CAD training across many different work groups cannot be underestimated. It is critical that management accurately assess these needs and calculate the associated training costs before choosing to adopt a CAD system.

Managers also need to factor training needs into the cost-benefit tradeoff. If the firm considers adopting a new CAD system that promises relatively modest benefits yet requires a great deal of employee training, it may be difficult to justify the new system. When the combined expense of the system and the training provides only incremental gains, managers may be wise to invest the money in more traditional manual methods.[36,42] On the other hand, firms experiencing market pressures such as changes in government or industrial standards, or customer demand for one-off products might *need* the system (and the associated training program) for survival. Thus, in such instances, the cost-benefit tradeoff is moot.

Small firms provide special cases for each of the issues raised above. Because of the limited resource availability in smaller companies, the issues often represent tighter constraints than those placed upon larger companies. For example, smaller firms frequently do not have the slack resources to provide a "buffer" for employees to learn the new technologies. As a result, the loss of productivity and the required training period that frequently accompany switching to a new technology create a loss of revenue that small firms may find difficult to overcome.[16,29,42] Small firms are also hurt by the high price of CAD equipment, which may consume a significant portion of their available funds, and limit their ability to develop an appropriate training program. These problems would seem to imply that the size of the firm determines of successful CAD use and should be considered in the pre-adoption decision. However, there are multitudes of small firms that have successfully implemented CAD systems and are doing very well financially.

Implementation Issues

Implementation issues are those confronted by the firm after it is already committed to the new technology. The main concern of training issues associated with the implementation process is how to most efficiently and effectively develop the workforce so that the system achieves its maximum potential. Some of the training issues associated with implementation are similar to the pre-adoption issues discussed above, while many issues are completely different. In the implementation phase, training issues involve not only the financial cost implied by a particular training strategy but also consideration of the effectiveness of the training methods, the development of desired skill sets, and the extent to which the value of design work has been enhanced through de-skilling. The training strategy involves a pattern of decisions regarding: (1) identity of the workers to be trained; (2) extent to which management will be trained; (3) location of the training sessions; (4) selection of the trainer; (5) timing of the training relative to installation; and (6) training methods employed. The issues inherent in a training strategy are discussed below.

When a new technology is brought into a firm, management must decide to what extent they are going to train the workforce. Specifically, the individual workers *within* specific functional groups that will receive training must be identified. Although much is being done in universities and technical schools to provide future workers with specialized technological skills,[1,2,18] almost all firms find the need to re-train a group of their existing employees to operate and work with a new system. From a financial standpoint, it is appealing to set up a specialized group within the firm to be trained, while the rest of the employees remain untrained. By doing so, the firm would limit its training expenses. This might be feasible if only *part* of the design work will be utilizing the new CAD system, while the remaining work will be done using existing methods.

Another tactic used to limit training costs restricts training only to new hires. Although these policies may sound appealing, in practice they often have led to counterproductive results.[18,22,27] Research shows that the deliberate exclusion of a portion of workers from a training program leads to perceptions of an "elite" group of workers; other untrained workers may feel they are being "put out to pasture." Such feelings frequently lead to poor morale and an erosion in worker/manager relationships.

Some firms avoid this training problem by hiring employees from other companies who are already trained.[22,27] In larger firms with established design teams, this has led to morale problems and results similar to those where workers were excluded from training. It is an interesting, but not surprising, finding to note that when companies choose to offer training only to select employees, there seems to be an age bias. Studies have found that although managers would not **say** that age entered into their decision process, the probability of being chosen to be a CAD user drops by 2% for each year of age.[22] For example, a six-year age difference would correspond to a 12% difference in the probability of a worker being selected for training. In this study, the average age of users was 39 years, while the average age of non-users was 48 years. This apparent bias may not be unfounded: Gist, Rosen and Schwoerer studied the ability of workers to learn new computer software and concluded that younger workers (under 45 years old) performed significantly better than their older counterparts on comprehensive examinations given after the training session.[20]

Training must also be considered for members of management. Research suggests that training multiple layers in the firm's hierarchy (e.g., both workers and managers) is beneficial, if not necessary, in many applications.[1,11,18] Brooks and Wells found that managers who are not familiar with the new CAD system may have trouble supervising workers because they fail to understand the implications of the changes in design work.[11] Often a skill differential arises between a trained worker and an untrained supervisor which creates conflict or a loss of status for the manager. An untrained supervisor may also experience difficulty in effectively planning and controlling the work flow since he or she has no basis from which to estimate drawing and alteration times. Evaluating the worker's progress and assessing performance of the individual also becomes difficult for the supervisor. The combination of these problems often leads to the untrained supervisor's "losing track" of the workers. The result can be a strained worker/supervisor relationship and a loss of productivity. The literature on Electronic Data Interchange (EDI) corroborates the above observations of CAD experiences. Carter, Monczka, Clauson and Zelinski note that management training in EDI is a major factor in increasing the probability that managers responsible for implementing EDI will succeed.[13] Thus, managerial training may well be a critical determinant to a successful CAD implementation.

The firm implementing a new technology must also select the method it will use to train the workers. In addition, management must decide whether the training will be conducted in-house using its own personnel or conducted externally by a vendor or an outside consultant. These decisions are often based on the firm's need for either a highly formalized or a more tutorial-oriented training program. Internal training programs can often be customized for the firm's specific needs relative to CAD and facilitate a more informal and open exchange of ideas than external programs. External education, on the other hand, is frequently more formalized and has the additional advantage of being less expensive. Often it is the only option for smaller firms that may not have the money, facilities, or personnel to conduct internal training.[29,42] Externally conducted training programs have the disadvantage of being more generic and, therefore, frequently less useful in meeting the unique needs of the firm.

Although researchers appear to agree upon the list of advantages and disadvantages regarding how training is conducted, there is not a consensus as to which type of training is better. Engleke, for example, recommends having the training done on-site, without vendors,[18] while Hubbard feels that the best training sessions are highly structured and relatively formal classes taught by professional instructors at off-site locations.[23] Majchrzak found that 45% of the firms she surveyed had in-house company-sponsored training programs.[29] She also found that larger firms generally chose to use the in-house, more focused training, while smaller firms did not.

Another type of training method is the informal transfer of training. This method allows some users to learn in the classroom and then return to the workplace and act as tutors in training other employees. Thus, training is transferred from the classroom to the work setting.[5] This method is effective only if the relevant knowledge is actually transferred. A critical factor in the successful transfer of training is the time interval that elapses between the classroom training and actual hands-on usage of the new technology. Beatty provides a case example of a company that trained its employees six months in advance of the implementation of a new CAD system.[6] Not surprisingly, most of the material taught in the training sessions had been forgotten in the period between training and actual usage. It is critical that a minimum amount of time elapse between CAD training and routine usage. Engelke estimates that the half-life for advanced software training is about two weeks if it is not used immediately.[18]

The timing dimension of CAD training does not have an obvious solution. If training is provided in advance of system installation, the worker is prepared to use the system once it is implemented, and, hence, the usual productivity loss due to "getting up to speed" is reduced. However, scheduling problems often arise. For example, the purchased equipment may arrive later than anticipated, or it may not run properly immediately after installation. In the meantime, the employees forget what they have learned. In contrast, some firms opt to wait and train the worker after the system has arrived and is functional, even though there will be unavoidable downtime while workers learn the system.

In addition to establishing the *timing* of the training program, it is also necessary to choose the appropriate training methodology. A thorough needs analysis should be undertaken to identify the design task, the personnel, and the desired type of training. The choice of the appropriate training method should consider learning objectives, trainee characteristics, current knowledge about the training process, and practical considerations such as constraints and costs in relation to benefits.[43] It is important that the selected method provide a good fit between the needs of the trainee relative to the design tasks which will be performed.[21]

Selection of training methods reinforces the development of different skill sets. Training programs can be classified as *formal* (e.g., classroom settings with lectures) or *informal* (e.g., tutoring, "hands-on" learning). Formal training reinforces the development of implicit skills which are generally acquired through formal engineering education or a technical training program. With respect to CAD, studies conducted by Beatty concluded that formal training is the prevalent method used in CAD environments.[6] Informal training, while less common in CAD environments, encompasses all other types of CAD knowledge transfer, including apprenticeships, tutorials, and location of workers who are knowledgeable of the CAD system with their less knowledgeable counterparts in the same work area. Some types of informal training tend to reinforce the development of tacit skills that largely are developed through experience. In addition, some types of informal training also have the benefit of being unencumbered by many of the environmental barriers to learning (e.g., fear of failure, boredom, negative reinforcement for asking questions, and lack of application to the work situation) which are often associated with formal methods.[10] In reallocating design work from professionals to less skilled employees, tacit skills diminish in relative importance when compared to implicit skills.[3] Although the quality of design work has traditionally been based on the tacit skills possessed by the designers, the acquisition of these skills may be de-emphasized through the choice of the training/education process.

Vendors of CAD software packages often provide training (usually formal) to firms that purchase their software, a feature which may play a significant role in making the training decision. Firms often assume that the vendor-supplied training that accompanies the software will suffice to bring their employees "up to speed" on the new system. Internal training (conducted by in-house personnel) is the norm for informal training; formal training may be conducted either internally or externally (using a vendor or an outside consultant to conduct the training). The advantages and disadvantages of both internally and externally conducted training programs in technology implementation are well documented in the literature.[5,17,22,27,32,33] However, it is the issue of formal vs. informal training in CAD environments that is the primary interest of the remainder of this chapter.

Finally, training issues must be balanced with the enhancement of the value added by design work. In an attempt to cut labor costs associated with using CAD, managers have tried to use less skilled workers to operate the new technology after its installation. With CAD, computer operators have been extensively substituted in the place of professional designers and draftsmen.[27,33,39] It has been suggested that the introduction of CAD systems causes a certain degree of de-skilling. The de-skilling process "results in the reduction of skills requirements by those using the new technology. In PCB design, a de-skilling process would involve a relative shift in work content, expanding routine tasks…."[39] Thus, this strategy attempts to achieve cost savings through CAD by allowing the firm to hire cheaper labor to operate the system. A similar strategy has been proposed by proponents of office automation for increasing the productivity in the office setting while decreasing the salaries of white collar workers.[24,40]

This de-skilling process has not, however, been successful in all applications within the CAD industry. A study of the West German mechanical engineering CAD industry examined firms that attempted to train a group of workers with no previous design experience to run the CAD system.[31] The study concluded that the firm's attempts to bypass the requisite design training and, instead, to emphasize computer literacy limited the success of the de-skilling process to only clerical tasks. Salzman[39] provides two examples of unsuccessful de-skilling attempts in which companies, one in the aerospace industry and the other in the electronics division of a large firm, attempted to assimilate workers trained only as computer operators into the CAD group. The study concluded that, in both cases, the attempts failed because the workers lacked the background knowledge to perform design work successfully. CAD, like many new technologies, is more than the automation of routine tasks; it requires the operator to possess a deep and thorough understanding for the design process to be successful. Attempts at de-skilling in several industries have produced equivocal results with some arguing that rather than enhancing the jobs of less skilled workers, de-skilling has actually degraded them.[4,8,9,44,46]

The literature states that, for many firms, the gains from implementation of systems such as CAD frequently come not only from savings in labor costs, but also from improvements in quality and speed of the design process.[29] In the competitive arenas of the electronics and telecommunications industries, cost, quality, and speed are no longer strategic tradeoffs in design work. Instead, a design group must be competitive in all three dimensions. If de-skilling through CAD is to be successful, it must simultaneously improve design performance in all three of these dimensions.

Organization Structure and Technology

Organizational issues constitute the third critical factor, after person and task, in the selection of a training program.[21] The organizational structure, whether organic or mechanistic, can help determine the CAD training strategy.[25,41] Burns and Stalker[12] discuss the importance of organizational structure to successful business ventures, showing that organic firms are more successful in innovative environments than mechanistic organizations. Organic firms are characterized by worker autonomy, a team orientation, a professional community, and decentralized decision making. Relationships between managers and workers tend to be less formal.[12,15,28] Davis and Wilkof[17] define an organic system as a professional organization which is bound together through formal and informal norms derived from a commonality of interest. It is this common interest that keeps the organization together. The rules tend to be less rigid, and the manager's spans of control are smaller. In organic environments, task accomplishment and innovation are moved to the most knowledgeable parties.[12]

Mechanistic structures, in contrast, are more formal and structured. They are characterized by less worker autonomy, centralized decision making, and more authoritative and hierarchical relationships between manager and worker.[12]

Link and Zmud[28] found that organic structures are the preferred environments for R&D activities. Their study concluded that organic structures encourage greater R&D efficiency. Covin and Slevin[16] reinforced the work of Burns and Stalker[12] with their results from a study of small firms. They found that an organic structure was more likely to be successful when the environment was highly competitive, while mechanistic structures were more successful in less competitive, stable environments. Davis and Wilkof,[17] while studying the transfer of scientific and technical information, observed the relationship between organizational structure and efficiency of information transfer. They concluded that one of the best ways to improve information transfer is to alter the organizational structure toward a more open organic system.

6.3   Issues of Organizational Fit in CAD

The formality of job tasks, type of decision making, and work environment are all part of the organizational context, and they can greatly influence the effectiveness of the training program. To explore these issues further, a study of the interaction of these constructs was undertaken.[33] To reduce the inconsistencies that might result if the investigation included such varied applications as CAD systems in architecture and automobile design, the scope of the study was limited to the use of CAD systems in the design and manufacture of printed circuit boards used in electronic devices. The CAD systems that were used by the firms constituting the sample were used for either schematic capture or single and/or multi-layered PC board layout. All firms were either in the electronics or telecommunications industries.

The firms participating in this study were selected based on the researchers' prior knowledge of their use of CAD systems, professional affiliations, and referrals from representatives of CAD software vendors. The sample was restricted to respondents who used different vendors and different CAD systems to prevent a systematic bias which might result from all respondents using the same CAD system. The firms varied in size from 75 to more than 20,000 workers. The unit of analysis was identified as the design work group. The size of the individual work groups selected varied from as little as one to as many as 100 CAD users. Studies were conducted in eight different design groups housed in seven different organizations. The two groups belonging to the same organization were located in different geographic locations, worked in diverse product lines, and used different CAD systems; thus, the work groups will be referred to as "firms." Two of the work groups were located in North Carolina (25%), five groups were located in Maryland (62.5%) and the remaining group was located in Texas (12.5%).

A multimethod design was used in gathering data for this study. A questionnaire was developed which used a 5-point Likert scale in questions designed to assess satisfaction with the CAD training methods employed, satisfaction with the CAD system, and benefits from CAD—including the extent to which design work had been de-skilled, realized cost savings, and quality improvements. In addition, interview questions were developed to provide richer insights into the issues and causal links in this exploratory study than the survey questions alone could have provided.[34] The interviews were conducted using a semi-structured format which allowed respondent elaboration on a number of issues, while keeping the flow of the interview along a predetermined course. This structure allowed respondents to elaborate on issues that they felt to be important or significant, while making sure all of the planned topics would also be discussed.

Given the interview format used in data collection, the size of the sample was intentionally kept small. Several authors have noted that smaller sample sizes are common for interview and case study research.[14,35,45] The person selected to be interviewed in each work group varied from firm to firm. Initially, it was the intention of the researchers to interview the manager of the design group. However, during the pilot study, it became apparent that the manager was not always sufficiently knowledgeable of the CAD system to answer the questions. Thus, the person interviewed in the study was selected by asking members of the design group (including the manager) who *they* felt had the most knowledge of the CAD system. As a result, interviews were conducted with designers, managers, and heads of internal training programs.

At the outset of the study, it was hypothesized that organizational context would be associated with distinct and unique types of training, which in turn, would provide specific advantages. The firms were broadly grouped as either organic or mechanistic; the classification methodology is discussed below. Training programs were classified as either formal or informal. The nature of the training program was determined from open-ended interview questions. The training program was determined to be formal if it involved primarily classroom settings and lecture format. Informal training programs were those where the workers were expected either to learn the system on their own *or* to receive unstructured one-on-one tutoring from a co-worker.

Each choice of training method might then be associated with its specific advantages, as partially suggested by Carter et al.[9] It should be noted, however, that while the Carter et al. study explored benefits associated with internal and external training programs, this research is examining benefits received as a function of the formality of the training program. In particular, de-skilling may be a benefit to firms using CAD as a core technology because of the potential savings that can be realized through the process of hiring and training inexperienced workers to perform the more routine aspects of design work. Based on a review of current research, the following definitions and indicators of the variables and constructs used in the study as related to the CAD environment are discussed below.

Organic Firm Structure

The organic nature of a firm using CAD is visible when workers are allowed influence over the decision process concerning work issues. Employees in organic environments exhibit more autonomy from their managers.[12] The level of autonomy is interpreted from interview questions that measure the amount of contact the worker has with his or her manager. Workers in independent, organic structures have less contact with their manager and receive less direction regarding work methods. In organic structures, the workers are assigned work and allowed the freedom and responsibility to complete it in ways they deem appropriate.[12,15] Another indicator of an organic firm is the workers' involvement in the decision process of acquiring a CAD system. In organic structures, consideration is given to the employees' opinion regarding which system to purchase. Although initially this variable seemed difficult to measure, simple questions such as, "Where is the CAD manager's office in relation to the CAD users" and "How frequently do you interact/speak with your manager/worker" are quite effective in revealing the work environment and its structure.

Mechanistic Structures

CAD firms with mechanistic structures are organized in a top-down fashion. Unlike organic structures, workers in mechanistic structures have little say in the day-to-day decisions that affect their occupations. Instead, management makes all decisions regarding the selection of the CAD system and how it is used. In general, management in mechanistic CAD firms makes itself more visible to the worker than in organic structures. Management is more likely to dictate rules and policy than to ask for the opinion and input of workers.

Formal Training

Formal CAD training programs are those with the traditional instructor/student relationship and are often taught in a classroom. They are scheduled in advance and may also include additional written supporting documents besides the instruction manual that accompanies the software. Formal training programs are often taught by outside consultants, such as CAD vendors. Large firms may maintain an in-house instructor. These instructors have frequently received their own training from outside sources. This method of CAD training was found to be prevalent in studies performed by Beatty.[6]

Informal Training

Informal CAD training encompasses most methods of transferring CAD system information within the actual work setting. For the purposes of this study, informal methods might include apprenticeships, tutorials, and location of knowledgeable and untrained workers in the same work area. Information transfer and learning can also take place during casual conversations.

Successful CAD Training

The success of any training program is ultimately dependent on user satisfaction with the system after the training is complete. Some (e.g., Adler[2]) have used financial measurements to calibrate the success of the CAD system. Although financial performance may be a valid success indicator, it can be highly susceptible to variations in the complexity and price of the components to be designed with CAD. Although this research was limited to only one type of CAD system (i.e., PCB design and layout), the pilot study revealed that differences in board complexity presented questions regarding the validity of comparisons of financial benefits. Furthermore, several of the younger firms had no basis upon which to compare the financial benefits reaped from the adoption of the CAD system; they had purchased their systems when the firm was founded.

Thus, there is no consensus regarding what constitutes the successful adoption of a technology. For the purpose of this study, successful CAD training was used synonymously with perceived satisfaction with the system. In addition, the extent to which the de-skilling process was used in the firm became a surrogate measure of the relative success of the training program. The presence of de-skilling was determined through open-ended interview questions. A firm was then classified as having allowed for de-skilling if they had *at least one* worker who had no previous design experience and was now using the CAD system.

6.4   Findings

In total, three of the firms observed were classified as mechanistic (based on the physical presence of management, level of control, and the role of the worker in CAD decision making processes), while five were classified as organic. Given the complex and dynamic competitive environment in both the electronics and telecommunications industries, it was not surprising that the organic structure was the predominant organizational form because organic structures tend to adapt better to unstable competitive environments. In total, five of the firms in the study exclusively used informal training methods, such as co-location, tutorials, and apprenticeships; two firms relied exclusively on formal instruction in a classroom setting, while the remaining firm used a combination of both methods. Six of the firms had achieved work de-skilling benefits, while two had not (see Tables 6.2 and 6.3).

The outcome of this analysis yielded a number of interesting observations. The first finding is the difference between the use of formal and informal training in mechanistic and organic firms. As can be seen, organic firms used more informal training methods, while mechanistic firms used formal techniques. Although this finding sounds trivial, it is not clear whether this is a true cause-effect relationship. In many cases, financial implications played a major role in the selection of the training method. Also, as previously noted, formal training methods have long been the predominant form of CAD training. Thus, both cost and competitive pressures may have created a situation in which the sample firms chose CAD training using informal methods.

Certain benefits were also associated with specific types of training programs. Table 6.3 displays the cross tabulation showing the relationship between the training method and the de-skilling process. Firms were classified as taking advantage of the de-skilling process if they had successfully trained one or more workers from a non-design related background to use the CAD system. Table 6.3 suggests that firms with informal training are able to take advantage of the de-skilling process. They can potentially gain financial

TABLE 6.2 Organization Structure vs. Training Method

	Formal	Informal
Organic		Firms B, D, E, F and H
Mechanistic	Firms A, C, and G	

TABLE 6.3 Training Method vs. De-skilling

	No De-skilling	De-skilling
Informal		Firms B, D, E, F and H
Formal	Firms A, G	Firm C*

*Only for one, special-case employee.

benefits as a result of hiring workers either with no previous CAD design experience or with non-technical backgrounds and then provide them with CAD training.

Firm C's experience is illustrative. The firm indicated that it spent in excess of $3,000 *per worker* on its formal CAD training program. However, it decided to initiate the de-skilling process by using **informal** training methods on *one* unskilled worker (out of a total of 20 CAD users). Unlike the rest of the training program, the informal nature of this worker's training meant there were no out-of-pocket costs. In addition, the one worker's training program was structured in a completely different manner. The otherwise highly structured training program was altered so that the worker was taught on a one-to-one basis. Whereas the formal program was of a predetermined length, the worker was allowed to pace herself through the instruction until she had mastered the system. Firm C's manager boasted that they had altered their training program so that this worker could be taught "from scratch." The manager further stated that if there was a similar need to train an unskilled worker in the future, he planned on using the same method again. When asked to comment on the formal training program, the manager stated that he felt that it had been of little value. His sentiments were not unique; other workers in firms that had conducted formal training programs felt that they had gained the majority of their system knowledge by learning it on their own.

In another illustration of the benefits of de-skilling, Firm E had trained a technician to use the system with a level of competency such that she could free engineers to do other work by assuming the smaller and less complicated projects. On the other hand, the firms with formalized training programs showed quite different results. Firms A and G had no experience in de-skilling their workforce, nor did they have any plans to attempt it.

Further, the survey revealed that there was almost universal satisfaction expressed by all firms in this study with their CAD systems. Regardless of the specific training program they used, all the firms interviewed were satisfied with the ease of system use, the time saved as a result of the system use, and the overall quality improvements. In regard to quality issues, respondents were most supportive of the gains in the ability to quickly reproduce highly legible copies of the original artwork. An interesting finding emerged when quality issues were further explored with the respondents. Although they were happy with the gains in quality, the universal response was that the CAD system did **not** reduce the number of design errors that occurred. Rather, the category of error just changed. The new system eliminated problems with illegibility of drawings, but greatly increased the number of "type-o's." The computerized systems still needed to be operated by humans, who inherently make errors.

Although there was universal satisfaction with CAD technology, satisfaction with the training program varied depending on the method used. The organic firms typically trained new workers through more informal means, such as co-location, informal tutoring and apprenticeships. In *all* cases, the respondents in organic firms were satisfied with the training procedures currently implemented and felt that they should remain in place.

A full seven of the eight firms were dissatisfied with the level of vendor support. Those firms (Firms A, C, and G) that received on-site, vendor-conducted training all complained about the poor value of their formal training programs. The most frequent complaint was the lack of system-specific training: vendor training was too generic. This result agrees with findings of the literature.[29] Firm C paid for a vendor training program and still was not satisfied. It then flew in a CAD consultant for a week, and *still* the workers claimed that they learned the most on their own. Seven of the firms, using both formal and informal training programs, felt that the after-the-purchase vendor help lines were a farce. The vendor was seldom, if ever, helpful in solving problems in a timely fashion. The comments of a respondent in Firm H were typical when he said, "I use them (vendor help lines) as a last resort, because they usually ask me to recreate my problem *exactly* in order, so that they can try to help me—and that takes forever. I'm better off just figuring it out myself, if I can." Even the firms most satisfied with the help lines only used them about twice a month. The majority of questions that arose regarding their CAD system were usually answered internally. Thus, satisfaction seemed to be inversely related to frequency of use.

Two of the three mechanistic firms (A and G) offering formal training utilized their own workers as instructors; these workers previously had received external training. The format for this instruction was highly structured with classrooms reserved for a block of a few days. In essence, these in-house instructors just "mimicked" the external training they received. The goal of this training policy was to save money on instruction directed at part-time users only. This in-house instruction was considered sufficient for workers who did not use the system full-time; full-time CAD users were still required to take CAD courses externally. These findings are similar to those found in the literature.[5] It is surprising, given the level of dissatisfaction that the firms expressed about the vendor-run programs, that these firms continued to rely on the vendor as a source of training. Perhaps the vendor, despite all the problems, was seen as offering certain instructional qualities that the firms felt they could not duplicate.

Prior research has argued that management familiarity with the CAD system is often correlated with its effective use. In our study, CAD managers in organic structures were not as familiar with the operation of the system as their counterparts in mechanistic structures. This may be partially due to the very nature of organic firms. As stated earlier, management in these firms operates in a more hands-off manner than do the managers in mechanistic firms. In addition, it may be easier to implement a management training program when there is a formal training program, rather than forcing the manager to learn the system from his workers—a situation which could be awkward and prone to creating conflict. The inability of these managers in organic firms to work with the new CAD system, however, is notable, especially in light of the findings of the literature.[1,6,11,29]

6.5 Discussion

The results of this study provide general support for a model that posits there is a direct relationship between the formalization of the training program and the organizational structure of the firm. It was the finding of this study that firms that had informal training programs were better able to take advantage of the opportunities for de-skilling. The de-skilling process frequently allows for financial savings in the use of CAD systems.[31,39] These potential savings can be considered a competitive advantage for the firms involved because they help to reduce costs and free up resources that the firm might be able to use better elsewhere (see Figure 6.1).

Mechanistic firms had the *apparent* advantage of being able to facilitate the establishment of CAD training programs for management. The word *apparent* was used above because it is not conclusive that management training is a necessity in *organic* firms. The interviews revealed that *none* of the CAD managers in firms with informal training programs was familiar with the operation of the system, while two of the three firms that utilized classroom instruction had managerial knowledge of the system. These results seem to conflict with previous research that has shown that management familiarity with a technology is often a prerequisite to the effective use of the technology.[1,7,13,29] Although this research has suggested that informal methods may pose barriers to CAD training for managers, time will tell if these

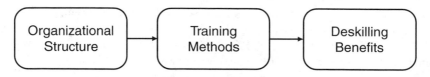

FIGURE 6.1　Observed training format and work de-skilling benefits.

firms will be penalized for not training their managers. It might be speculated that the inability of management to truly understand the CAD system may create long-term compromises in the firm's ability to find innovative ways of using CAD or in the ability of management to manage CAD workers effectively. Thus, de-skilling benefits in the short term may imply strategic costs in the long term.

The results of this study further suggest that firms concerned with gaining specific competitive advantages through CAD can *choose* the method of training to achieve their goals. Clearly, firms that wish to leverage the de-skilling process should select more informal training techniques, while firms concerned with training *all* levels of the firm should opt for a more formal program. Thus, the choice of the training strategy becomes a deliberate decision aimed at achieving strategic objectives, rather than a predestined selection, determined only by the nature of the organizational structure.

6.6   Conclusions

The implementation of a new CAD system presents many issues which the firm must confront if the true potential of the system is to be realized. The discussion here provides general empirical support for the argument that the type of training program chosen by a firm is closely related to the organizational structure. For many firms, the choice of training program involves determining the best fit to the specific organizational structure, as well as considering the required level of resource commitment. The cost-benefit analysis for a CAD system often focuses on financial requirements during the pre-adoption stage and on efficiently achieving the benefits of CAD during the implementation stage. De-skilling has been cited as one benefit that has financial implications for the firm as well as the potential to enhance value added by design work.   These findings suggest that *if* firms are particularly concerned with benefits associated with the de-skilling process, then it might be in their best interest to use more informal methods to train their workers.

One mechanistic firm in the sample adopted informal training methods in order to take advantage of the de-skilling process. It might be suggested that other mechanistic firms with similar goals alter their training format to do the same. By allowing for informal, less-constrained flows of information, these mechanistic firms might be able to achieve the same quality of CAD education that their organic counterparts seem to enjoy. Organic firms that are concerned with management training, on the other hand, might be advised to pursue more formal methods of educating CAD managers. As noted above, this training may promote the effective management of the CAD system as well as the development of a strategic vision for the use of CAD.

Finally, the apparent difference between training programs for mechanistic and organic CAD firms suggests many interesting questions. For example, would it be advantageous for mechanistic firms to adopt more organic *structures*, at least in their CAD design groups, or is the simple adoption of different training *methods* enough? Is it ever cost-effective for a firm to use a hybrid training program consisting of both formal and informal methods, as this research seems to suggest? Are there other CAD training strategies which would emerge if the study were broadened to encompass different industries? These questions merit future research.

References

1. S. A. Abbas and A. Coultas, Skills and Knowledge Requirements for CAD/CAM in *CAD/CAM*, in *Education and Training: Proceedings of the CAD ED 83 Conference*, P. Arthur (Ed.), Anchor Press, 1984.
2. P. S. Adler, CAD/CAM: Managerial Challenges and Research Issues, *IEEE Trans. Eng. Mgmt.* Vol. 36, No. 3, pp. 202–215, 1989.
3. P. S. Adler, New Technologies, New Skills, *California Management Review*, Vol. 29, No. 1, pp. 9–28, 1986.
4. P. Attewell, The De-skilling Controversy, *Work and Occupations*, Vol. 14, No. 3, pp. 323–346, 1987.
5. T. T. Baldwin and J. K. Ford, Transfer of Training: A Review and Directions for Future Research *Personnel Psychology*, Vol. 41, pp. 63–84, 1988.

6. C. A. Beatty, Tall Tales and Real Results: Implementing a New Technology for Productivity, *Business Quarterly*, Vol. 51, No. 3, pp. 70–74, 1986.

7. C. A. Beatty and J. R. Gordon, Barriers to the Implementation of CAD/CAM Systems, *Sloan Management Review*, Vol. 29, No. 4, pp. 25–33, 1988.

8. P. Botsman and P. Rawlinson, Trade Unions and New Technology: Talking to Pelle Ehn, *Work and People*. Vol. 12, No. 1, pp. 8–10, 1986.

9. H. Braverman, *Labor and Monopoly Capital: The Degradation of Work in the 20th Century*, New York: Monthly Review Press, 1974.

10. R. D. Bretz and R. E. Thompsett, Comparing Traditional and Integrative Learning Methods in Organizational Training Programs, *Journal of Applied Psychology*, Vol. 77, No. 6, pp. 941–951, 1992.

11. L. S. Brooks and C. S. Wells, Role Conflict in Design Supervision, *IEEE Trans. Eng. Mgmt.*, Vol. 36, No. 4, pp. 271–282, 1989.

12. T. Burns and G. M. Stalker, *The Management of Innovation*, London: Tavistock Press, 1961.

13. J. R. Carter, R. M. Monczka, K. S. Clauson, and T. P. Zelinski, Education and Training for Successful EDI Implementation, *Journal of Purchasing and Materials Management*, pp. 13–19, 1987.

14. T. D. Cook and D. T. Campbell, *Quasi-Experimentation: Design and Analysis Issues for Field Settings*, Houghton Mifflin Company, 1979.

15. J. A. Courtright, G. T. Fairhurst, and L. E. Rogers, Interaction Patterns in Organic and Mechanistic Systems, *Academy of Management Journal*, Vol. 32, No. 4, pp. 773–802, 1989.

16. G. C. Covin and D. P. Slevin, Strategic Management of Small Firms in Hostile and Benign Environments, *Strategic Management Journal*, Vol. 10, pp. 75–87, 1989.

17. P. Davis and M. Wilkof, Scientific and Technical Information Transfer for High Technology: Keeping the Figure on its Ground, *R&D Management*, Vol. 18, No. 1, pp. 45–58, 1988.

18. W. D. Engelke, *How to Integrate CAD/CAM Systems: Management and Technology*, New York: Marcel Dekker Inc., 1987.

19. J. Forslin and B. M. Thulestedt, Computer Aided Design: A Case Strategy in Implementing a New Technology, *IEEE Trans. Eng. Mgmt.* Vol. 36, No. 3, pp. 191–201, 1989.

20. M. Gist, B. Rosen, and C. Schwoerer, The Influence of Training Method and Trainee Age on the Acquisition of Computer Skills, *Personnel Psychology*, Vol. 41, pp. 255–265, 1988.

21. I. L. Goldstein, Training in Work Organizations, *Annual Review of Psychology*, Vol. 31, pp. 229–272, 1980.

22. P. S. Goodman and S. M. Miller, Designing Effective Training through the Technological Life Cycle, *National Productivity Review*, Vol. 9, No. 2, pp. 169–177, 1990.

23. S. W. Hubbard, *CAD/CAM: Applications for Business*, Phoenix, AZ: Oryx Press, 1985.

24. K. Hughes, Office Automation: A Review of the Literature, *Industrial Relations*, Vol. 44, No. 3, pp. 654–677, 1989.

25. D. F. Jennings and S. L. Seaman, Aggressiveness of Response to New Business Opportunities Following Deregulation: An Empirical Study of Established Financial Firms, Vol. 5, No. 3, pp. 177–189, 1990.

26. G. L. Lee, Managing Change with CAD and CAD/CAM, *IEEE Trans. Eng. Mgmt.* Vol. 36, No. 3, pp. 227–233, 1989.

27. J. K. Liker and M. Fleisher, Implementing Computer Aided Design: The Transition of Nonusers, *IEEE Trans. Eng. Mgmt.* Vol. 36 No. 3, pp. 180–190, 1989.

28. A. N. Link and R. W. Zmud, Organizational Structure and R&D Efficiency, *R&D Management*, Vol. 16, No. 4, 317–323, 1986.

29. A. Majchrzak, A National Probability Survey on Education and Training in CAD/CAM, *IEEE Trans. Eng. Mgmt.*, Vol. 33, No. 4, pp. 197–206, 1986.

30. A. Majchrzak and H. Salzman, Social and Organizational Dimensions of Computer-Aided Design, *IEEE Trans. Eng. Mgmt.*, Vol. 36, No. 3, pp. 174–179, 1989.

31. F. Manske and H. Wolfe, Design Work in Change: Social Conditions and Results of CAD Use, in Mechanical Engineering, *IEEE Trans. Eng. Mgmt.* Vol. 36, No. 4, pp. 282–297, 1989.

32. W. McGehee and P. W. Thayer, *Training in Business and Industry.* New York, Wiley, 1961.
33. C. McDermott and A. Marucheck, Training in CAD: An Exploratory Study in Methods and Benefits, *IEEE Trans. Eng. Mgmt.,* Vol. 42, No. 10, pp.410–418, 1995.
34. A. McKinney and A. Marucheck, The Effect of Environmental, Technological, and Human Resource Factors on the Success of EDI: An Exploratory Study, Kenan-Flagler School of Business Working Paper, Chapel Hill, NC, 1990.
35. J. Meredith, N. L. Hyer, D. Gerwin, S. R. Rosenthal, and U. Wemmerlov, Research Needs in Managing Factory Automation, *Journal of Operations Management,* Vol. 6, No. 2, pp. 203–218, 1986.
36. P. G. Raymount, New Directions in Training, in *CAD/CAM in Education and Training, Proceedings of the CAD ED 83 Conference.* Paul Arthur (Ed.), Anchor Press, 1984.
37. H. Rolfe, In the Name of Progress? Skill and Attitudes towards Technological Change, *New Technology, Work and Employment,* Vol. 5 No. 2, pp. 107–121, 1990.
38. L. M. Saari, T. R. Johnson, S. D. McLaughlin, and D. M. Zimmerle, A Survey of Management Training and Education Practices in U. S. Companies, *Pers. Psychology,* Vol. 41, pp. 731–743, 1988.
39. H. Salzman, Computer-Aided Design: Limitations in Automating Design and Drafting, *IEEE Trans. Eng. Mgmt.,* Vol. 36, No. 4, pp. 252–261, 1989.
40. P. Sassone, Survey Finds Low Office Productivity Linked to Staffing Imbalances, *National Productivity Review,* Vol. 8 No. 2, pp. 147–158, 1992.
41. B. Schneider and A. Reichers, On the Etiology of Climates, *Personnel Psychology,* Vol. 36, pp. 19–39, 1983.
42. D. M. Schroeder, C. Gopinath, and W. Congden, New Technology and the Small Manufacturer: Panacea or Plague? *Journal of Small Business Management,* Vol. 27, No. 3, pp. 3 10, 1989.
43. S. I. Tannenbaum and G. Yukl, Training and Development in Work Organizations, *Annual Review of Psychology,* Vol. 43, pp. 399–441, 1992.
44. S. P. Vallas, New Technology, Job Content, and Worker Alienation: A Test of Two Rival Perspectives, *Work and Occupations,* Vol. 15, No. 2, pp. 148–178, 1988.
45. R. K. Yin, *Case Study Research: Design and Methods,* London, U.K. Sage Publications, 1989.
46. G. Zicklin, Numerical Control Machining and the Issue of De-skilling, *Work and Occupations,* Vol. 14, No. 3, pp. 452–466, 1987.

7

Intelligent Techniques for the Planning, Design, and Manufacture of Progressive Dies

A.Y.C. Nee
National University of Singapore

B.T. Cheok
National University of Singapore

7.1   Introduction

Nowadays, metal stampings are used in almost every mass-produced product. This is because metal stamping is a flexible production process that can be used to produce complex, accurate, strong, and durable articles cheaply and quickly. In metal stamping, successful products are the results of good product and tool design. Modern CAD/CAM technology and new ideas in design and construction of tools, coupled with increased speed and rigidity of the presses, have contributed to the continuous use of metal stamping production processes to manufacture increasingly more sophisticated products.

Progressive dies are high precision tools used for producing metal stampings. In a progressive die, the workpieces are advanced from one station to another. At each station, one or more die operations, such as piercing, notching, blanking, lancing, shaving, drawing, extruding, embossing, coining, blanking, and forming are performed on the sheet metal strip. The result is a finished component deposited on the stack at every stroke of the press.

7.2   Planning, Design, and Manufacture of Progressive Dies

The design and manufacture of a progressive die begin with the task of planning for the sequence of die operations that need to be performed to manufacture a part. This planning process is called the strip design. This is a very important process as it affects the optimum operation of a progressive die and has a direct bearing on the cost of fabricating the die and on the cost of metal stampings produced. This planning process includes selecting the best layout of the flat pattern to optimize stock material usage(nesting), selecting and ordering a combination of punches that will stamp out the required part, selecting piloting holes to guide the strip as it progresses along the die, and ensuring that the workpieces are attached to the strip by means of a carrier web until the final cut off at the last station.

Based on the sequence of operations decided during the strip design, the die is designed. The die is designed such that the workpieces can be accurately fed and located at each station. It has to cater to the removal of slugs, ejection and stripping of parts, precision guidance for its punches, ease of replacement of punches for sharpening or replacement, safety of operation, and robustness of the die to withstand high speed cyclic stamping forces. The die is constructed by assembling together die components such as punches, punch plate, die block, pilots, strippers, stock guides, guide bushings, guide posts, and die sets located by dowel pins and held together by fastening screws.

There are two categories of components in a progressive die: (1) standard die components such as standard punches, pilots, guide bushings, guide posts, dowel pins, fastening screws, and die sets which can be purchased from die components manufacturers; and (2) non-standard components such as punch plate, die blocks and non-standard punches that need to be custom-made in the tool room. The manufacturing plan for fabricating the non-standard components will be developed first. This includes the selection of cost-effective setups, selection of fixturing requirements, selection of machine and cutting tools, generation of NC tool-paths, and heat-treatment processes. Thereafter these components are machined and treated. Finally, all the components are assembled together to form the progressive die.

7.3   Factors Influencing the Use of Intelligent Techniques for the Planning, Design, and Manufacture of Progressive Dies

The processes involved in the planning, design, and manufacture of a progressive die are shown in Figure 7.1. They involve many inter-related tasks requiring the support of multi-disciplinary engineering and tooling skills based on theories and principles, design codes, heuristic procedures, personal expertise, and experience. There is a wide variety of research on the use of intelligent techniques in solving design problems. It is essential to apply the most appropriate artificial intelligence techniques to handle these design and tool-making tasks when developing a computer-aided, integrated planning, design, and manufacturing

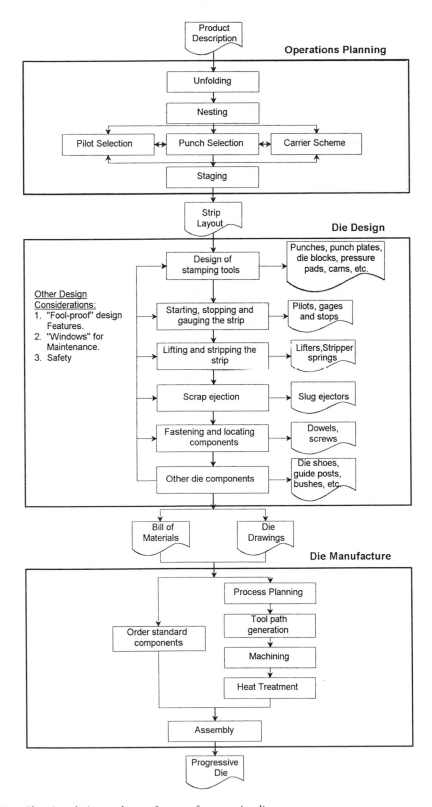

FIGURE 7.1　Planning, design, and manufacture of progressive dies.

system for progressive dies. The following factors are to be considered when developing an intelligent system to assist the die maker:

- Die making is an iterative, feedback process. The system should allow the user to generate initial trial designs quickly and the resulting information should be used to direct the search for a solution. Each trial design (iteration) adds information (i.e., provides feedback to the user) thus progressively moving towards the final solution. Here, the role of the computer should be restricted to what it can do best, i.e., evaluation of functions, execution of procedures, search data, and presentation of information to the designer in an appropriate form. During the iteration process, the die maker should be given adequate control over the computer in order to exercise his personal experience and judgment to handle ambiguity, redefine design objectives, relax constraints, exercise common sense, apply knowledge from domains not programmed into the computer, and exercise his creativity.

- Die making is a complex process and should be structured into smaller, inter-related, manageable subtasks. The system should be able to control the planning and design process so as to allow the user to combine a series of sub-optimal solutions at each of the subtasks into the most preferred solution. While a computer can be programmed to achieve an optimum solution for some of the die making tasks, the progressive die produced by integrating the optimum solutions from these subtasks is not always the best solution. For example, the nested arrangement that provides the best material utilization may not necessarily be the preferred solution when the additional constraint of accommodating the indirect piloting holes is considered. In another case, a strip layout that requires the least number of stations may not necessarily be the preferred solution as it may be difficult or impossible to accommodate all the tools within the limited die area, or the resulting die may be structurally too weak or difficult to maintain.

- The die making process involves the processing and manipulation of different types of data and knowledge by the computer. First, interactive graphical aids need to be provided for the user to create, view, and manipulate the plans, designs, and engineering models. Second, a knowledge-based system is required to provide the framework to store and manipulate the objects and rules associated with die making. Third, numerical routines are needed to process, transform and manipulate the geometrical data to support the various spatial planning, pattern processing, and recognition tasks. Finally, a solid modeler is required to provide the 3-D representation of the progressive die.

- Die design is the synthesis of a semi-custom product. It involves the synthesis of the progressive die (engineering model) from the strip layout (plan model) by selecting standard tooling components and generating non-standard tooling components, modifying them, and then assembling these components together.

- The die maker usually builds a progressive die by adapting old designs to meet the manufacturing requirements of the new product. In other words, the die is not always generated from first principles.

We will now examine how the various intelligent techniques can be applied to develop a computer aid for the planning, design, and manufacture of progressive dies.

7.4   Metal-Forming Features to Formalize the Description of the Product

Machining features have been used to formalize the description of a workpiece when developing computer-aided process planning and automated fixturing systems for Computer Numerical Control (CNC) machining. Similarly, metal-forming features can be used as a language to formalize the description of metal stampings in a knowledge-based system.

A metal stamping can have the following features:

Stamped features: holes, notches, cutouts, slots, etc.
Formed/bend features: bends, flanges, hems, seams, drawings, embossing, etc.
Formed/stamped features: countersinks, etc.

Each of these features can be associated with a metal stamping operation. They have attributes in the form of their geometrical and topological descriptions and technical specifications (e.g., tolerancing and finishing properties). A product can be described by connecting these features using feature relationships such as "connect_to" and "part_of." The feature representation of a workpiece is shown in Figure 7.2. There are several advantages associated with the use of metal-forming features to describe a workpiece in an intelligent system:

- The feature representation provides an "intelligent" description of the product.
- The feature representation provides a convenient means for representation as objects in a knowledge-based system.
- As each feature is associated with a stamping operation, it facilitates the generation of the process plan and the generation of the die. For example, when we are staging the die operations, we can build the rules to ensure that the sides of the wall of a bend feature must be notched before it is bent. Thereafter, the punches can be configured from the notch features and the tools associated with bending (i.e., bending punch, bending block, pressure pads, etc.) can be configured from the bend feature.
- The "intelligent" association between a feature and the stamping operation required to manufacture it makes it possible to develop product design tools that provide product designers with tooling advice. For example, Mantripragada et al. (1996) developed an interactive design tool for box-type sheet metal parts that can be used to alert the designers to potential production problems, defects, and failures; it can also provide them with information that can be used to explore alternative design, evaluate trade-offs, and arrive at optimal design for the given process conditions.
- The features and the associated "feature-relationship tree" can be used as a criterion to index the product in a design data base. The design data base is a data base of products and their associated progressive dies manufactured by a company. A die designer usually builds a progressive die by adopting an old design to meet the manufacturing requirements of the new product. Hence the features and the feature-relationship tree of a new product can be used to search through the design data base and retrieve the past design of the nearest product manufactured by the company for use as a reference to manufacture the new product. The use of features to index a product can be considered as the basis for the development of a case-based planning system for progressive dies. We will discuss the application of a case-based planning approach for progressive die design in another section.

7.5 Pattern Processing and Recognition Tasks Associated with Operations Planning

In essence, the operations planning task to develop the strip layout is a spatial planning exercise involving pattern processing and shape recognition skills best performed by human beings. Very little research work has been done to solve this problem and the operations planning task remains a major stumbling block in the development of an integrated system for the manufacture of progressive dies. Computer-assisted pattern processing and shape recognition tasks are computationally intensive in nature. If the computer takes too long to perform these tasks, it would render the interactive, iterative feedback planning system ineffective. Therefore, there is a need to reduce the spatial planning tasks into smaller and simpler pattern processing and shape recognition subtasks done by the computer, while leaving the more complex decision making tasks to the user.

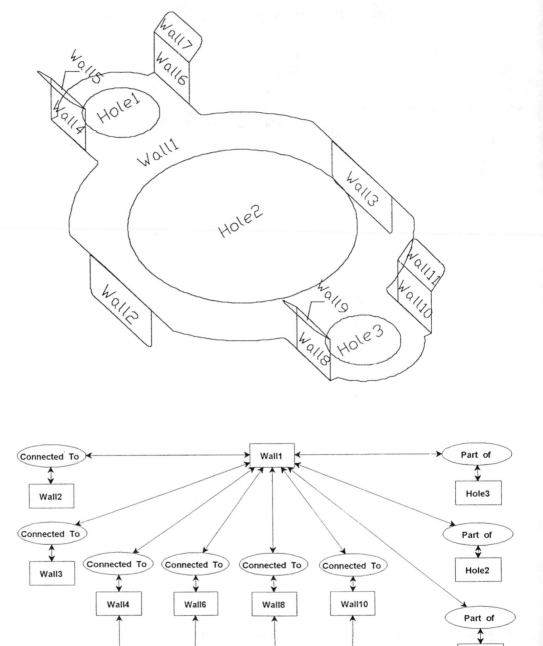

FIGURE 7.2 Features of a workpiece and the schematic representation of the feature relationship.

(a)

(b)

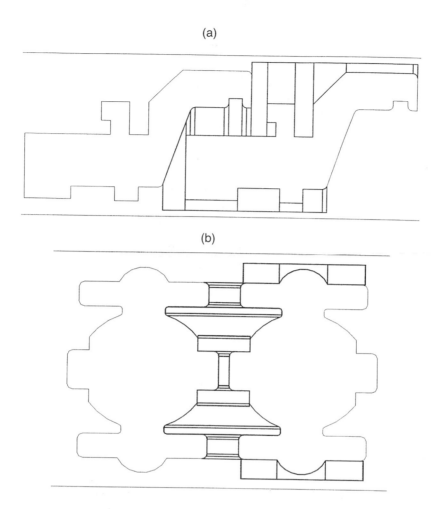

FIGURE 7.3   Strips used to notch out the external profiles of a workpiece. (a) Strips generated by projecting edges vertically. Thin strips of web are provided at both sides of strip for carrying the workpiece, (b) strips generated by project edges horizontally.

A "Strip" Approach for Punch Shape Decomposition

One of the key tasks in operations planning for progressive dies is to select an optimum combination of punches to notch out the external profile of the workpiece progressively. This is achieved by decomposing the scrap areas between successive workpieces into smaller areas. This simplifies the punch construction and makes it easier to remove the resulting smaller scrap pieces from the die.

A strip approach can be used to achieve this objective. This is done by projecting strips from each edge of one workpiece either horizontally or vertically to the edge of another workpiece or to the edge of the strip (Figure 7.3). The following rules can be used to control the development of the strips:

- When projecting strips vertically from the edge of the workpiece to the edge of the strip, the punch will overlap the edge by the overlap allowance if a center web carrier is used. Otherwise, it will leave a thin strip of web at the side to carry the workpiece through the die.

- Punches to notch away concave features on the external profile of the workpiece are selected first. Thereafter, these concave features are closed for the generation of strips.

- Filleted corners are not used for strip generation. Instead they are tagged and temporarily ignored. The new corner of the edge used for projection is redefined as the meeting point of the projections

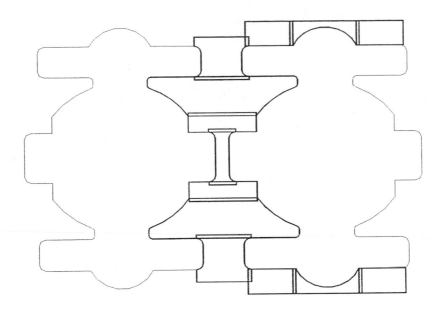

FIGURE 7.4   Final punch shapes selected by modifying the strips in Figure 7.3(b).

of the two adjacent edges to the fillet. The strips are generated for the new edges. Finally, the filleted corners are re-introduced to the punch shape.

- The edge of the punch that is adjacent to the edge of a workpiece is the cutting edge. The punch clearance is automatically applied to this edge.
- Overlap allowance is automatically applied to the non-cutting (unbounded) edge to prevent blurr formation.

These punch shape modification commands can be used to select the final punches required to stamp out the external profile of a workpiece. Figure 7.4 shows how the strips initially generated to stamp out the part shown in Figure 7.3a can be modified into practical punch shapes.

The strip approach has the tendency to produce too many notching punches. In addition, it does not take into consideration the relative location of the center of pressure of the notch contour to the shape of the punch. Hence, there is a need for the user to retain control in the final selection of the punch shapes. This can be achieved by providing a set of punch shape modification commands for the user to modify the strips to meet his requirements. Some of the functions of punch shape modification commands required are

- Automatic calculation and display of the center of pressure of a notch contour relative to the shape of the punch
- Combination of two or more neighboring shapes into one
- Stretching selected edges of a shape by an offset distance
- Spiltting a larger shape into two smaller ones
- Automatic generation of a shape based on boundaries defined by the users.

The complex task of punch shape decomposition is reduced by dividing the process into smaller, manageable subtasks where the computer concentrates on the number-crunching geometry processing tasks while the designer concentrates on making decisions to guide the system towards an optimum solution. This generate-then-modify approach appears to be an efficient tool to solve the progressive die design automation problem. It helps to eliminate the tedious construction and calculation chores associated with the traditional approach and hence allows the designer to concentrate on selecting the optimum shapes.

A Skeletal Approach for the Recognition of Punch Shapes

The design of progressive dies involves the process of matching the profiles of the punch designed by the user with a catalogue of standard punch shapes supplied by the die component's manufacturer. There are two advantages associated with using standard components: first, they are usually cheaper than custom made component, and second, they are readily available, hence reducing the production lead time. When a die designer is developing the strip using the traditional approach, he is continually taking off dimensions from the product drawings to develop the various punch shapes. Hence, the manual task of matching punch profiles with the standard punches provided in the catalogue becomes a natural extension of the measuring and construction tasks. However, in an automated environment, the punch profiles are derived directly from the geometry of the respective features stored in the knowledge base. It would be a tedious exercise for the user of the die design automation system to pick off the dimensions of the punch profiles developed by the system and manually select the standard punch shapes from the catalogue. Furthermore, it would defeat the main objective of providing the die design automation system, i.e., to allow the designer to concentrate on the important task of making design decisions by relieving him from the mundane and tedious tasks of measuring, checking, and flipping through catalogues.

The use of the skeleton as a descriptor of shape has been adopted by many researchers to simplify image processing and recognition problems. Several efficient "skeletonization" schemes have also been developed. One group of researchers (Wu et al., 1994a, Wu and Chen, 1994b) introduced the use of the simplified line skeleton for the classification of 2-D workpieces. They defined the simplified skeleton of the polygon as the set of points where the firings of two non-neighboring edges meet in their advancing paths. They divided the simplified skeleton into real and virtual links. They also proved that the simplified skeleton correctly represents the global shape information of a rectilinear contour.

The Enhanced Simplified Line (ESL) skeleton was developed at the National University of Singapore specifically for the recognition of punch shapes. The ESL skeleton is a modification of Wu and Chen's simplified line skeleton. In addition to the real and virtual links, extension links are introduced to represent the local shape information at the corners and ends of a profile. In addition, the ESL skeleton can be applied to non-rectilinear contours. If a contour consists of circular segments, they are approximated by straight lines; its ESL skeleton is derived regardless of whether the sides are non-rectilinear. The steps required for the construction of the ESL skeleton for an L-shaped contour are illustrated in Figure 7.5. In the final skeleton, FG and DE are the real links. FG has a shrinkage grade of (d1+d2) while DE has a shrinkage grade of d1. GD is the virtual link while the dashed lines are the extension links.

A commercial die components catalogue consists of many standard punch shapes. For example, there are 78 basic and special punch shapes in the Face '88–'89 Misumi Press Die Standard Components and System Technical Specifications (Misumi, 1988). After detailed analysis of the ESL skeletons of the standard punch shapes, it was discovered that the following skeletal features can be used to help recognize

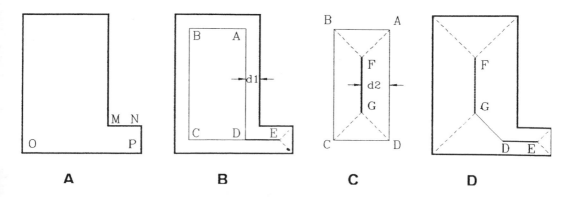

A   **B**   **C**   **D**

FIGURE 7.5   Steps taken to generate the ESL skeleton of an L-shaped contour.

standard punch shapes offered in the Face '88–'89 catalogue. They are

- Number of real links (RLs)
- Number of virtual links (VLs)
- Number of open nodes in the simplified skeleton (ONs)
- Real links orthogonal to each other? (OR?)
- Real links symmetrical about shape center? (SY?)
- Number of extension links (ELs)
- Number of vertices among all extension links with non-zero shrinkage grade (EFs)

Of these features, RLs, VLs, ONs, ELs, and EFs are derived directly from the ESL skeletal data. OR? and SY? are topological descriptors of the ESL skeleton and require additional processing for their derivation. On further analysis, the following findings are noted:

1. All the standard punch shapes have five or fewer real links. In other words, contours with 5 or more real links can be immediately classified as non-standard.
2. More than half of these standard shapes can be distinguished by considering only four of these features, namely, RLs, VLs, ELs, and EFs.
3. The rest can be distinguished using all seven features, except for seven pairs of shapes that have identical skeletal features but differ in minute dimensional or topological details.
4. For these seven pairs, additional conditions must be fulfilled before such contours can be positively identified. The confirmation checks to distinguish between two sets of standard shapes that share identical skeletal features are explained in Figure 7.6.

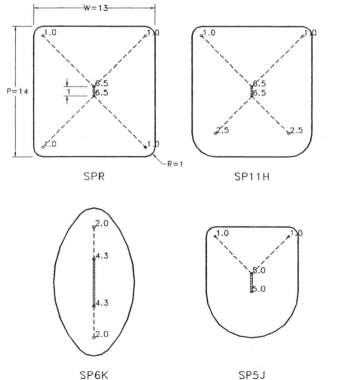

SPR and SP11H can be distinguished by checking the shrinkage grades of the extension links.

SP6K and SP5J can be distinguished by comparing the topology of the extension links.

FIGURE 7.6   Additional confirmation checks required to distinguish between two standard shapes having similar ESL skeletal features.

These findings can be used to develop a hierarchical search method which attempts to use the least features (and hence least computer resources) to classify a shape. If a decision cannot be made, it will move to the next level of search where more skeletal features of the input contour need to be extracted and matched. At the end of the search, a contour will be classified as a standard punch shape or as non-standard if it fails all the tests. The algorithm for the hierarchical search method used to match an input contour with standard shapes provided by the Face '88–'89 Catalogue is shown as a flowchart in Figure 7.7. Once a contour is recognized as a standard shape, the skeletal data can be used to calculate its principal dimensions.

The ESL skeleton and the seven features are independent of contour scale, position, and orientation. This approach therefore offers an efficient and complete solution to the task of recognizing a bounded set of standard shapes. In addition, the skeletal approach would help simplify the task of decomposing complicated non-standard punch profiles into smaller, standard/non-standard punch shapes because every real link essentially represents a trapezoidal region. By removing a real link, we are effectively decomposing the affected profile into a simpler shape and a trapezoidal contour.

7.6   Some Rules for Operations Planning

Before the strip layout can be developed, the die maker must decide on the piloting scheme and the carrier scheme to guide and carry the strip as it progresses through the die.

Piloting Schemes

Pilots are used to guide a strip into position before the die operations at a station are executed. Four different piloting schemes are used in progressive dies. The rules for the selection of piloting holes under the various schemes are described below.

Direct Piloting

Direct piloting consists of piloting in holes pierced in the workpiece at an earlier station. This is the preferred piloting scheme as it is most economical in terms of die construction and strip material usage. Pilot holes are always pierced first and should be made available at every station up to the last (or next-to-last) station. Because of the continuous insertion and removal of the pilot, there is a tendency for the shape of the hole to be distorted. Hence, the following conditions must be satisfied before a hole on the workpiece can be selected for use as a pilot hole:

1. It is circular in shape.
2. The specified tolerance is not high.
3. It is big enough for use as a pilot hole.
4. It does not lie on the folded portion of the workpiece.
5. It is not too close to the edge of the workpiece.
6. It is not too close to another hole on the workpiece.

From the list of holes that satisfy the above conditions, the most suitable pilot holes are selected based on the following priority:

1. If only one hole is available, it will be selected in the first instance.
2. If there are a number of holes, select the two largest holes (which are equal in diameter); the distance between them in the direction perpendicular to the feed direction must be greater than a minimum preset distance. The holes must be located on opposite sides of the part.
3. Select the two largest holes (the diameters of which are within a preset percentage of each other) that satisfy the earlier conditions.
4. Select the hole that is nearest to the centroid of the unfolded portion of the workpiece.

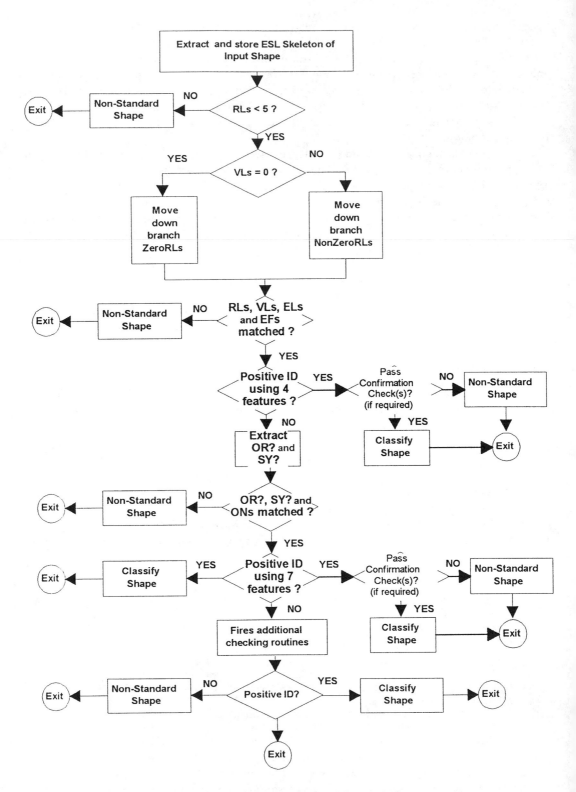

FIGURE 7.7   Flowchart illustrating hierarchial search method used for punch shape recognition.

Semi-Direct Piloting

In cases where a workpiece has closely toleranced holes which invalidate the use of direct piloting, a semi-direct piloting scheme may be used. The operation of a semi-direct piloting scheme is similar to a direct piloting scheme. The difference is that the pilot hole diameters will now be made slightly smaller than the actual holes to be pierced. The actual hole size will be pierced over the piloting hole in the last station. In this way, the distortions to the hole size caused by the actions of the pilot will not affect the shape of the final hole. Progressive dies using semi-direct piloting schemes cost more to construct as there is a need for additional punches at the last station. The shortlisting and selection criteria for semi-direct pilot holes are similar to direct pilot holes except that closely toleranced holes can now be shortlisted for selection.

Indirect Piloting Using Holes Located on Scrap Sections

Indirect piloting by locating the pilot hole in the scrap section formed by the notching operations is adopted when there is no suitable hole on the workpiece for use as a direct or semi-direct pilot hole. Progressive dies using indirect piloting schemes are more complex in construction than those using direct piloting schemes.

Scrap areas satisfying the following conditions are shortlisted for use to accommodate the indirect pilots:

1. Scrap areas which are adjacent to folded portions of the workpiece are not suitable. This is because they are always notched in the earlier stations in preparation for the subsequent bending operation.
2. Scrap areas must be big enough to accommodate the pilot with sufficient clearances at the sides.

The final location of the indirect pilot holes will be the two scrap areas which are furthest apart in the direction perpendicular to the feed direction.

Indirect Piloting Using Holes Located on the Side Carrier Webs

When it is not feasible to locate pilots on the scrap sections formed by the notching operations, indirect pilots can be located on the side carrier webs. In this case, the strip width will need to be increased to accommodate the pilot holes. The selection of indirect pilot holes on the side carrier webs will depend on the shape and location of the punches relative to the edges of the strip. To reduce strip width, it may also be necessary to recess some of the punches next to the selected piloted holes. It would be very difficult to develop routines to process the spatial information required to locate the pilot holes. Hence, the pilot holes should be selected interactively by the user.

Carrier Schemes

Another decision to be made by a die maker is to determine the method of carrying the workpieces as they progress through the progressive die. There are two types of carrier schemes: center carrier and side carrier.

Center Carrier

In this case, the workpieces are connected and carried by scrap sections at or near the center of the strip. These scrap sections will be parted in the last station to allow the workpiece to be severed from the strip. This carrier scheme is preferred as it requires the minimum strip width. The scrap sections and the respective parting punches are selected from scrap sections which satisfy these conditions:

1. Scrap areas that connect the two neighboring workpieces
2. Scrap areas that are not adjacent to folded portions of the workpiece
3. Scrap areas of sizes greater than a minimal value to provide sufficient strength to carry the workpieces.

Side Carrier

In cases where a center carrying scheme cannot be used, a side carrier approach can be used. Here, the workpiece will be carried by scrap webs at one or both sides of the strip.

When a side carrier scheme is used, the punch decomposition process will generate punch strips such that thin strips of web are provided at the sides of the strip to carry the workpiece. The parting punches will be selected from the scrap sections which satisfy these conditions:

1. Scrap areas that connect the edge of the strip to the workpiece
2. Scrap areas that are not adjacent to folded portions of the workpiece
3. Scrap areas with sizes greater than a minimal value to provide sufficient strength to carry the workpieces.

7.7   Spatial Planning Techniques for Staging the Die Operations

The final planning task in the development of the strip layout is the staging of the die operations such that when all the stamping operations have been completed on the strip in the progressive die, the required part is produced in the last stage. To automate the staging process, it is necessary to estimate the amount of die area each of the stamping operations requires. This can be achieved by developing rules and functions to derive the envelope area required to mount the respective tools in the die. For example, the amount of die area required by a punch will depend on the following factors:

1. The size and shape of the punch
2. The type of mounting arrangement at the punch plate
3. The minimum distance between holes on the punch plate required to maintain structural strength and also to provide room for the operator to access the punch during assembly and replace the punch during operation. This can be represented by a user-defined die area expansion factor used to expand the envelope area derived using the first two factors.

The functions to calculate the envelope areas for the stamping operations can be stored as monitors in the object representation of the features in the knowledge base. When the staging task is performed, these monitors will be fired and the envelope areas calculated when needed.

The strip layout can be derived automatically by staging the die operations in the following order:

1. Piercing operations on pilot holes are staged first.
2. Piercing operations on other internal holes are staged next.
3. Notching operations of external profiles are staged next.
4. Notching operations of external profiles used to accommodate indirect pilot holes (if any) are staged next.
5. Bending operations are staged next.
6. Cam operations (on precise holes on bend features of the workpiece) are staged next.
7. Finally, the cut-off operations and internal holes used as semi-piloting holes (if any) are staged.

During the staging process, the die operations are represented by their envelope areas and laid on the strip according to their priority. For envelope areas having the same priority, they are ordered according to their relative distance from the center of the strip. Those envelope areas further away from the center will be laid ahead of those nearer to the center. If an envelope area overlaps an existing one on the station, it will be carried to the next station. The staging process will continue until the envelope areas of all the die operations are laid on the strip. To reduce the time taken by the computer to check for the interference of envelope areas, the shapes of envelope areas are simplified to either rectangular or circular.

The strip layout produced using the above heuristics will always attempt to achieve the smallest die area possible. The system can provide facilities for the user to control the final strip layout. First, the tightness of the die can be controlled by changing the die area expansion factor. If the factor is a larger value, the die operations will be spaced further apart. Second, interactive commands can be provided to allow the user to insert idle stations and to move a die operation from one station to another.

7.8　Iterative Feedback Approach for Operations Planning

We have described the intelligent techniques that can be used to assist in the various tasks involved in the operations planning in progressive die design. These tasks are interdependent and decisions made in one task will affect the outcome in the others. For example, the actual punch shapes selected will depend on the piloting schemes used. On the other hand, the actual pilot holes (for an indirect piloting scheme) cannot be selected until the shapes of the notching punches are known. In other words, an effective computer aid must provide the following facilities for the user to manage the iterative feedback design process:

1. Generate a "first-guess" solution for each of the subtasks quickly.
2. Provide interactive aids to modify the solution.
3. Use the modified solution as a starting point to generate the solutions for the other subtasks.
4. At any time, move from one subtask to another to study how the solutions affect each other, and, if necessary, modify them to ensure that the constraints in all the subtasks are satisfied.

The iterative feedback design procedure can be achieved by integrating an interactive graphics user interface (preferably in the form of a CAD system) with a knowledge-based system. The CAD system will provide interactive graphics aids for the user to visualize and manipulate the plan as it is developed and stored in the knowledge base. Commands can also be provided via the CAD interface to let the user define break points in a task and to step from one task to another.

Using the techniques described so far, a prototype design system can be developed to generate the strip layout from the feature-based description of a product. As an illustration, the staging and strip layout required to manufacture the product shown in Figure 7.2 is shown in Figure 7.8. In addition, Cheok et al. (1995) described how these techniques are used to generate the plans to manufacture a different workpiece using a prototype progressive die design system developed at the National University of Singapore.

7.9　A Model-Based Reasoning (MBR) Approach for Die Synthesis

The operations planning process converts the features of a product into a plan model describing the stamping operations at each station required to manufacture the product. The next stage in progressive die design is to generate the engineering description of the die from the plan model. Once the engineering model is described in the knowledge base, a 3-D solid modeling description of the die can be produced in the CAD system for visual inspection and verification and down-stream manufacturing. The symbolic relationships between entities in the CAD data base, the feature tree, the plan model and the engineering model are shown in Figure 7.9.

Dym and Levitt (1991) have described an architecture and some methodological ideas for building knowledge-based engineering systems using the model-based reasoning (MBR) approach. The MBR approach can be used as the framework to automate the synthesis of a progressive die from strip layout. The features of the MBR approach for die synthesis are as follow:

1. The die structure is represented by a hierarchy of its components classified into branches of subassemblies and by topological links which are automatically deduced using spatial reasoning techniques. The generic hierarchy of die components represented in the knowledge base is shown in Figure 7.10.
2. The die components can be classified into two synthesis models:
   a. *Components or subassemblies whose descriptions are directly dependent on the product features and the strip layout*—The rules and formulation to describe these components are specially coded and stored in a library of standard components in a component hierarchy based upon standard naming conventions given in Figure 7.10. Rules or procedures that reference additional attributes of named components then generate multiple abstraction links for components. Examples of

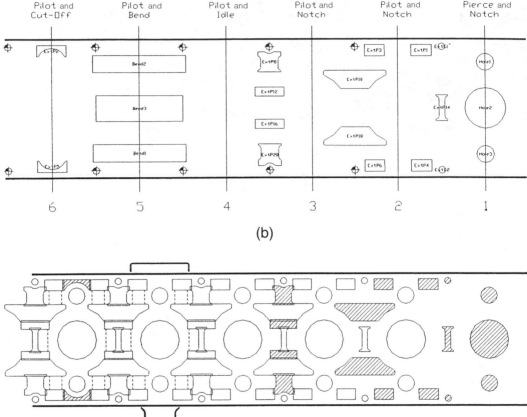

FIGURE 7.8   The staging and the strip layout for the manufacture of the part shown in Figure 7.2. (a) The stamping operations selected at the various stations, (b) the strip layout.

such components and subassemblies are the piercing, notching, and blanking punches, pilots, and bending punch subassemblies.

b. *Components whose descriptions are deduced by reasoning about their geometrical and topological relationship with reference to the standard components described above*—The punch plate is such a component. Its size and topological description will depend on the tools it supports.

The schema for the description of some of the die components will be detailed later. The relationship between the component abstraction and the product hierarchy is illustrated in Figure 7.11. Functions and rules to derive the respective values of the slots of the objects in the product hierarchy can be implemented in the following manner:

- Functions and rules that do not require the resolution of conflicts will be stored as monitors in the respective slots of the object in the component abstraction and will be executed when an object is created in the product hierarchy. For example, the function to calculate the screw size will be stored as a "when needed" monitor in the size slot of the screw object representation in the component abstraction. When a new screw needs to be added to the die assembly, the monitor will be fired to calculate its size.

- Rules that require the resolution of conflicts will be represented as rules external to the object. Additional rules may be implemented to help resolve conflicts. For example, the locations of the

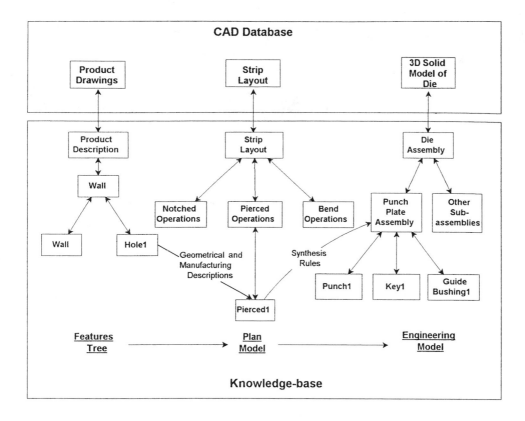

FIGURE 7.9 Symbolic relationship between entities in the CAD data base and the knowledge base.

screws and dowels as specified by the respective rules may conflict with the predefined positions of the punches and the pilots, as well as the rules on distance between holes and distance between holes and edge of plate. As the latter takes precedence, rules are introduced to shift the offending screws or dowels.

3. The knowledge base is supported by a library of standard catalogue die components, each identified by a part identification. The parametric dimensions of these standard components are retrieved from a library of data files and stored as multivalued slots in the respective objects in the component abstraction.

7.10   Schemas for the Description of Some Die Components

Piercing, Notching, and Blanking Punches

Piercing, notching or blanking punches, P, can be described by the following function:

$$P = f(S,M,L)$$

where S = punch shape
     M = mounting method
     L = length of the punch

The derivation of the punch shape was described earlier. The length of the punch is a function of the shut height and various components' thickness, which affects the vertical height of the die. To help simplify the problem, only three mounting methods are adopted. The rules for the selection of mounting method are

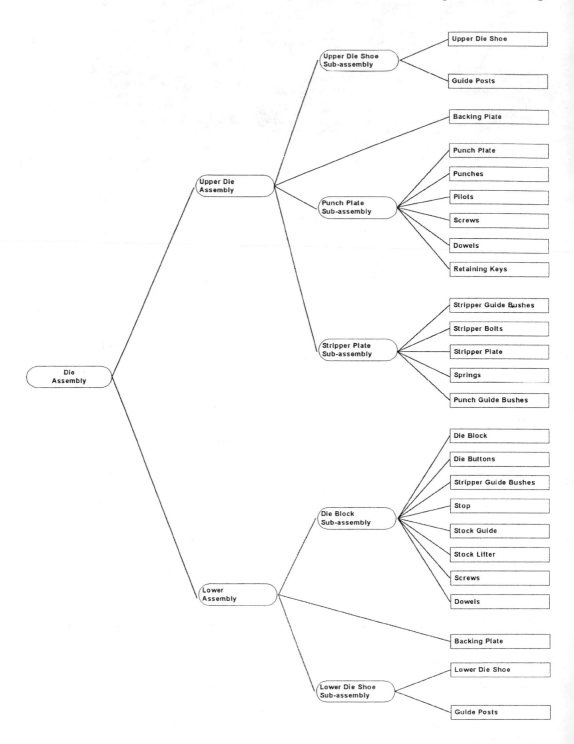

FIGURE 7.10   Generic hierarchy of die components. (From Cheok, B.T. et al. 1994. With permission.)

- If the punch is UNCLASSIFIED, then a STRAIGHT mount is used.
- If the punch is CLASSIFIED and the ENVELOPE_TYPE is CIRCULAR, then a BLOCK mount is used.
- If the punch is CLASSIFIED and the ENVELOPE_TYPE is RECTANGULAR, then a SHOULDER mount is used.

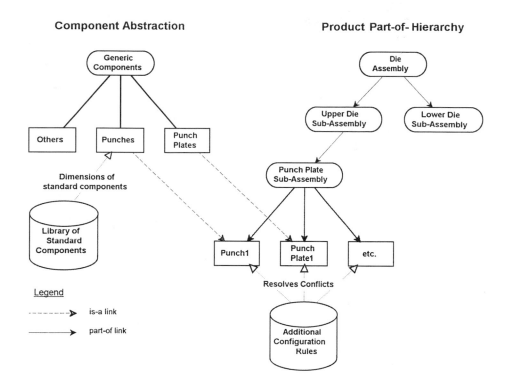

FIGURE 7.11   Relationships between component abstraction and product hierarchy.

The shape of the envelope depends on the aspect ratio of the punch shape. If the aspect ratio is near the value 1, a circular envelope area is used. The envelope areas are used to stage the die operations. The relationship among the punch shape, envelope type, and mount type is illustrated in Figure 7.12.

Bending Punches

To simplify the problem of generating the bending punches, the bending operations required to form a feature can be classified into any of the four predefined bending configurations shown in Figure 7.13. The dimensions of the various components can be derived from the various empirical formulas provided in die design textbooks. These dimensions can also be used to derive the envelope area to represent the spatial requirement to support the bending operation during strip development.

Punch Plates

Stamping punches and bending punches are components and subassemblies whose descriptions are directly dependent on the product features and the strip layout. Punch plates, die blocks, and strippers are some of the components whose descriptions can be deduced by reasoning about their geometrical and topological relationship with reference to the tools that they support. For example, the schema used to generate the description of a punch plate is given in Table 7.1.

7.11   Rules for the Placement of Locators and Fasteners

The various components of a die assembly are held together by fastening screws and placed in position by dowels. Screws and dowels must be placed such that they can effectively serve their intended purposes without interfering with the actual stamping tools and affecting the strength of the die. In other words,

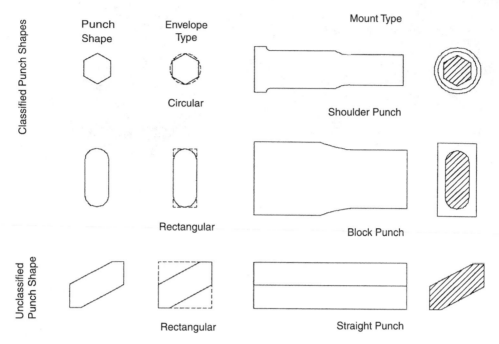

FIGURE 7.12 Relationship between punch shape, envelope shape, and mount type. (From Cheok, B.T. et al. 1996. With permission of the Council of the Institution of Mechanical Engineers.)

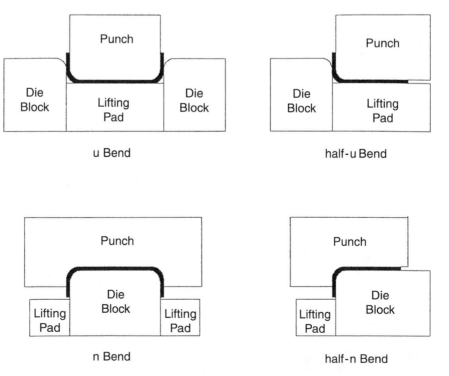

FIGURE 7.13 Basic bending configurations to support the various bend features (bend angle not necessarily 90°). (From Cheok, B.T. et al. 1996. With permission of the Council of the Institution of Mechanical Engineers.)

TABLE 7.1   Schema for the Generation of Punch Plates

Component Name: Punch plate
Function: Hold and support punches
Requirements: 1. Provide sufficient area to mount punches, fasteners, locators, pilots,
                    and other components.
               2. Provide sufficient strength to bear stripping forces.

Geometrical and Topological Features	Governing Rules and Functions
Plate Thickness	Function of punch sizes.
Plate Length and Width	Spatial arrangement of punches as defined by strip layout.
	Location and size of fasteners, locators and pilots.
	Rules on minimum distance between holes.
	Rules on minimum distance between holes and plate edges.
Plate Location	Below backing plate.

Rules governing size and location of retaining holes for:
Punches: spatial arrangement of punches as defined by strip layout.
Pilots: spatial arrangement of pilots as defined by strip layout.
Dowels: rules on locators.
Screws: rules on arrangement of fasteners; size is a function of stripping forces.

the screws and dowels should be placed in locations so that they are sufficiently far apart to provide the fastening strength and positional accuracy required. At the same time, they cannot be too close to the other holes and pockets on the plates so as to adversely affect the strength of the plates they are holding and locating. The following steps can be used to place screws and dowels on plates:

1. The envelope areas of the holes and pockets on the plate are derived. The envelope area of a hole is the circumscribing circle whose diameter is equal to the diameter of the hole enlarged by a factor to satisfy die design rules controling minimum distances between holes.
2. The envelope rectangle circumscribing the envelope areas of all the holes and pockets is derived. This represents the minimum plate area required to mount the tools (on punch plate) or accommodate the holes (on die block).
3. Thereafter the envelope areas of the holes to accommodate the screws and dowels are placed at the corners of the envelope rectangle according to the predefined configurations, some of which are shown in Figure 7.14.
4. The arrangement in which the locators are farthest apart is the preferred solution and will be accepted immediately if it satisfies the die space constraint.
5. Otherwise, the arrangement which requires the minimal die space will be accepted.
6. The envelope area including the holes for the screws and dowels is derived. This represents the minimum plate area required by the punch plate.
7. The actual plate size will be selected by matching the dimensions of the envelope area with the next bigger plate provided in the standard catalogue component library.
8. Finally, the position of the screws and dowels is adjusted to be further away from the center of the plate to take advantage of the bigger plate area.

7.12   3-D CAD Representation of the Progressive Die

The MBR approach can be used to generate a complete description of the progressive die assembly in the knowledge base. This has to be translated into CAD models for visualization and manufacturing purposes. Traditionally, die designers have used 2-D drawings to represent the die assembly. However, in an automated design environment, 2-D drawings are no longer adequate or efficient. There are several

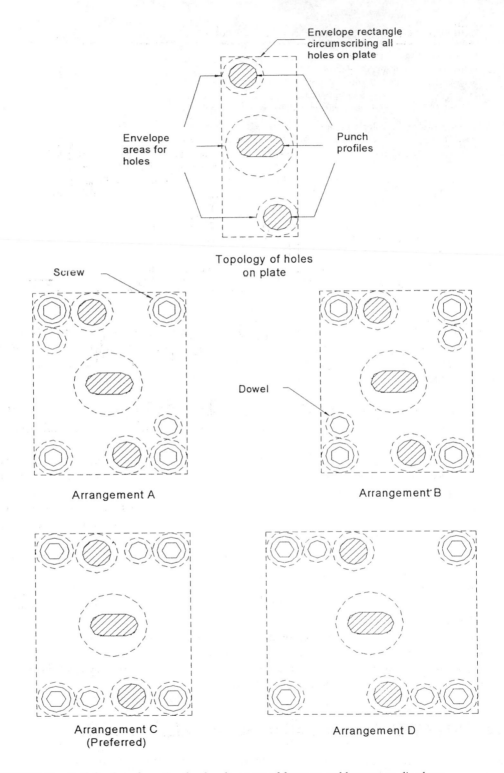

FIGURE 7.14   Predefined configuration for the placement of fasteners and locators on die plates.

strong reasons to support a move to 3-D CAD representation of progressive dies:

1. In an automated design environment where most of the design tasks are left to the computer, it is important to provide the designer with an effective tool to perform the final inspection of the die assembly. A 3-D model is the ideal visualization aid for the designer to use to check for and correct any mistakes made by the design automation system.
2. Similarly, a 3-D model of the strip can be laid on die assembly both in the open and shut positions. This provides a fast and effective means for the visual confirmation of the total solution. Any operational problems that may be caused by the interference of the strip with the die can be picked up quickly.
3. An accurate CAD representation of the non-standard components can be used directly to program the NC codes required for their fabrication.
4. Many CAD/CAM systems provide aids to produce 2-D drawings from 3-D models. Hence, traditional die drawings can still be produced very quickly.

3-D CAD models can be produced from the engineering model in several ways, depending on the capabilities of the knowledge-based system and the CAD system. For example, functions can be written in the knowledge-based system to scan the engineering model and translate it into neutral file descriptions. Macros can be written in the CAD system to read these neutral files and generate the 3-D CAD model. Alternatively, the knowledge-based system can fire functions to generate solid modeling kernel data files for each and every component. These files can then be read by a CAD system.

Figure 7.15 shows the exploded view of a punch plate subassembly generated using the MBR approach. Figure 7.16 shows a partial assembly of a two-stage piercing and blanking die generated using some of the techniques described.

7.13   Feature-Based Process Planning for the Manufacture of Progressive Dies

There have been many attempts to automate the process planning function for the manufacture of engineering components. Invariably, these systems involve a machining feature extraction function to transform the geometrical and topological information of a product model held in a CAD system into relevant machining features such as holes, slots, fillet, chamfer, etc. This approach involves the use of complex algorithms that consume a considerable amount of computer resources.

There is no need for the feature extraction task if one adopts the MBR approach for die synthesize. This is because the machining features can be extracted directly from the engineering model of the die stored in the knowledge-based system. For example, the following machining features can be deduced from the model-based description of the punch plate:

- Dowel holes are straight thru circular holes.
- Holes for screws on punch plate are threaded counterbore holes.
- Retaining holes for straight punches are non-circular thru holes.
- Retaining holes for block punches are rectangular thru holes.
- Retaining holes for circular shoulder punches are counterbore holes.
- Retaining holes for non-circular shoulder punches (i.e., with keys) are complex features consisting of a thru hole and an irregular pocket.
- Retaining holes for pilots are counterbore holes.

Similarly, formulas for the calculation of dimensional and geometrical tolerance values for these machining features can be programmed as methods in their object representations. In other words, the MBR approach is able to generate sufficient information to initiate the automated process planning task. Hence, the framework identified in this chapter can be extended to perform the process planning task if

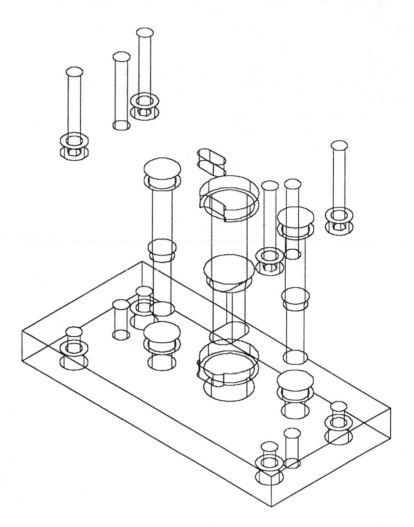

FIGURE 7.15   Exploded view of a punch plate subassembly consisting of a plate, 3 pierce punches, 4 screws, and 3 dowels.

the rules associated with the various processes planning tasks can be programmed in the knowledge-based system. Automated process planning is itself an extensively researched topic and is not within the scope of this chapter. Lee et al. (1993) demonstrated a knowledge-based process planning system for the manufacture of progressive dies. It is believed that the process planning knowledge encapsulated in their system can be adopted for use in the framework proposed in this chapter.

7.14   Future Developments: Designing from Past Experience Using a Case-Based Approach

The intelligent techniques described thus far explain how a knowledge-based approach can be used to develop an integrated system for the design and manufacture of progressive dies. This system is able to take the feature-tree description of a product and develop the strip layout, generate the engineering model of the die, and develop the process plans required to manufacture the die components. By itself, such a system would not only greatly reduce the lead time required to produce the progressive die needed to mass produce a product, but also greatly reduce reliance on the expertise of experienced die and tooling

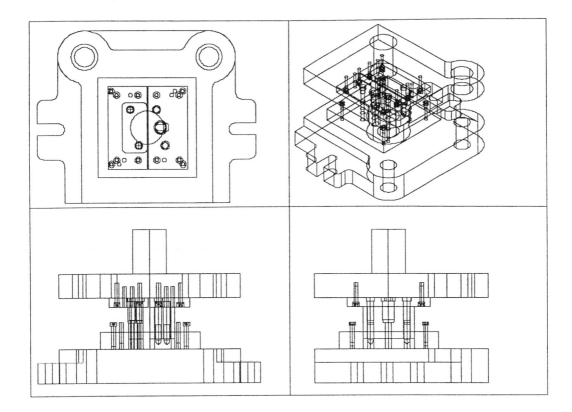

FIGURE 7.16 3-D CAD model of a partial assembly of a two-stage pierce and blank progressive die (showing only the die shoes, punch plate assembly, and die block). (From Cheok, B.T. et al. 1996. With permission.)

designers. Most importantly, it is believed that the framework proposed will provide the basic knowledge for the development of an even more efficient intelligent die design and manufacture system using a case-based approach.

It has long been recognized that many decisions made by human beings are not reasoned from first principles. Instead, when confronted with a decision making task, humans first try to relate the problem at hand to the closest situation they have experienced in the past. They will then try to adapt their past experience to solve the problem at hand. This case-based approach is also commonly used by the die designer. When designing a die to mass produce a product, the die designer first examines the product carefully and then tries to recall whether he has worked on a similar type of product. If he has, he will retrieve the drawings and use them as a reference for the new design. If we are able to develop a computerized die design system based on the case-based approach, it will provide the following benefits:

1. The design, which is based on an earlier working design, will have a very high chance to work successfully.
2. The design will probably be one of the best solutions. This is because only good past designs will be stored as cases in the design data base.
3. There is a better chance that components used in the earlier design may be recycled for use in the new design.
4. The experience of a die designer will not be restricted to designs he has worked on in the past; instead, he will have access to all the good designs developed by the company.

In the 1980s, Schank (1986), Kolodner and Riesbeck (1986), and others pioneered basic research in case-based reasoning (CBR). Today, industry research laboratories and graduate students at several universities have demonstrated CBR in a wide variety of domains including law, cooking, and American football. Schank's student, Hammond (1989), described the architecture and the implementation of a case-based planner for the generation of Szechwan cooking recipes. Some key components of a CBR system are

- A means to store and index past cases.
- A means to retrieve the nearest case.
- A means to modify the old case to meet the goals of the new case.

Today, the implementation of these components has been restricted to relatively "simple" domains such as generation of cooking recipes, criminal sentencing, legal reasoning in patent law, etc. In such domains, it is possible to describe the criteria used to index the cases and describe the goals as textual statements for storage and manipulation in a program.

One of the greatest challenges facing the development of a case-based system for die design is to find a way to describe the product (which is basically geometrical in nature) in simple descriptive statements such that it can be indexed and stored in a library of cases. Now, it is believed that the feature-tree of a product can be used to index a metal-stamping workpiece for storage as a library of cases. The feature tree can also be used to search and retrieve the nearest case for adaptation. However, the feature-tree of a product can be topologically rather complicated and the pattern matching process required to locate a similar tree or the nearest tree from a library of products can be tedious and computationally rather complex. A complete treatment of a case-based approach for die design is outside the scope of this chapter. Furthermore, there are still many problems that need to be addressed. However, we believe that the techniques described in this chapter provide some of the initial building blocks to achieve the ultimate objectives.

7.15 Conclusion

The tool and die industry is facing three severe challenges: First, the increasingly competitive market has forced companies to push out new products more frequently to attract consumers. This has put a tremendous amount of pressure on product and tooling designers who are required to work under the constraints of increasingly shorter lead times. Second, as consumer expectation becomes more sophisticated, the high product quality demanded invariably requires the use of new materials with high standards of accuracy and finishing. The tools and dies required to make these products need to be more precise and sophisticated, stretching the expertise of experienced tool and die designers to their limits. Third, in the newly industrialized and industrializing countries, young people are reluctant to take up craftsmanship-related courses that require long apprentice training programs. Hence, the tool and die industry will find it very difficult to maintain the pool of well-trained, experienced personnel it requires to sustain the growth needed to keep pace with the continuous demands for new and innovative products by the consumer.

The use of conventional CAD/CAM systems has helped to automate the drafting, NC programming work associated with design and tool making. This has partly helped to alleviate the problems posed by the challenges identified above. However, it has not replaced the experience-based, trial-and-error approach. A longer term solution is to introduce intelligent CAD/CAM practices to the die making industry to help it meet with these challenges.

Further Information

The authors were awarded funding from the National Science and Technology Board (NSTB) of Singapore on November 1, 1996 to develop an Intelligent Progressive Die (IPD) system, a knowledge-based software for the planning, design, and manufacture of progressive dies.

References

Cheok, B.T. et al. (1994). Some aspects of a knowledge-based approach for automating metal stamping die designs, *Computers in Industry*, Vol. 24, 81–96.

Cheok, B. T., Foong, K. Y., and Nee, A. Y. C. (1996). An intelligent planning aid for the design of progressive dies, *Proceedings of the Institution of Mechanical Engineers*, Vol. 210, 25–35.

Dym, C. L. and Levitt, R. E. (1991). An architecture for model-based reasoning, In *Knowledge-based Systems in Engineering*, McGraw-Hill, New York, 179–190.

Hammond, K. J. (1989). Case-based planning: viewing planning as a memory task, *Perspectives in Artificial Intelligence*, Acadamic Press, Boston, MA.

Kolodner, J. L. and Riesbeck, C. K. (1986). *Experience, Memory and Reasoning*, Lawrence Erlbaum Associates, Hillsdale, NJ.

Lee, I. B. H., Lim, B. S., and Nee, A. Y. C. (1993). Knowledge-based process planning system for the manufacture of progressive dies, *International Journal of Production Research*, Vol. 31, No. 2, 251–278.

Mantripragada, R., Kinzel, G., and Altan, T. (1996). A computer-aided engineering system for feature-based design of box-type sheet metal parts, *Journal of Materials Processing Technology*, Vol. 57, 241–248.

Misumi. (1988) Face '88–'89, Misumi Press Die Standard Components and System Technical Specifications, Misumi Shoji Co. Ltd., Tokyo, Japan.

Schank, R. C. (1986). *Explanation Patterns: Understanding Mechanically and Creatively*. Lawrence Erlbaum Associates, Hillsdale, NJ.

Wu, M. C., Chen J. R., and Jen S. R. (1994a). Global shape information modelling and classification of 2-D workpieces. *International Journal of Computer Integrated Manufacturing*, Vol. 7, No. 5, 261–275.

Wu, M. C. and Chen J. R. (1994b). A skeleton approach to modelling 2-D workpieces, *Journal of Design and Manufacturing*, Vol. 4, 229–243.

Computer-Aided Design (CAD) Technique for Flexible Manufacturing Systems (FMS) Synthesis Utilizing Petri Nets

D.Y. Chao
National Cheng Chi University

Abstract

The knitting technique provides a set of simple synthesis rules to construct a large Petri net (PN), thus avoiding time-consuming verification. We show that the synthesized nets are well behaved and form a new class called synchronized choice nets. In order to determine the applicable rules and detect rule violations, we present the concept of temporal matrix that records relationship (concurrent, exclusive, sequential, ... etc.) among processes in the PN. Upon each generation, the algorithm will consult and update the matrix. The complexity for the algorithm is $O(\alpha^2)$ where α is the total number of processes. This algorithm has been incorporated into our X-Window-based tool for design, analysis, simulation, testing, and synthesis of communication protocols, etc. The rules and algorithm have been extended to include the once forbidden generations between exclusive transitions and arcs with multiple-weights, i.e., general Petri nets (GPNs).

8.1 Introduction

Modern automated manufacturing systems face increasing pressure to improve efficiency and cost effectiveness and to ensure a quick return on large investments in new equipment.[27–29] They are also characterized by the fast dynamics of such systems that may be disturbed by various factors such as urgent orders, possible shortages of tools and materials, energy constraints, and unexpected failures of the equipment including machines, instruments, and computers. It is obvious that decisions have to be made in a short time within the environment of highly automated and interrelated manufacturing systems of the present day. The use of computers in the design and control of manufacturing systems thus becomes mandatory.

Over the past 20 years, a new type of manufacturing system, flexible manufacturing system (FMS), has emerged. An FMS is a large and complex system consisting of many interconnected components of hardware and software. Typically an FMS consists of several machines interfaced with automated material handling and computer control. The main tasks in designing an FMS include process routing, the selection of a sequence of operations, and scheduling the assignment of time and resources.

PN theory has been applied to specifications, validation, performance analysis, control code generation, and simulation for manufacturing systems.[4,8,9,27–29,31,32,37,42,52,53,55,56,62–64] The first step toward these applications is modeling (or synthesis) of PNs for FMS.[32,53] A PN model is constructed for an FMS. The analysis of this PN model is conducted and system properties are claimed. The PN must satisfy three properties: boundedness, liveness, and reversibility.[3,6] These properties are critical for an FMS to operate in a stable, deadlock-free, and cyclic way.

Advantages of PN modeling include:

1. Specifications can be expressed in a highly formal way offering more promise for automated analysis.
2. The model has a natural graphical, as well a textual, representation.
3. Automated analysis is possible for properties, including deadlock-free, mean throughput, mean delay, buffer overflow, and resource contention.
4. Much research has been done on the use of timing with PNs, e.g., Molloy.[35] There are programs available to support automated analysis of timed Petri nets. Deterministic timing models are available for specifying real-time systems.[17,23] Stochastic PNs (SPN) are preferable for performance analysis in systems without real-time constraints. There are simulation programs available for cases where analytic analysis is not practical because of space or time complexity.
5. PNs are able to express solutions to asynchronous events and blocking.
6. It is possible to implement step-wise refinement of system specifications.
7. Some research has been done on the automatic translation of extended PN models into code.

PN, although a good model for FMS, has its problems. As the system grows in complexity, the modeling process becomes crucial, the analysis is no longer an easy problem, and it takes a very long time to get the analysis results. It has been shown that the complexity of the reachability problem (also called the state explosion problem) of the Petri nets is exponential. To solve the complexity problem, two approaches have been proposed: reduction[24–26,30,33,36,44] and synthesis. Reduction is to reduce a PN to a smaller one while retaining its properties. Analysis is then performed on the smaller PN with much less time required. Two different techniques have been implemented. Reduction is of limited value because modification and re-analysis may have to be conducted if the analysis methods have detected some undesired properties. Synthesis provides the designer a set of guidelines or rules to construct a large PN without logical errors; hence, no analysis is needed.

Simulation and animation are time consuming, but are useful to observe evolution of states; thus, they are useful to verify the modeling correctness, to study system transient behavior, and to check analytical results and theories. Analysis provides only steady-state performance figures such as minimum cycle time and identifies bottlenecks. Because FMS operations may be distributed over a large area, it is difficult to have a global view of the complete system. To simplify the design, debugging, and maintenance, it is necessary to visualize the operations in a central location; there is an urgent need for the graphical modeling and animation of FMS.

Our tool XPN-FMS[4] integrates modeling, simulation, animation, analysis, and synthesis in one software package. It can model and analyze the performance of an FMS cell based on Petri net theory. Using X-window graphical interface and animation, this software allows modeling and analysis visualization and is easy to understand and manipulate. It lets a user draw the factory layout of an arbitrary FMS cell on screen using the tool. A corresponding Petri net model is deduced (not automatically) from the FMS cell and can be displayed on the screen. It can also display the animated overall operating process of the FMS corresponding to the Petri net model. By experimenting with different inputs and comparing the output in a timed period, a user can quickly determine how to improve the efficiency, reduce the cost, and pin down where the bottleneck is. This software is of practical value for industrial engineers and manufacturing managers.

For the PN models of FMS that are decision-free, we exend the theory and algorithm of a unique matrix-based method to search for subcritical loops and to support scheduling and dealing with transition periods. This has been applied to find the optimal input sequence control. Recently, we have simplified the tool to eliminate the codes for animation with slight modifications. A designer can design factory layout using regular "graph" buttons and perform animation and simulations. This helps reduce the effort necessary for the designer to learn to use the tool.

Various synthesis approaches[1,2,6,8–16,18–22,27–32,34,36,40,41,47,48,50,51,54,55,58–64] have been proposed. According to Jeng,[28] two dominant synthesis approaches are bottom-up and top-down. Bottom-up approaches start with decomposition of systems into subsystems, then construct subnets for subsystems and merge these subnets to reach a final PN by sharing places,[1,42] transitions,[29,31,52] and elementary paths[31,32,51] or by linking subnets.[19] Top-down approaches begin with a first-level PN and refine it to satisfy system specifications until a certain level is reached. The stepwise refinement technique by Vallette[51] is a top-town approach where transitions are replaced by well-formed blocks. Suzuki and Murata[50] generalized this technique, but with the disadvantage of the need to analyze the blocks. Our knitting technique supplements this approach by synthesizing the blocks without analysis.

To automate the synthesis, rules must be developed as in Esparza and Silva[22] and Datta and Ghosh[18,19] to synthesize free-choice (FC) and extended FC (EFC). But they are unable to synthesize asymmetric-choice nets (AC), and one needs to analyze whether the subnet is reducible.[19] Futhermore, they do not have explicit algorithms and the associated complexity. Thus, most techniques do not deal with CAD explicitly. We have devised some simple and yet effective rules to guide synthesis, e.g., for communication protocols[58] and automated manufacturing systems.[3,46,48,58] They required no analysis and the synthesized nets constitute an new class of nets that are more general than FC, EFC, and AC.[6] This new class of nets is referred to as the Synchronized-Choice (SC) nets.

SC nets possess some interesting properties as they satisfy the following two requirements:

1. If a circuit Γ has a PT-handle γ, γ is bridged to Γ through a TP-bridge B.
2. If a circuit contains a TP-handle, then this TP-handle is bridged to the circuit through a PT-bridge.

The knitting technique by the author[6,8–15] is a rule-based interactive approach. A net N^1 is expanded to N^2 by inserting new paths bearing some physical meaning such as increasing concurrency, alternatives, etc. The generations are performed in such a fashion that all reachable markings in N^1 remain unaffected in N^2; hence all transitions and places in N^1 stay live and bounded, respectively. N^2 is live and bounded by making the new paths (NP) live and bounded. This notion is novel compared with other approaches because it aims to find the fundamental constructions for building any PNs. There are two advantages: (1) reduction of the complexity of synthesis as an interactive tool, and (2) providing the knowledge from which the construction of the class of nets may be built. It therefore opens a novel avenue to PN analysis. Due to the simplicity of the rules, it is easily adapted to computer implementation for allowing the synthesis of PNs to be performed in a user-friendly fashion.

A temporal matrix (T-Matrix) is proposed in this chapter to record the relationships (concurrent, exclusive, sequential, ... etc.) among the modeled processes. Tracking all the processes and the relationships among processes in large PNs is a difficult task and thus makes the automatic tracking of rule applicability upon generation and the automatic updating of the T-Matrix desirable.

The knitting rules are useful for analysis and reduction. For a given PN, we construct its T-Matrix and check mutual relationships among processes against the rules. Any violation spots potential bad designs. The reverse process of removing processes (reduction), according to the rules, should preserve the properties of the PN. Rather than reducing modules to transitions, we remove paths to reduce the PN. We have implemented a reduction algorithm based on the rules; the code is very simple, containing less than 100 lines.

In an earlier paper,[6] we proved that the synthesized nets are bounded and conservative based on a linear-algebra technique. This chapter will show that the structural relationships among processes in a PN are identical to the temporal ones. Based on this, we are able to prove the properties of live, reversible, and marking monotonic, but will omit the proof for boundedness. This chapter develops such an algorithm and its X-Window implementation. In addition, the T-Matrix can record self-loops and find maximum concurrency with linear time complexity, which helps for processor assignments.

The T-Matrix, however, cannot synthesize certain classes of nets. For instance, there are no ordering of firings among a set of transitions that are exclusive to each other. Sometimes, these transitions must execute one by one. In addition, if the synthesized net is initially safe, it stays safe for any reachable marking. It is marking monotonic, i.e., it will not evolve into a deadlock by adding more tokens. The synchronic distance between any two transitions in a synthesized net is either one or infinite. Certain classes of net will evolve to become unsafe, marking nonmonotonic, and to allow any positive synchronic distance.

The above limitation springs from the fact that some generations are forbidden. Allowing the forbidden generations would produce[34] new classes of nets. To maintain well-behavedness, new generations must accompany the forbidden generations. Based on this, this chapter develops a set of new rules to generate new classes of nets. First, we remove the restriction of forbidding the generation of new TT-path between two exclusive transitions. We add another new TT-path such that the above two exclusive transitions are sequentialized with synchronic distance of one. These two transitions are both exclusive and sequential to each other—a new temporal relationship, sequentially exclusive, denoted as "SX" or $\uparrow$. Second, we extend the rules to general Petri nets where arcs may carry multiple weights.

None of the existing tools integrate drawing, file manipulation, analysis, simulation, animation, reduction, synthesis, and property query in one software package. Furthermore, because PNs model discrete-event systems, the tool finds applications in communication protocols, flexible manufacturing systems, (extended) finite state machines, expert systems, interactive parallel debuggers,[11] digital signal processing (DSP),[5,7,11,65] etc.

We have enhanced the tool to include not only PNs but also state diagrams and data flow graphs (DFGs) with few code changes. Thus a designer can choose the model with which he is familiar. For instance, DSP professionals do not know PNs well. They can, however, draw DFGs and obtain iteration bounds, critical loops, rate-optimal scheduling, and others by just clicking a button.[5,7,11] Section 8.3 formalizes the TT rule and PP rule after the preliminaries in Section 8.2. Section 8.4 deals with the T-Matrix for PNs. Section 8.5 outlines the algorithm for updating the matrix and its time complexity. Section 8.6 discusses the X-Window implementation. Section 8.7 extends the rules to allow forbidden generations between exclusive transitions and GPNs. Its application to an FMS is presented. This section also briefly summarizes other features of the CAD tool. Section 8.8 concludes the approaches presented in this chapter. Notations are provided in the Appendix.

8.2   Preliminaries

Terminology

A Petri net[38–39,43] is a directed bipartite graph consisting of two types of nodes: *places* (represented by circles) and *transitions* (represented by bars). Places represent conditions or the presence of raw materials, and transitions represent events. Each transition has a certain number of input and output places indicating the preconditions and postconditions of the event. The holding of the condition or a raw material in a place is indicated by a *token* (represented by a dot) in the place. The system status is represented by the holding of a pattern of tokens in places, which is called a *marking*.

Definition: Let

$$P = \{p1, p2, ..., pa\}, T = \{t1, t2, ...tb\}, \text{ with } P \cup T \neq \phi \quad \text{and} \quad P \cap T = \phi$$

$$I: P \times T \to \{0, 1\}$$

$$O: T \times P \to \{0, 1\}$$

$$M_0: P \to \{0, 1, 2, ...\}$$

then $N = (P, T, I, O, M_0)$ is an ordinary marked Petri Net (OPN).

In this definition, pi $(1 \leq i \leq a)$ is called a place, ti $(1 \leq i \leq b)$ a transition, I an input function defining the set of directed arcs from P to T, O an output function defining the set of directed arcs from T to P, and M_0 an initial marking whose ith component represents the number of tokens in place pi. Note the functional values of both I and O are restricted to 0 and 1. If other positive integer values are allowed, then the N is a general Petri net (GPN).

Definition: The firing rules are

- A transition $t \in T$ is enabled if and only if the marking at p, $m(p) > 0, \forall p \in P$ such that $I(p, t) = 1$;
- An enabled transition t fires at marking M (with components $m(p)$), yielding the new marking M' (with components $m'(p)$,

$$m'(p) = m(p) + O(p, t) - I(p, t), \forall p \in P.$$

The Marking M' is said to be reachable from M. Given N and its initial marking M_0, the reachability set $R(N, M_0)$ (abbreviated as R) is the set of all markings reachable from M_0.

Definition: A marked Petri net N is B-bounded if and only if $m(p) \leq B, \forall p \in P$ and $M \in R(N, M_0)$ where B is a positive integer. If $B = 1$, N is safe. N is live if and only if $\forall t \in T$, and $\forall M \in R(N, M_0)$, $\exists$ a firing sequence σ of transition to lead to a marking that enables t. N is reversible if and only if $M_0 \in R(N, M)$, $\forall M \in R(N, M_0)$. N is *well-behaved* if N is live, bounded, and reversible. $M_s = M_c(A) - M_d(B)$ is defined as: $\forall p \in A$, if $\neg (p \in B)$, then $m_s(p) = m_c(p)$, else $m_s(p) = m_c(p) - m_d(p)$. The operation $+$ is defined similarly. The synchronic distance between $t1$ and $t2$ in N is defined as $d_{12} = \text{Max}\{\sigma(t_1) - \sigma(t_2), \sigma \in L(N, M)\}$, where $\sigma(t)$ is the number of times t appears in σ and $L(N, M)$ is the set of all firing sequences from M. $d(W, V)$ is the maximum number of firings of transitions in set W without any transition firing in set V.

The synchronic distance is a concept closely related to the degree of mutual dependence between two events in a condition/events system.[49]

Definition: A node x in $N = (P, T, I, O, M_0)$ is either a $p \in P$ or a $t \in T$. The post-set of node x is $x \bullet = \{y \mid \exists$ an arc $(x, y)\}$ and its pre-set $\bullet x = \{y \mid \exists$ an arc $(y, x)\}$. A directed elementary path (DEP) in N is a sequence of nodes: $[n_1 n_2 ... n_k]$, $k \geq 1$, such that $n_i \in \bullet n_{i+1}$ $1 \leq i < k$ if $k > 1$, and $n_i = n_j$ implies that $i = j$, $\forall 1 \leq i, j \leq k$. An elementary cycle in N is $[n_1 n_2 ... n_k]$, $k > 1$ such that $n_i = n_j, 1 \leq i \leq j < k$, implies that $i = 1$ and $j = k$.

Definition: An A-path is a DEP whose places initially have no tokens.

Most new path generations are A-paths. When the new path generation (from a transition to a transition) forms a new circle, then we need to add tokens to the new A-path.

Definition: A basic process is defined as elementary cycle $[n_1 n_2 ... n_k]$ in a PN where n_1 is a place holding tokens.

The synthesis always starts from a basic process that is well-behaved. We add new paths in such a fashion that the N^2 remains well-behaved.

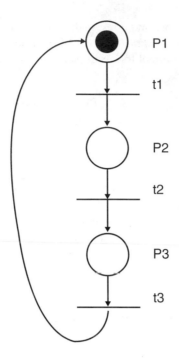

FIGURE 8.1   A basic process. (From Reference 2a. With permission.)

Definition: Let $t_a \in T$ be enabled in M_0, $\forall p_h \in \bullet t_a$, p_h is a *home place*. The set of home places is denoted as $H = \{p_h | p_h$ is a home place$\}$.

Home places actually form the home states of the synthesized net.

Definition: A pseudoprocess (PSP) in a PN is a DEP $[n_1 n_2 ... n_k]$, $k \geq 2$ such that
$|\bullet n_i| = |n_i \bullet| = 1, 2 \leq i \leq k - 1$, or
$k = 2$. The PSP is termed a *virtual PSP* (*VP*).

$n_g(n_j) = n_1(n_k)$ is defined as the *generation point* (*joint*). $\dot{n}_g(\dot{n}_j) = n_2(n_{k-1})$ is defined as the *next generation point* (*next joint*).

Figure 8.1 shows a basic process with $p1$ as a home place. In Figure 8.2(3), $[t1\ p2\ t2\ p3\ t3]$ ($[p1\ t1\ p2]$) is a PSP whose $n_g = t1(p1)$ and $n_j = t3\ (p2)$, respectively. Note the VP in Figure 8.3 (Π_1) contains only n_g and n_j. n_g, n_j, n_g and n_j are also used in new-path generations.

Definition: Let NP be a set of new paths generated from a set of nodes called generation points (n_g) in N^1 to another set of nodes called joints (n_j) in N^1 to produce $N^2 \cdot n_g(n_j)$ is a node $n_k \in NP$ and $n_k \in n_g \bullet (\bullet n_j)$.

A PN consists of a number of PSPs; each is considered a process. This may save the time involved in the synthesis compared with that which considers each individual node. When we generate a new PSP from n_g to n_e, we generate a new process or a new PSP. Mutual relationships among processes may be sequential, concurrent, or exclusive as will be defined below.

Definition: If Π_1 and Π_2 in a PN are in the same elementary cycle, then

- $\Pi_1 \leftrightarrow \Pi_2$, i.e., they are sequential (SQ) to each other, if the cycle has tokens, if the DEP from Π_1 to Π_2 is an A-path, then $\Pi_1 \rightarrow \Pi_2$; i.e., Π_1 is sequential earlier (SE) than Π_2; otherwise $\Pi_1 \leftarrow \Pi_2$; i.e., Π_1 is sequential later (SL) than Π_2. A special case of "SE" ("SL") is "SP" ("SN") if the n_j (n_g) of Π_1 equals $n_g(n_j)$ of Π_2.
- otherwise, if the cycle contains no tokens, then $\Pi_1 o \Pi_2$; i.e., they are cyclic (CL) to each other.

Definition: Let DEP1 $= [n_{ps} n_2 ... n_k \Pi_1]$ and DEP2 $= [n_{ps} n_{2'} ... n_{k'} \Pi_2]$, where DEP1 $\cap$ DEP2 $= \{n_{ps}\}$, Π_1 and Π_2 are two PSPs, and there are no DEPs (called bridges) from a node ($\neq n_{ps}$) DEP1 to a node

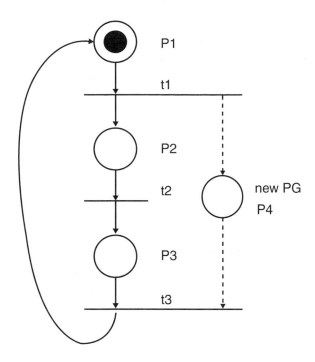

FIGURE 8.2 Create a pure-generation (dashed line) by using TT rule. (From Reference 2a. With permission.)

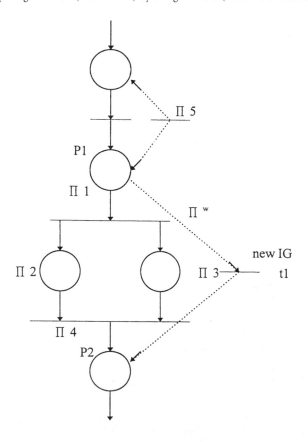

FIGURE 8.3 New interaction-generation (IG) connection Π_1 and Π_4. (From Reference 2a. With permission.)

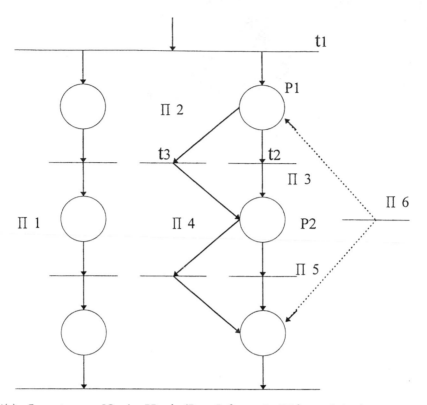

FIGURE 8.4(a)   Generate a new IG using PP rule. (From Reference 2a. With permission.)

	1	2	3	4	5	6	
1		CN	CN	CN	CN	CN	CN: Concurrent
2	CN		EX	CL	CL	CL	EX: Exclusive
			SE	SE			
3	CN	EX		CL	CL	CL	SE: Sequential
				SE	SE		Earlier
4	CN	CL	CL		EX	CL	SL: Sequential
		SL	SL				Later
5	CN	CL	CL	EX		CL	CL: Cyclic
		SL	SL				
6	CN	CL	CL	CL	CL		

FIGURE 8.4(b)   An example of the matrix to record the relationship between PSPs. (From Reference 2a. With permission.)

($\neq n_{ps}$) DEP2 and vice versa. n_{ps} is called a **prime start node**. If $n_{ps} \in P$, then $\Pi_1 | \Pi_2$. $\Pi_1 \| \Pi_2$ if none of $\Pi_1 | \Pi_2$, $\Pi_1 \leftrightarrow \Pi_2$, and $\Pi_1 \text{ o } \Pi_2$ hold.

Note $\Pi_1 \| \Pi_2$ implies $n_{ps} \in T$ or Π_1 and Π_2 are in two separate components of a PN that is not strongly connected. Also, in any synthesized PN using the TT and PP rules, Π_1 cannot be both concurrent and exclusive to Π_2 using the above definition. This is, however, not true for PNs that cannot be synthesized using the TT and PP rules.

The above definition of *sequential, earlier,* and *later* seem to be also *marking related.* However, in the synthesized net, any cycle contains at most a place marked with tokens; thus the marked place can be considered a *structure.* In Figure 8.3, the new PSP Π^w is Π_1 and $\Pi_2 | \Pi^w$. In Figure 8.4(a), $\Pi_1 \| \Pi_4$. Π_3 and Π_6 are on an elementary cycle containing Π_3, Π_5, and Π_6; hence $\Pi_3 \text{ o } \Pi_6$ in this figure.

The structural relationship between any two PSPs is recorded in a matrix. An entry A_{ik} corresponding to row i and column k represents the relationship between Π_i and Π_k. For example, if Π_i is concurrent

(a)

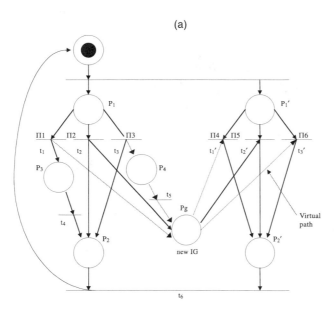

FIGURE 8.5(a) An example of the application of completeness rule 1. (From Reference 2a. With permission.)

	1	2	3	4	5	6
1		EX	EX	CN	CN	CN
2	EX		EX	CN	CN	CN
3	EX	EX		CN	CN	CN
4	CN	CN	CN		EX	EX
5	CN	CN	CN	EX		EX
6	CN	CN	CN	EX	EX	

FIGURE 8.5(b) Original T-Matrix (excluding the new IG). LEX $(\Pi_2, \Pi_5) = \{\Pi_1, \Pi_2, \Pi_3\}$, LEX $(\Pi_5, \Pi_2) = \{\Pi_4, \Pi_5, \Pi_6\}$. (From Reference 2a. With permission.)

to Π_k, then $A_{ik} = A_{ki} = CN$, where CN stands for *concurrent*. Such a matrix is termed a **Temporal-Matrix** (T-Matrix) because the structural relationship between two PSPs resembles their temporal relationship as explained below; see also Lemma 1.

It is easy to see that if we execute a PN starting from the initial marking involving Π_1, then $\Pi_1 \rightarrow \Pi_2$ implies that Π_1 is executed earlier than Π_2; $\Pi_1|\Pi_2$ implies that there is no need to execute Π_2 to proceed; $\Pi_1\|\Pi_2$ implies that both of them need to be executed to proceed. Intuitively, $\Pi_1 \leftrightarrow \Pi_2$, if they are subject to an intra-iteration precedence relationship; $\Pi_1\|\Pi_2$, if they can proceed in parallel; and $\Pi_1|\Pi_2$ if it can return to its initial marking with only one of them executed.

Complementing the prime start node is the prime end node defined as follows:

Definition: Let DEP1 $= [\Pi_1 n_k \dots n_2 n_{pe}]$ and DEP2 $= [\Pi_2 n_{k'} \dots n_2 \cdot n_{pe}]$, where DEP1 $\cap$ DEP2 $= \{n_{pe}\}$, Π_1 and Π_2 are two PSPs, and there are no DEPs (called bridges) from a node ($\neq n_{pe}$) in DEP1 to a node ($\neq n_{pe}$) of DEP2 and vice versa. n_{pe} is called a **prime end node.**

The definitions of prime (start and end) nodes can be extended to a pair of nodes rather than PSPs. Note in an SC, the n_{ps} and n_{pe} of any pair of nodes may not be both transition or both places. In such cases, however, the corresponding two handles must have bridges across them. An example is shown in Figure 8.5(a) where nodes t_4 and t_5 are mutually exclusive since their prime start node is p_1, a place, but their prime end node is t_6, a transition. The corresponding two PT handles $[p_1 t_1 p_3 t_4 p_2 t_6]$ and $[p_1 t_3 p_4 t_5 p_g t_1' p_2' t_6]$ have TP bridges $[t_3 p_2]$ and $[t_1 p_g]$ across them. Also note that on the two handles from the prime start node p_1 to the prime end node t_6, there are other prime end nodes p_2 and p_g, respectively. There are no bridges between the two PP handles $[p_1 t_1 p_3 t_4 p_2]$ and $[p_1 t_3 p_2]$. Such a pair of nodes p_1 and p_2 is defined as a **pure pair of prime nodes.**

Note if the pair of nodes are both output (input) nodes of a prime start (end) node, then the corresponding prime end (start) node must be the same type as the prime start (end) node; i.e., they must both be transitions or both be places. For instance, t_1 and t_3 (p_2 and p_2') are both output (input) nodes of $p_1(t_6)$; their prime end is also a place (transition) p_2 or p_g (t_1 or t_2 or t_3). Thus, in an SC, the two prime nodes of a pure pair must both be transitions or places.

Definition: If $\exists \Pi$, $n_1 \in \Pi$ and $n_2 \in \Pi$, then $n_1 \leftrightarrow n_2$. If $n_1 \in \Pi_1$ and $n_2 \in \Pi_2$, $\Pi_1 \neq \Pi_2$; the structural relationship between n_1 and n_2 is A_{12}.

Independent, uncoordinated action may cause an N to be unbounded and nonlive. A single generation may thus necessitate that additional generations maintain coordination and the associated searching of additional n_g and n_j as in Rules TT.4 and PP.2 below. This searching motivates the following two definitions.

Definition: A local exclusive set (LEX) of Π_i with respect to Π_k, X_{ik}, is the maximal set of all PSPs that are exclusive to each other and are equal or exclusive to Π_i, but not to Π_k. That is, $X_{ik} = \text{LEX}(\Pi_i, \Pi_k) = \{\Pi_z | \Pi_z = \Pi_i \text{ or } \Pi_z | \Pi_i, \neg \ \Pi_z | \Pi_k), \forall \Pi_{z1}, \Pi_{z2} \in X_{ik}, \Pi_{z1} | \Pi_{z2}\}$.

Definition: A local concurrent set (LCN) of Π_i with respect to Π_k, C_{ik}, is the maximal set of all PSPs which are concurrent to each other and are equal or concurrent to Π_i, but not to Π_k, i.e., $C_{ik} = \text{LCN}(\Pi_i, \Pi_k) = \{\Pi_z | \Pi_z = \Pi_i \text{ or } \Pi_z || \Pi_i, \neg (\Pi_z || \Pi_k), \forall \Pi_{z1}, \Pi_{z2} \in C_{ik}, \Pi_{z1} || \Pi_{z2}\}$.

Examples of LEX and LCN appear in Figures 8.5 and 8.6, and Rules TT.4 and PP.2, respectively, employ them.

The definition of LEX (LCN) between two PSPs can be extended to the LEX (LCN) between two transitions (places). Thus $\text{LEX}(t_i, t_k)$ [$\text{LCN}(p_i, p_k)$] is a set of transitions (places), instead of PSPs. Let $G(J)$ denote the set of all $\Pi_g(\Pi_j)$ involved in a single application of the TT or PP rule. To avoid the unboundedness problem, it is necessary to have a new directed path from each PSP in X_{gj} to each PSP in X_{jg}. Both X_{gj} and X_{jg} can be determined from the T-Matrix even though $d_{gj} = \infty$, which is, therefore, of no use to the synthesis.

Since synchronic distance depends on the marking, the determination of synchronic distance between all pairs of transitions may take exponential time complexity. It is easier, however, to determine the synchronic distance due to the structure of the net; such a synchronic distance is called the *structure synchronic distance* (d^s) as defined below. Since we construct most new generations based on the net structure, rather than on the marking, of interest is the structure of synchronic distance, rather than the synchronic distance itself. We may find the structure synchronic distance of a synthesized net by examining the synchronic distance of the same net with each home place holding only one token; i.e., by making the net safe. However, not all such synchronic distances are useful to the synthesis. For instance, we will observe that when $t_g \| t_j$ and $d(X_{gj}, X_{jg}) = 1$, the information of $d_{gj} = \infty$ is of no use and we define that $d_{gj}^s = 1$. On the contrary, when $d(X_{gj}, X_{jg}) = \infty$, the TT-generation from t_g to t_j may cause unbounded places in the TT-path. This problem cannot be detected by checking the T-Matrix only. The information of $d_{gj} = \infty$ is indeed useful here, and the value of d_{gj}^s is assigned to be ∞. Hence we have following definition:

Definition (Structure Synchronic Distance): The structure synchronic distance between transitions t_g and t_j is $d_{gj}^s = 1$ if $d(X_{gj}, X_{jg})$ is finite and $d_{gj}^s = \infty$ if $d(X_{gj}, X_{jg}) = \infty$.

8.3   The Synthesis Rules

We first present possible types of generations followed by the definitions of the rules.

In order for the rules to be complete, all possible generations must be considered. The types of generations depend on n_g and n_j in two factors: (1) whether they are transitions or places and (2) their structural relationship. For factor 1, there are four types, **TT, PP, TP, and PT** generations, defined as follows:

Definition: The n_g and n_j of a TT (PP, TP, and PT) generation are transition (place, transition, place) and transition (place, place, and transition), respectively.

For factor 2, n_g and n_j can be in the same or different PSPs. For the latter case, there are five possibilities: (1) sequential earlier (SE), (2) sequential later (SL), (3) concurrent (CN), (4) exclusive (EX), and (5) cyclic (CL).

Definition: Let Π_g (Π_j) denote the PSP that contains the generation point (joint). Pure generation (PG) generates paths within a single PSP (Figure 8.2), i.e., $\Pi_g = \Pi_j$. Interactive generation (IG) generates paths between two PSPs (Figure 8.4), i.e., $\Pi_g \neq \Pi_j$.

Definition: If $n_g \rightarrow n_j$ prior to the generation, then it is a forward generation; otherwise it is a backward generation.

In Figure 8.2, [*t*1 *p*4 *t*3] is a forward generation; $\Pi6$ in Figure 8.4(a) is a backward generation. A backward TT generation needs the addition of tokens to the NP to avoid the resulting N being nonlive. This constitutes Rule TT.2.

The idea of constructing the rules is simple. An eligible generation upon N^1 to produce N^2 should not alter the marking and firing behavior of N^1. Furthermore, the new PSP must be live and bounded.

One of the following three actions is taken depending on the type of generation: (1) forbidden, (2) permitted, or (3) permitted but need more generations. For instance, a TP generation causes an extra token to be injected into the original net N^1 whenever the NP gets a token. This alters the marking behavior of N^1 and the net is unbounded. A PT generation robs a token from N^1 and causes N^1 to be nonlive, hence changing its firing behavior. To avoid this problem, add more generations or forbid such generations. The rules are complete in the sense that all possible generations have been considered.

The following definitions for TT and PP rules have considered all possible structural relationships between n_g and n_j. Some generations are forbidden, some require the addition of tokens, while others require further generations.

Definition: TT Rule:
For an NP from $t_g \in \Pi_g$ to $t_g \in \Pi_j$ generated by the designer,

(0) TT.0 If $t_g \mid t_j$ or only one of them is in a cycle, which was solely generated using Rule PP.1, then signal "forbidden, delete the Π^w " and return.

(1) TT.1 If $\Pi_g = \Pi_j$, signal "a pure *TT generation*;" otherwise signal "an interactive *TT generation.*"

(2) TT.2 If $t_g \leftarrow t_j$, signal "forming a new cycle."
If, without firing t_j, there does not exist a firing sequence σ to fire t_g, then insert token in a place of NP.
If $\Pi_g = \Pi_j$, return and the designer may start a new generation.

(3) TT.3 If the structure synchronic distance $d_{gj}^s = \infty$ then
(a) TT.3.1
Apply Rule TT.4.
(b) TT.3.2
Generate a new TT-path path to synchronic t_g and t_j such that after step 3.c, d_{gj}^s changes from ∞ to 1.
(c) TT.3.3
Go to step 2.

(4) TT.4
(a) TT.4.1
Generate a *TP-path* from a transition t_g of each Π_g in X_{gj} to a place p_k in the NP.
(b) TT.4.2
Generate a *virtual PSP*, a *PT-path*, from the place n_j to a transition t_j of each Π_j in X_{jg}.

Definition: PP Rule:
For an NP from $p_g \in \Pi_g$ to $p_j \in \Pi_j$ generated by the designer,

TABLE 8.1 Summary of the Rules

Conditions of Generations	PP Rule	TT Rule
A. cycles created	—	TT.2
B. no cycles created	—	—
B.1 only one of n_g and n_j in a cycle from PP.1	—	TT.0
B.2 exclusive	PP.2	TT.0
B.3 concurrent	PP.0	—
B.3.1 neither n_g nor n_j in a cycle from PP.1	—	TT.4
B.3.2 both n_g and n_j in a cycle from PP.1	—	TT.3
B.4 sequential or cyclic	PP.1 (PG) PP.2 (IG)	TT.1 (PG) TT.4 (IG)

(0) PP.0 If $p_g \parallel p_j$, then signal "forbidden, delete the Π^w" and return.

(1) PP.1 if $\Pi_g = \Pi_j$, signal "a pure *PP generation*;" otherwise signal "an interactive *PP generation*."

(2) PP.2

    (a) PP.2.1 Generate a *TP-path* from a transition t_k of NP to a place p_j of each Π_j in C_{jg}.

    (b) PP.2.2 Generate a virtual PSP, a *PT-path*, from a place p_g of each Π_g in C_{gj} to the transition $\dot{n}_g$ of NP.

Note that some generations may require the application of more than one rule. For instance, a backward IG may need both Rules TT.2 and TT.4. Note also the partial dual relationship between the TT and PP rules. Replacing transitions with places (and vice versa) reversing each arc in Rule TT.4, we obtain Rule PP.2. The rules are summarized in Table 8.1.

Section 8.6 demonstrates the application of these rules to an automated manufacturing system. Note that during each application of a rule, more than one new PSP may be generated; we call such a generation a *macro generation*. Otherwise, it is a *singular generation*.

In the remainder of this chapter, all nets mentioned refer to nets synthesized using the rules presented above.

Below, we explain the rules starting with the rules that forbid generations. Forbidden TT and PP generations come from synchronic and concurrency mismatches, respectively, which cause uncoordinated actions. For a TT generation, a single parameter, the synchronic distance between t_g and t_j, can unify all cases of the structural relationships between n_g and n_j.

Definition: Synchronic Mismatch: A TT generation causes a synchronic mismatch of $d_{gj}^1 \neq d_{gj}^w$, where d_{gj}^1 (d_{gj}^w) denotes the synchronic distance between t_g and t_j and N^1 (NP).

Definition: Concurrency Mismatch: A PP generation causes a concurrency mismatch if $(p_g \parallel p_j)$ and $\exists$ a $p \in P$, $p \parallel p_g$, and $\neg (p \parallel p_j)$.

There are three possible synchronic distances: 1, ∞, and the integers in between. A TT generation should not alter the synchronic distance between t_g and t_j. But a TT-path implies a synchronic distance of 1 between t_g and t_j. Thus, if the synchronic distance between t_g and t_j is 1 prior to the generation, the TT generation will not alter the reachable markings of N^1. If t_g (t_j) can fire infinitely often without firing t_j (t_g), then the NP will get unbounded tokens (become nonlive due to deficiency of tokens).

In terms of structural relationship, $d_{gj} = \infty$ for the following cases: (1) $t_g | t_j$, (2) t_g and/or t_j is in a cycle generated earlier using the PP rule, (3) $t_g \leftrightarrow t_j$ or $t_g \parallel t_j$, $\exists t_x$ such that $t_g | t_x$, and/or $t_j | t_x$, and t_g, and (4) t_g and t_j are in two separate PNs. For cases (1) and part of (2) where only one of t_g and t_j is in a cycle, the generation is chosen to be forbidden to make Rule TT.0. The rest of the cases need additional generations to make Rules TT.3 and TT.4.

(a)

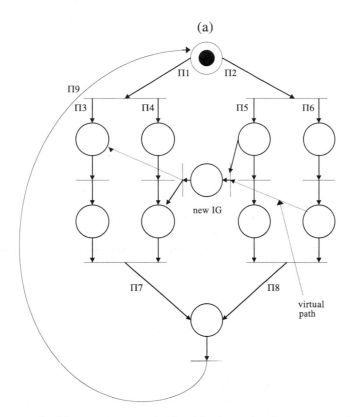

FIGURE 8.6(a) An example of the completeness rule 2 involving interactions between two exclusive PSPs using PP rule. LCN (Π_4, Π_5) = {Π_3, Π_4}, LCN (Π_5, Π_6) = {Π_5, Π_6}. (From Reference 2a. With permission.)

	1	2	3	4	5	6	7	8	9
1		EX	SP	SP	EX	EX	SE	EX	SE
2	EX		EX	EX	SP	SP	EX	SE	SE
3	SN	EX			EX	EX	SP	EX	SE
4	SN	EX	CN		EX	EX	SP	EX	SE
5	EX	SN	EX	EX		CN	EX	SP	SE
6	EX	SN	EX	EX	CN		EX	SP	SE
7	SL	SL	SN	SN	EX	EX		EX	SP
8	EX	SL	EX	EX	SN	SN	EX		SP
9	SL	SL	SL	SL	SL	SL	SN	SN	

FIGURE 8.6(b) The T-Matrix of Figure 8.6(a). (From Reference 2a. With permission.)

Rule PP.0 forbids concurrent mismatch. Since $\Pi_g \| \Pi_j$, they need to cooperate to complete the task. But the NP diverts the token away to break the coordination and induces a deadlock; hence, Rule PP.0 forbids a PP generation between concurrent PSPs.

Rules TT.4 and PP.2 model the messages exchange between two parties and the context switching between two processes, respectively. They are called completeness rules 1 and 2, respectively. To satisfy the completeness requirement[1] in communication protocols, each PSP in X_{gj} (X_{jg}), if executed, must send (receive) a token by firing the relevant t_g (t_j).

In the context switching, all local contexts in C_{gj} must switch to those in C_{jg}. See Figures 8.5 and 8.6 for the application of these two rules.

The correctness of these rules is established in the sequel. We first show the correctness for a synthesized net where each home place holds eactly one token. Let N^a denote such a net, N^b the net after adding tokens to N^a, and N^{1a} and also N^a. We then show that the synthesized net is **marking monotonic**; that is, it remains well-behaved by adding tokens to places in the synthesized net.

Rules TT.1, TT.2, and PP.1 do not require further generations. They are simple and have been dealt with in much of the literature using the concept of reductions and expansions. The rest rules are established to satisfy the following guidelines for adding NPs.

1. No disturbance to N^1: inhibiting any intrusion into normal transition firings of N_1 prior to the generations. This guarantees that neither unbounded places nor dead transitions would occur in N_1 since it was live and bounded prior to the generations.
2. Well-behaved NP: inhibiting any dead transitions and unbounded places in NP.

There are two kinds of intrusions. One is to chnage the marking of N_1; the other is to eliminate some reachable markings. In order for NP to be live, it must be able to get enough tokens from p_g or by firing t_g. When tokens in NP disappear, the resultant marking of the subnet N_1 in N_2 must be reachable in N_1 prior to the generation of NP. Furthermore, the marking of NP must be reversible in N_2.

The second intrusion may occur when the joint is a transition that may not be enabled in case of backward TT generation, hence, causing some markings in N_1 to be unreachable. Thus if the joint is a transition, NP must be able to get tokens from a generation point within each iteration.

Based on the concept of *no intrusion*, the rules are constructed according to the following guidelines:

1. From M_0, each t_g must be potentially able to be fired (always fires or has fired $\forall \sigma$ in N_1 enabling the joint that is a transition).
2. These tokens must disappear from NP before it gets unbounded tokens, and return to an N_1 to reach a marking M_2. The marking of the subnet N_1 in N_2, $M_2(N_1)$ of all possible sets of such M_2 is identical to those in N_1 without the new paths.

Theorem 1: Any PN resulting from a singular application of Rules TT.1 and PP.1 to an N^{1a} synthesized from a basic process is well-behaved.

Proof: This theorem can be proved by repetitively applying reduction and expansion.[30,34,36]

Theorem 2: Any PN resulting from a singular application of Rules TT.2 to a N^{1a} synthesized from a basic process is well-behaved.

Proof: This theorem can be proved by repetitively applying reduction and expansion.[30,34,36]

Lemma 1: For an N^2, any pair of PSPs cannot be both concurrent and exclusive to each other.

Proof: If one assumes to the contrary, then there are two n_{ps}, one a $t \in T$ and another a $p \in P$. This can only happen when they are PT or PP generations between concurrent PSPs or TP or TT generations between exclusive PSPs. These four types of generations are forbidden.

This lemma implies that any time start node of a pair of PSPs must be either transition or place and cannot be both. Thus, similar to an SC, the two prime nodes of a pure pair must both be transitions or places. This lemma leads to the equivalence of structural and temporal relationship, which is marking- or time-related, rather than structurally related. Two transitions (places) are temporally concurrent to each other if they can fire (have tokens) simultaneously in a safe PN. Without this lemma, p_1 and p_2 may not (may) be able to hold tokens simultaneously, even if $p_1 \| (|) p_2$ since there may be two DEPs from a $p_{ps} (t_{ps})$ to p_1 and p_2, respectively.

Lemma 2: For an N^2, let $M_\alpha^2 \in R^2$ (the R for N^2), where all Πs in C_{12} are marked, then $M_\beta^2 = M_\alpha^2 - \mu(C_{12}) + \mu(C_{21})$ is a reachable marking, where $\mu(C_{12})$ $(\mu(C_{21}))$ denotes that each Π in C_{12} (C_{21}) holds exactly one token.

Proof: After each Π in C_{12} fires once, there exists a σ for one of C_{21} to gain a token. By the equivalence of temporal and structural relationship, each Π in C_{21} will gain a token. Hence the lemma.

This lemma ensures that the application of Rule PP.2 will not alter the reachable markings of N^1. The following lemma also comes from the equivalence of temporal and structural relationship.

Lemma 3: Using Rule TT.4, if one t_j in X_{jg} gets a token in each of its input places except the one from $\dot{n}_j$, then any subsequent firing sequence, if long enough, will inject a token to n_j and fires the above t_j.

Proof: This lemma comes from the fact that, prior to the generation, $d(X_{gj}, X_{jg}) = 1$, since $X_{gj} \leftrightarrow$ or $\| X_{jg}$.

This lemma, along with Lemma 4, ensures that no tokens get accumulated in the NP indefinitely. Note: if $t_j \rightarrow t_g$, then the NP is marked by Rule TT.2. t_j can fire without the token injected by firing a t_g. But subsequent firing will always lead to the firing of a t_g to restore M_0^w.

The above firing sequence, denotd as σ^w, including both t_g and t_j and $\forall\, t \in \sigma^w$, $t \in T^w$ (the set of transitions in the NP), is called a *complete firing sequence* of the NP. Lemma 4 shows that after a σ^w, the marking of N^{1a} in N^{2a}, $M^{1/2}$ will be a reachable one in N^{1a}. The correctness of this lemma is based on the following:

Observation 1: $\forall t \in T^w$, if $t \neq n_g$, $t \neq \dot{n}_j$, and $t \neq \dot{n}_g$

1. $|\bullet t| = 1$.
2. $\forall$ pair of t_1 and $t_2 \in T^w$, $\neg\,(t_1 \| t_2)$ using Rule TT.4, and $\neg\,(t_1 | t_2)$ using Rule PP.2.
3. Any TP (PT) generation occurs between exclusive (concurrent) PSPs.
4. Any pair of PSPs Π_1 ad Π_2, joining at a node n_k in the NP, then $\Pi_1 \|(|) \Pi_2$, if $n_k \in T^w (P^w)$.

Proof: $(1) |\bullet t| > 1$ only for transition joints within the NP. But this occurs only under Rule PP.2.1 where the joint is an $\dot{n}_g$. (2) Consider Rule TT.4 first. Any pair of new PSPs corresponding to the TP-path using Rule TT.4.1 are mutually exclusive since they have the same p_{ps} as the two corresponding Π_g. The same result applies to Rule TT.4.2 since the n_{ps} is a place p_j. The case for Rule PP.2 can be proved in a dual fashion. Cases 3 and 4 can be similarly proved.

The observation derives from the fact that there are only TP and PT generations beyond the first generation using Rules TT.4 and PP.2. From (2) of Observation I, any pair of PSPs Π_1 and Π_2 using Rule TT.4 (PP.2) is eitther $\Pi_1 \leftrightarrow \Pi_2$ or $\Pi_1 |(\|) \Pi_2$. Thus, a complete firing path through the NP is a single DEP using Rule TT.4 and includes all transitions in the NP using Rule PP.2.

Lemma 4: For the N^2 after applying Rule TT.4 (PP.2), after injecting a token (tokens) into the NP by firing only one t_g (the $\dot{n}_g$), the subsequent firing sequence will lead to $(M_\alpha^{1a}, M_0^w)[\sigma^w \rightarrow (M_\beta^{1a}, M_0^w)$, where both M_α^{1a} and $M_\beta^{1a} \in R^{1a}$.

Proof: We first show that the injected tokens will eventually disappear from the NP due to the following facts: 1. There are no internal TP-paths whose $p_j = n_j$ and 2. Injected tokens can flow freely inside the NP since every t inside the NP has only one input place by Observation 1. We then show that upon the disappearing of the injected tokens from the NP, the resultant M^2 contains a $M^{1/2} \in R^{1a}$. This is true for Rule TT.4 since the subfiring sequence in $N^{1/2}$ leading to the firings of t_g and t_j is exactly the same as that in N^{1a} prior to the generation. For Rule PP.2, the token disappearing is equivalent to the switching of tokens from C_{gj} to C_{jg}, which leads to a submarking $M^{1/2} \in R^{1a}$ by Lemma 2.

Lemma 5: $\forall \sigma$, such that $M_0^2[\sigma M_0^2$, either $\sigma^w \subseteq (\sigma$ or $\forall\, t \in \sigma, \neg)(t \in T^w)$.

Proof: This lemma comes directly from Lemmas 3 and 4.

Theorem 3: Any PNs resulting from the singular applications of Rules TT.4 and PP.2 to a N^{1a} synthesized from a basic process are well behaved.

Proof: Prove by induction. It is easy to see that a basic process is live, bounded, and reversible. Then, assuming the N (i.e., N^{1a}) from the $(N-1)$th synthesis steps is bounded, live, and reversible, the proof needs to show that after each full synthesis step using Rule TT.4 or PP.2, the resulting N^{2a} remains so. $\forall \sigma$ that does not include any node in the NP, it will fire in exactly the same manner as that in N^{1a} and leads to the same reachable marking. Hence this case need not be considered.

Reversible: We need to show that a σ exists such that $(M_\alpha^{1a}, M^w) [\sigma > (M_0^{1a}, M_0^w)$. There are two cases: (a) $M^w = M_0^w$ and (b) $M^w \neq M_0^w$.

(a) Two cases if σ (1) does or (2) does not include any $t \in T^w$. For case (2), the subnets that involve σ are exactly those in N^{1a}. Hence, the σ exists. For case (1), Lemma 5 dictates that a $\sigma^w \subseteq \sigma$. Let $\sigma = \sigma_c \sigma^w \sigma_d$, then $(M_\alpha^{1a}, M_0^w) [\sigma_1 \sigma^w > (M_\beta^{1a}, M_0^w)]$, where $M_\beta^{1a} \in R^{1a}$. The problem then reduces to case (2) which has been proved.

(b) $M^w \neq M_0^w$, which implies a partial σ^w. By Lemma 5, all sbsequent firing sequences, if long enough, can complete the rest of σ^w to restore M_0^w. The case then is similar to case (a) and can be proved similarly.

Live: Assume to the contrary, then $\exists$ a $R^{2a} = (M_\alpha^{1a}, M^w)$, $\exists t \in T^2$; some of its input places, which are mutually concurrent, can never get tokens. But this is impossible by the equivalence between temporal and structural relationship (which ensures that all input places of t are possible to get tokens) and the requirement that the NPs in Rule TT.4.2 and PP.2.2 should be virtual (which ensures that the above possibility should be definite) by preventing uncoordinated movement of tokens in a set of concurrent places. When a token enters a place with more than one output transition, it may fire any of its output transitions and thus it may result in uncoordinated movement of tokens.

That is, the two tokens generated by firing a transition may move to the input places of two exclusive transitions, respectively. Since they are mutually exclusive, none of them will be able to be fired; the N hence is not live. The virtual PSP generations in Rules TT. 4.2 and PP.2.2 prevent such uncoordinated movements.

One of the above two tokens moves in N^{1a}; the other in the NP. Uncoordinated movements of these two tokens intrude the marking behavior of N^{1a}; this is impossible by Lemma 5, which states that guideline (2) is satisfied.

Note that Guideline (1) can only be revoked when the a t_g has input places from the NP. Then, by Rule TT.2, the NP will have a token in its initial marking, which allows t_g to be potentially able to be fired from M_0^{2a}.

Thus both guidelines are satisfied. Based on this, we can show that the t is potentially able to be fired. If (1) $M^w = M_0^w$, then by Lemma 4, the presence of NPs does not change the markings of N^{1a}; hence, if t is not in the NP, t is potentially able to be fired. If t is in the NP, then by Observation 1 and Guideline (1), t is also potentially able to be fired. Now if (2) $M^w \neq M_0^w$, then also by Lemma 4 there exists a subsequent firing sequence to drive M^w back to M_0^w and then $M^{1/2}$ is a reachable marking in M^{1a}. The case then degenerates to case (1).

Theorem 4: Any PN resulting from a singular application of Rule TT.3 to an N^{1a} synthesized from a basic process is well-behaved.

Proof: Lemma 5 does not hold for Rule TT.3.1 because Lemma 3 does not either. This is because t_g may fire infinitely often relative to t_j. Now Rule TT.3.2 prevents such infinite firings of t_g with respect to t_j and forces $d_{gj} = 1$. Lemma 3, and hence Lemma 5 also hold for Rule TT.3. The rest of proof follows that for Rule TT.4 in Theorem 3.

Theorem 5: Any PNs resulting from the applications of the TT and PP rules to a basic process are bounded, live, reversible, and marking monotonic.

Proof: Prove by induction. It is easy to see that a basic process is live, bounded, and reversible. Then, assuming the N from the $(N-1)$th synthesis steps is bounded, live, and reversible, the proof needs to show that after each full synthesis step using a certain TT or PP rule, the resulting N^{1a} remains so. The correctness of this is established by Theorems 1–4. Note that we have been assuming that there is only one token in every home place. Now prove the property of marking monotonic by showing that N^b is well-behaved.

Reversible: Let $M_\alpha^{2a} \in R^{2a}$ in N^{2a}, then

$$M_\alpha^{2a} = M_0^{2a} + Ax^{2a}$$

where A is the incidence matrix and x^{2a} a firing vector. Since N^{2a} is reversible, $\exists$ a x^{2a} such that $Ax^{2a} = 0$ and $M_\alpha^{2a} = M_0^{2a}$. Now add some tokens to some places in N^{2a} to result in N^b. The relevant equation is

$$M_\alpha^b = M_0^b + Ax^b$$

The above firing vector x^{2a} also makes $M_\alpha^b = M_0^b$ since $Ax^{2a} = 0$. Now choose the least positive integer u such that $x = ux^{2a} - x^b \geq 0$; i.e., every component of x is nonnegative. We have

$$M_\alpha^b + Ax = M_0^b + A(ux^{2a}) = M_0^b$$

Thus N^b is reversible.

Live: The proof of N^b being live is exactly the same as that in Theorem 3.

The bounded property can also be proven by deriving the P-invariants[15] for the synthesized nets.

Theorem 6:   The class synthesized PN belongs to the class of synchronized-choice (SC) nets.

Proof: A basic process is obviously an SC. Then assuming N^1 is an SC, we need to prove that N^2 is also an SC. Prove for the TT-rule first. The proofs for PP rules are similar. In order to satisfy the first (second) requirements, we need to show that for any pair of PT (TP) handles, there are two TP bridges across them. If all nodes and arcs of the two handles are in N^1, then by assumption they satisfy the two requirements. We need only consider the case where only part of the handles are in the new paths. Such handles are called mixed handles.

Consider the generations by TT-rules first. The case of PP-rules is similar. We need not consider the pure-TT generation, because the new path forms a PP-handle to N^1. For a generation by Rule TT.3, inside the new paths, there are only place joints formed by TP-path generations using Rule TT.3.1. But there are no pairs of TP-handles containing these TP-paths. This is because the TP-paths start from t_gs and all n_{ps} of LEX1 are places.

The only pairs of fixed PT-handles have their $n_{pe} = t_g$ and $n_{ps} = $ the p_{ps} of LEX1. One handle of the pair is in N^1; part of the other is in the new path. Two of the TP-paths generated using the Rule TT.3.1 form the two bridges between the above two TP-handles, thus satisfying the first requirement.

Note that an SC may not be synthesized; this happens when the SC is not well-behaved. For instance, when the PT-path (TP-path) generated using Rule TT.4.2 (PP.2.2) is not a virtual one, the N^2 is an SC but with deadlocks.

8.4   Temporal Matrices for Petri Nets

The rules should be implemented as an interactive and visual-aid tools. The designer can view the PN model being designed on the screen and use a mouse to input paths according to the design specification. Each time a path is generated, the system should check the new path generation against the rules. If some rule is violated, a warning should be issued to the designer to request either deletion of this new path or addition of more PSPs. In order to automate this process, a mechanism is needed to

- Record the relationship between any two PSPs,
- Determine the applicable rule upon a path generation, and generate additional paths and entries, if necessary, or, if no rules are applicable, delete this path.

We use matrix to perform the first function. Instead of recording connectivity between places and transitions, the matrix in our algorithm records structural relationship between PSPs. Before a path is generated, the T-Matrix needs to be consulted to verify that no rule is violated. Whenever a new path is created, a PSP may be separated into several new PSPs, and relationships between some *PSP*s are changed. Therefore, the total number of *PSP*s needs to be updated, as do the matrix entries.

Example: Figure 8.4(a) shows a PN model with its matrix shown in Figure 8.4(b). Originally there are five *PSP*s in solid paths Π_1 to Π_5; Π_6 in a dashed path is to be added by IG. Π_1 is *concurrent* to all other *PSP*s. Therefore, all entries in the first column and the first row are *CN*. Π_2 and Π_3 are *exclusive* to each other, as are Π_4 and Π_5. Therefore,

$$A_{23} = A_{32} = A_{45} = A_{54} = \text{EX}.$$

Notice Π_6 returns to form cycles; therefore, some entries need to be updated. Prior to Π_6 generation, Π_3 is structurally *sequentially earlier* (SE) to both Π_4 and Π_5. After the addition, Π_6, Π_4, and Π_5 become SE to Π_3, implying the formation of cycles.

We have described how the T-Matrix records the relationship. The next section presents the algorithm and the associated time complexity for updating the T-matrix.

8.5   The Algorithm and Its Complexity

The algorithm consists of two steps:

1. Create new entries in the T-Matrix for each NP.
2. Determine the applicable rule; generate additional NPs if necessary and update the T-Matrix accordingly. If no rule is applicable or some rules are violated, delete the path.

For example, if a PP-path is generated between p_g and p_j and $p_g \parallel p_j$, then Rule PP.0 dictates that this path should be deleted. It is easy to find the structural relationship between the new *PSP* Π^w and the *PSP*s directly involved with the generation, but not obvious as to the structural relationship between *PSP*s far away from (no direct connection with) the Π^w. In the following, we will develop techniques for new entries for PG and IG, respectively.

Determination of Entries for Pure-Generation

After the PG of a Π_l from Π_g, Π_g is split into three new *PSP*s: $\Pi_{g_1} \rightarrow \Pi_{g_2} \rightarrow \Pi_{g_3}$. The entries related to Π_g should be deleted. The values for the new entries are

$$A'_{ik} = A_{gk}, \quad A'_{ki} = A_{kg}$$

where Π_k is any *PSP* other than Π_g, Π_l, Π_{g_1}, Π_{g_2} and Π_{g_3}, and Π_i is one of the newly created *PSP*s. Note that Π_{g_1} and/or Π_{g_2} may be empty. The other new entries are the relationships among the newly created *PSP*s:

$$A'_{g_2 g_1} = A'_{g_3 g_1} = A'_{g_3 g_2} = \text{SL},$$
$$A'_{g_1 g_2} = A'_{g_2 g_3} = A'_{g_1 g_3} = \text{SE}$$

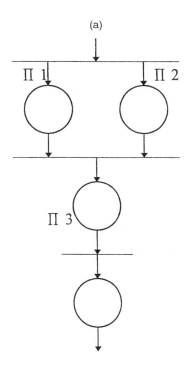

FIGURE 8.7(a) Original Petri nets with 3 *PSP*s. (From Reference 2a. With permission.)

	1	2	3	
1		CN	SP	SP: Sequential Previous
2	CN		SP	SN: Sequential Next
3	SN	SN		

FIGURE 8.7(b) The T-Matrix of the net in Figure 8.7(a) prior to the pure-generation. (From Reference 2a. With permission.)

and

$$A'_{lg_2} = CN \qquad \text{for PG } using \; TT \; rule$$

$$A'_{lg_2} = EX \qquad \text{for PG } using \; PP \; rule$$

Therefore, no modifications for existing entries are necessary for PG. There are four new PSPs. Each new entry takes constant time to enter; the corresponding time complexity is $O(n)$, where n is the number of PSPs before the generation. We add a PG in Figure 8.7 using the PP rule with the T-Matrix and PN shown in Figure 8.8. There are four new entries: Π_3, Π_4, Π_5, and Π_6, respectively.

Determination of Entries for Interaction-Generation

In the following, we consider only the case where $\Pi_g | \Pi_j$ or $\Pi_g \| \Pi_j$; the case where $\Pi_g \rightarrow \Pi_j$ or $\Pi_{j_1} \rightarrow \Pi_g$, can be treated similarly.

For an IG of Π_l from Π_g to Π_j, Π_g (Π_j) is split into $\Pi_{g_1} \rightarrow \Pi_{g_2}$ ($\Pi_{j_1} \rightarrow \Pi_{j_2}$). Note that $\Pi_g \rightarrow \Pi_l \rightarrow \Pi_{j_2}$. The structural relationship between the new PSP Π_l and existing PSPs must be determined. Also some structural relationships among existing PSPs are changed. For instance, independent of the structural relationship between Π_g and Π_j prior to generation, after generation $\Pi_{g_1} \rightarrow \Pi_l \rightarrow \Pi_{j_2}$. It is easy to determine the structural relationships among PSPs directly involved in the generation, but rather difficult to determine them between PSPs far away from the generation and PSPs directly involved in the generation.

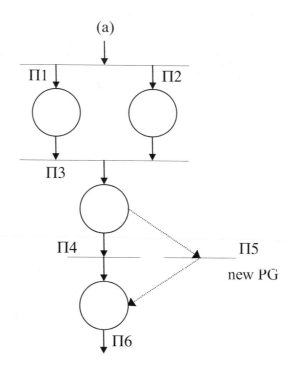

FIGURE 8.8(a)   After adding a pure-generation, there are 6 PSPs in the Petri net. (From Reference 2a. With permission.)

	1	2	3	4	5	6	
1		CN	SP	SE	SE	SE	SP: Sequential Previous
2	CN		SP	SE	SE	SE	SN: Sequential Next
3	SN	SN		SP	SP	SE	
4	SL	SL	SN		EX	SP	
5	SL	SL	SN	EX		SP	
6	SL	SL	SL	SN	SN		

FIGURE 8.8(b)   Updated T-Matrix. (From Reference 2a. With permission.)

The new matrix (A') entries are determined by the following lemma.

Lemma 6:   For an IG of Π_l from Π_g to Π_j, then update A to A' as follows:

(1) $A'_{ik} =$ SE and $A'_{ki} =$ SL, where $ik = g_1g_2, j_1j_2, g_1l, lj_2, lq, g_1j_2, rl, rj_2$, and $rq,$

(2) $A'_{mz} = A_{mz}$ and $A'_{zm} = A_{zm}$, where $mz \neq rq, qr$ $(m, z \in \{r, q, u, v\})$, and

(3) $A'_{tz} = A_{sz}$ and $A'_{zs} = A_{zt}$, where $st = gg_1, gg_2, jj_1, jj_2$ $z \in \{r, q, u, v\}$, and $tz \neq j_2r, g_1q$

where Π_r, Π_q, Π_u, and Π_v should be old PSPs not belonging to C_{gi} or X_{gi} such that $\Pi_r \rightarrow \Pi_g$, $(\Pi_j \rightarrow \Pi_q$, $\neg \; \Pi_u \rightarrow \Pi_g)$ $(\Pi_u$ not *sequentially earlier* to $\Pi_g)$, $\neg (\Pi_u \rightarrow \Pi_j)$, $\neg (\Pi_v \rightarrow \Pi_g)$, and $\neg (\Pi_v \rightarrow \Pi_j)$.

Proof: Case (1) is obvious when neither $i = r$ nor $k = q$. For $i = r$, the presence of Π_l affects the following entries: (a) $A_{rq} = A_{gj}$ and (b) A_{hq}, where $\Pi_g \rightarrow \Pi_h$, $\neg (\Pi_h \rightarrow \Pi_q)$ and $\neg (\Pi_q \rightarrow \Pi_h)$ (note $\Pi_j \rightarrow \Pi_q$).

(a)

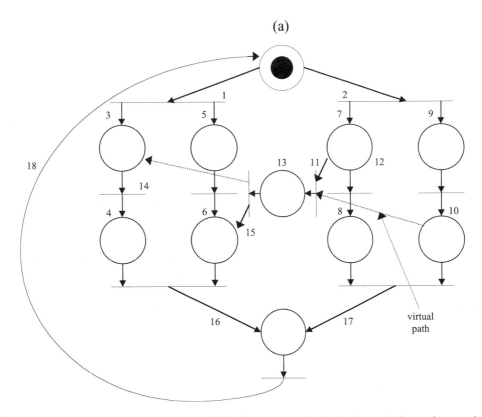

FIGURE 8.9(a) After the interaction-generation (IG), there are 18 PSPs in Petri nets. Each number stands for a PSP. (From Reference 2a. With permission.)

For case (a), prior to the generation, no path passes through both Π_r and Π_q (otherwise $\Pi_r \to \Pi_q$); after the generation, there is a path passing through Π_r, Π_l, and Π_a. Hence, $A_{rq} = $ SE; this completes the proof for case (1). For case (b), note the two paths $\Pi_{g_2} \dots \Pi_h$ and $\Pi_l \Pi_{j_2} \dots \Pi_q$ intersect at the generation point of Π_l; this might change the relationship between Π_h and Π_q. However, it is easy to see that $A_{hq} = A_{gj}$ and from the structure definitions of "|"and "||", we have $A'_{hq} = A_{hq}$. Hence, except for $mz = rq$, $A'_{mz} = A_{mz}$ where $m, z \in \{r, q, u, v\}$. Hence case (2) proved.

Case (3) can be proved similarly to cases (1) and (2).

The corresponding time complexity is $O(n^2)$ mainly incurred by the operation $A'_{rq} = $ SE. To complete the step of the single application of a TT or PP rule, additional TP- or PT-paths may have to be generated from/to the NPs. The matrix update after each such generation can be performed in a similar fashion to that in Lemma 6 with the same time complexity.

Figure 8.9 shows the PN after the IG and the updated T-Matrix (refer to Figure 8.6). Any alphabetic number in Figure 8.9(a) stands for a PSP. Π_3 in Figure 8.6 has been separated into two PSPs, Π_3 and Π_4, respectively. Note this PN does not belong to the class of asymmetric-choice nets; note also that all pairs of exclusive (concurrent) PSPs have places (transitions) as their n_{ps} and n_{pe}; i.e., the PN belongs to the class of synchronized-choice nets. The synthesized PN in Figure 8.5(a) also belgnls to the class of synchronized-choice nets.

Complexity of the Algorithm

The complexity for the algorithm consists of the follwing components:

1. Determining which rule is applicable and checking the rule violation.
2. Updating matrix entries.

	1	2	3	4	5	6	7	8	9	10	11	12	13	14	15	16	17	18
1		EX	SP	SE	SP	SE	EX	EX	EX	EX	EX	EX	EX	EX	EX	EX	EX	SE
2	EX		EX	SE	EX	SE	SP	SE	SP	SE	SE	SE	SE	SE	SE	SE	SE	SE
3	SN	EX		SP	CN	CN	EX	EX	EX	EX	EX	EX	EX	EX	EX	SE	EX	SE
4	SL	SL	SN		CN	CN	SL	EX	SL	EX	SL	SL	SN	SN	CN	SP	EX	SE
5	SN	EX	CN	CN		SP	EX	EX	EX	EX	EX	EX	EX	EX	EX	SL	EX	SE
6	SL	SL	CN	CN	CN		SL	EX	SLL	EX	SL	SL	SL	CN	SN	SP	EX	SE
7	EX	SN	EX	SE	EX	SE		SP	CN	CN	SP	CN	SE	SE	SE	SE	SE	SE
8	SE	SL	EX	EX	EX	EX	SN		CN	CN	EX	CN	EX	EX	EX	EX	SP	SE
9	EX	SN	EX	SE	EX	SL	CN	CN		SP	CN	SP	SP	SE	SE	SE	SE	SE
10	EX	SL	EX	EX	EX	EX	CN	CN	SN		EX	EX	EX	EX	EX	EX	SP	SE
11	EX	SL	EX	SE	EX	SE	SN	EX	CN	EX		CN	SP	SE	SE	SE	EX	SE
12	EX	SL	EX	SE	EX	SE	SL	EX	SN	EX	CN		SP	SE	SE	SE	EX	SE
13	EX	SL	EX	SP	EX	CN	SL	EX	SL	EX	SN	SN		SP	SP	SE	EX	SE
14	EX	SL	EX	CN	EX	SP	SL	EX	SL	EX	SL	SL	SN		CN	SE	EX	SE
15	EX	SL	EX	CN	EX	SP	SL	EX	SL	EX	SL	SL	SN	CN		SE	EX	SE
16	SL	SL	SL	SN	SL	SN	SL	EX	SL	EX	SL	SL	SL	SL	SL		EX	SE
17	EX	SL	EX	EX	EX	FX	SL	SN	SL	SN	EX	EX	EX	EX	EX	EX		SP
18	SL	SL	SL	SL	SL	SL	SL	SL	SL	SL	SL	SL	SL	SL	SL	SN	SN	

FIGURE 8.9(b)   The new T-Matrix of the net in Figure 8.9(a). (From Reference 2a. With permission.)

For factor 1, it takes $O(1)$ time to consult the T-Matrix to find A_{gj}. For factor 2, the complexity to update entries has been shown to be $O(n)$ for PG and $O(n^2)$ for each PSP generation during an IG.

Thus each PSP generation takes $O(n^2)$ time complexity. Let n_i be the number of PSPs prior to the ith generation. The time complexity for the ith generation is $O(n_i^2)$. After a PG, one PSP is deleted, while at most four new PSPs are created. Thus after the PG, the number of PSPs increases at most from n_i to n_i+3.

For an IG of an NP between two PSPs, two PSPs are deleted and at most five new PSPs are created. Hence after the IG of a single PSP, there are at most $n_i + 3$ PSPs. The time complexity for the next generation is bounded by $O([n_i + 3]^2)$.

Let α be the total number of PSPs at the end of a design. If we add the above time complexity for all steps, the total time complexity would be $O(\alpha^3)$. However, this is overly estimated by noting that every entry is updated at most twice because the entry can only be updated when it is "CN" or "EX." In other words, it can be updated from "CN" or "EX" to "SQ," but not vice versa. Since we have α^2 entries, the total time complexity is therefore $O(\alpha^2)$.

Note that the T-Matrix does not include any virtual PSPs as they have no internal nodes (i.e., neither n_g nor n_j) to serve as an n_g or an n_j. Thus any PSP in the T-Matrix must contain at least one internal node and the total number of PSPs in the T-Matrix are less than the total number of nodes, β, in the PN. Therefore, the total time complexity is $O(\beta^2)$. When many PSPs contain a number of internal nodes, the number of nodes outnumbers that of PSPs. Operations on the T-Matrix should result in significantly more reduction in time for synthesis than if we consider each node individually.

8.6   X-Window Implementation

We have implemented the above algorithm by incorporating it into a multi-function Petri net graphics tool based on X-Window (Motif version). This software package[4,11] is a user-friendly CAD tool for designing, verifying, simulating, querying (including cycle time, invariants, incidence matrix, inputs/outputs of neods, etc.), reducing, and synthesizing PN. The structure of this tool is shown in Figure 8.10.

This tool can draw, erase, copy, move, pan, and zoom in/out objects such as places, transitions, states, rectangular and elliptical objects (filled and nonfilled), and texts (with a variety of fonts). The user can

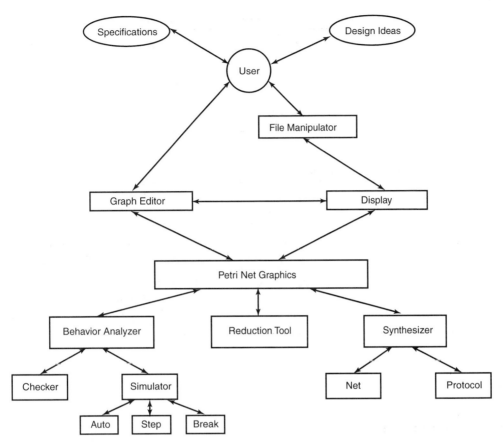

FIGURE 8.10   Structure of the CAD tool for designing, analysis, and synthesis of protocols. (From Reference 2a. With permission.)

construct a PN either using the "DRAW" button to draw transitions, places and arcs, or using the "FILE" button to input a PN file. Buttons "Move," "Pan," and "Zoom," to allow large PNs to be drawn. Clicking the "Pan" button centers the graphics at a chosen location. In addition, the vertical (horizontal) scrollbar allows the graphics to be moved up and down (left and right). After the PN is drawn or displayed on the screen, graphical interconnection among places and transitions are translated into internal representations for further manipulations.

A user can initiate the synthesis by clicking the "Petri" button of the "Synthesis" menu. Afterwards, no further clicking is necessary for further generations except for clicking the "T-Matrix" button to display the T-Matrix in the bottom window (called "text_w") after a new generation. An example of this process to the synthesis of an automated manufacturing system (Figure 8.11) is shown in Figure 8.12. Shown in the bottom is a message window "text_w." After each generation step, this window displays the kind of generation, and signals any rule violations or whether tokens must be added to places in the NPs.

Upon an interaction generation, a new window pops up that displays an arrow linking two sets of nodes and invites the designer to pick one node in the left-hand (right-hand) set as $n_g(n_j)$. Each time a node is picked, the tool will call a filtering procedure to eliminate nodes in the set that are sequential to the node just picked and then redisplay the window. This process continues until both sides are empty and no more NPs need to be generated for this specific IG.

In the displayed T-Matrix, each structural relationship is expressed by a single letter (e.g., "E") different from the two letters (SE) presented earlier. The correspondence is shown in Table 8.2. In addition, $A_{ik} =$ 'Y' if $A_{ik} = $ 'X' and Π_i is in a cycle that has a place with more than one input transition; $A_{ik} = $ 'Z' if

TABLE 8.2   Structure Relationship Correspondence

Sequentially earlier	E	SE
Sequentially later	L	SL
Sequentially next	N	SN
Sequentially previous	P	SP
Exclusive	X	EX
Concurrent	U	CN
Cyclic	C	CL

TABLE 8.3   Major Components

Components	Number	Function
Entries	2	Inputs two types (A and B) of raw materials
Exits	2	for final A parts and B parts, respectively.
Robots	R1	Serves machines, M1, M2, and M4 with no priorities. Randomly choose one ready machine with raw material.
	R2	Loads machine M3 and M5.
		From Machines M1 and M2
AGV Systems		
AGV1		Deliver intermediate A parts to M3.
AGV2		Deliver intermediate B Parts to M5.

$A_{ik} = `U$' and Π_i is in a cycle that has a place with more than one input transition. We now apply the tool to the synthesis of an automated manufacturing system (Figure 8.11).

Description of a Manufacturing System

This automated manufacturing system consists of the following major components (Table 8.3): two entries, two exits, five machines, two robots, two automatic guided vehicles (AGV), and related conveyors. It can produce two types of products. An unlimited source of raw materials is assumed. Once machines, robots, or AGVs start any operation, they cannot be interrupted until the work is complete. We now build up a well-behaved PN model.

Modeling and Synthesis Process

First, synthesize the production portion from Entry 1 to Exit 1, followed by that from Entry 2 to Exit 2. For the first portion, pick a basic process by identifying a sequence of operations of loading by R1, machining by M1, unloading via AGV1, loading by R2, and machining by M3 modeled as shown by the solid cycle [p1 t1 p2 t2 p3 t3 p4 t4 p5 t5 p6 t6 p1] in Figure 8.12(a). In this basic process, p1 models the availability of A-raw materials.

Upon this basic process, generate a new PSP [p2 t7 p7 t8 p4] shown as the dashed line in Figure 8.12(a) using Rule PP.1 since $n_g = p2$ and $n_j = p4$; both are from the same basic process. The new PSP models another alternative to produce A-parts by Machine 2. Now there are three unique PSPs, i.e., Π_1 (or psp1) = [p4 t4 p5 t5 t5 p6 t6 p1], Π_2 (or psp2) = [p2 t2 p3 t3 p4], and Π_3 (or psp3) = [p2 t7 p7 t8 p4]. The tool also displays the T-Matrix after the PP generation in the bottom window. It is easy to see that $\Pi_2 | \Pi_3$ and Π_1 is sequential to both Π_2 and Π_3.

Now since the operation in p3 requires Robot 1, generate a backward TT-path from t2 (in Π_2) to t1 (in Π_1), i.e., [t2 p8 t1]. In addition, a token in place p8, standing for Robot 1, using Rule TT.2. Furthermore, since the local exclusive set LEX (Π_2, Π_1) = {Π_2, Π_3}, generate a TP-path [t7 p8] via Rule TT.4.1 from a transition in Π_3 to p8. Since LEX(Π_1, Π_2) = {Π_1}, nothing more needs to be added via Rule TT.4.2.

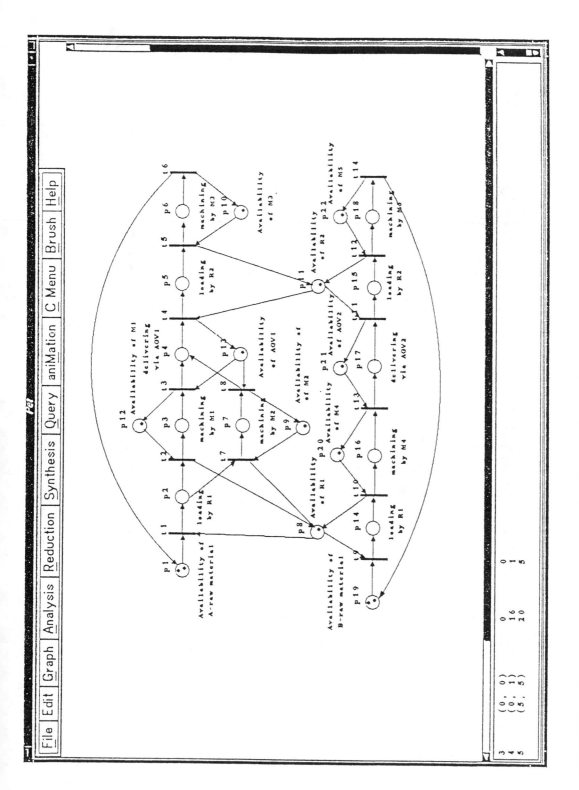

FIGURE 8.11 A Petri net model of an automated manufacturing system. (From Reference 2a. With permission.)

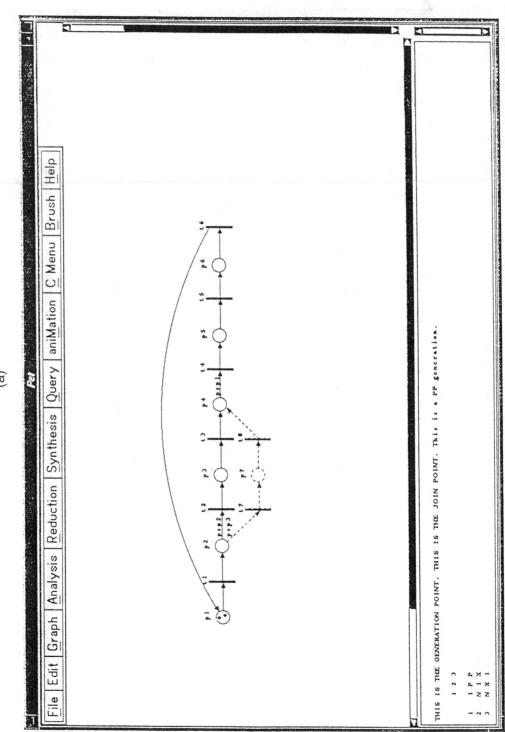

(a)

FIGURE 8.12(a)   Add [p2 t7 p7 t8 p4] to the basic process via PP1. (From Reference 2a. With permission.)

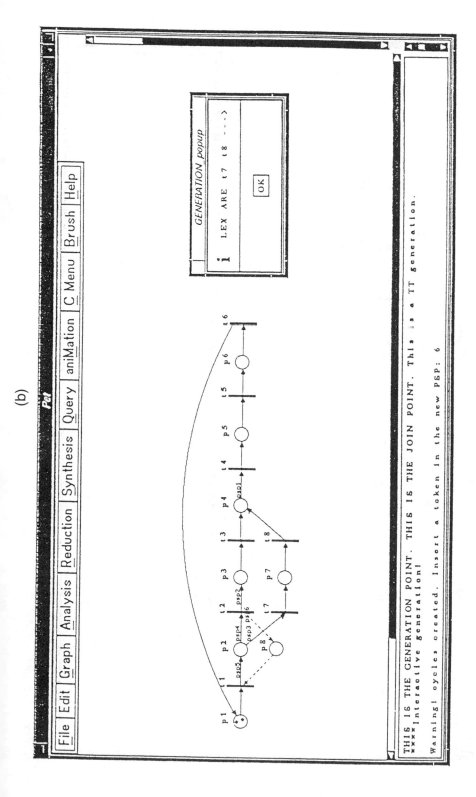

FIGURE 8.12(b)   Add [t2 p8 t1]. (From Reference 2a. With permission.)

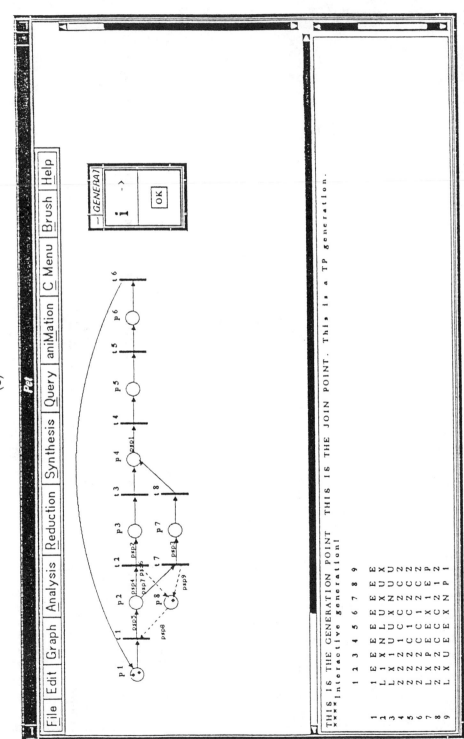

FIGURE 8.12(c)   Add a token to *p8* via TT.2, add a TP-path [*t7 p8*] via TT.4. (From Reference 2a. With permission.)

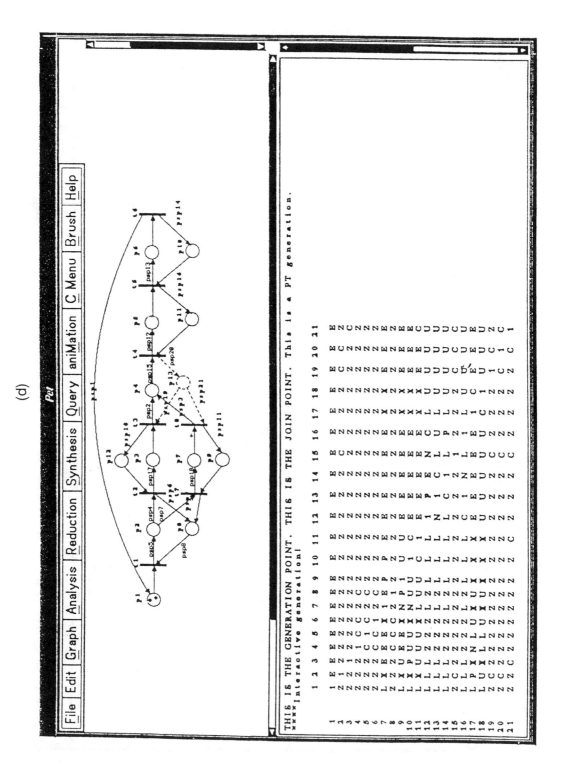

FIGURE 8.12(d) Add [*t8 p9 t7*], [*t6 p10 t5*], [*t5 p11 t4*], and [*t3 p12 t2*]; add [*t4 p13 t3*] and [*p13 t8*] via TT.2 and TT.4. (From Reference 2a. With permission.)

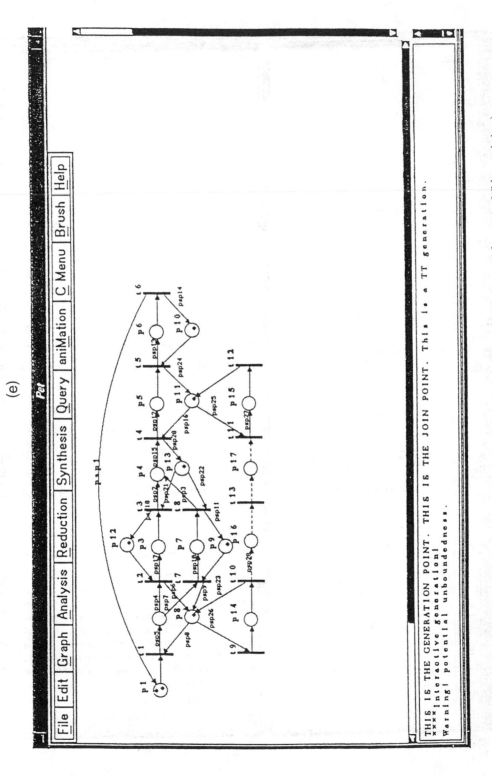

FIGURE 8.12(e)   Add [*p8 t9 p14 t10 p8*] and [*p11 t11 p15 t12*] via PP.1 add [*t10 p16 t13 p17 t11*]. (From Reference 2a. With permission.)

Note that there are two transitions, $t7$ and $t8$, in Π_3, we can pick either one as the t_g of the new TP-path. The tool conveys this information to the designer by popping up a new window, inside which is displayed an arrow linking two sets of transitions: $\{t7, t8\}$ and—(an empty set). Picking $t7$ results in the dotted lines in Figure 8.12(c). The tool then pops up another window linking two empty sets and signaling the completion of this step of synthesis. The set of home places now is $H = \{p1, p8\}$.

Machines 1, 2, 3, and R2 are required to start their corresponding operations. Thus, using Rules TT.2 and TT.4, generate the new paths: $[t3\ p12\ t2]$, $[t8\ p9\ t7]$, $[t5\ p11\ t4]$, and $[t6\ p10\ t5]$ (Figure 8.12(d)) where $p12$, $p9$, $p11$, and $p10$ model the availability of M1, M2, R2, and M3, respectively.

Considering the AGV1, first generate a TT-path between $t4$ (in $\Pi = [p4\ t4]$) and $t3$ (in $\Pi' = [t3\ p4]$), i.e., $[t4\ p13\ t3]$. Then add a token to $p13$ using Rule TT.2. Since LEX($[p4\ t4]$, $[t3\ p4]$) = ϕ, add nothing, according to Rule TT.4. LEX($[t3\ p4]$, $[p4\ t4]$) can be any one of $\{[t8\ p8]\}$, $\{[t7\ p7\ t8]\}$, and $\{[p1\ t6]\}$. Using either $\{[t8\ p4]$ or $\{[t7\ p7\ t8]\}$, add the virtual PT-path from $p13$ to $t8$ as the dotted line in Figure 8.12(d). This implies that once Machine 2 completes its processing, AGV1 delivers the part. Theoretically, one may use $\{[p2\ t7]\}$ and then a PT-path $[p13\ t7]$ would be added. $H = \{p1, p8–p13\}$.

Having synthesized the partial system, i.e., the production portion from Entry 1 to Exit 1, now synthesize the rest of the portion, from Entry 2 to Exit 2, to involve Machines 4 and 5, AGV2, and Robots 1 and 2. For $[t7\ p8\ t1]$ and $[t5\ p11\ t4]$, applying Rule PP.1 leads to the two cycles $[p8\ t9\ p14\ t10\ p8]$ and $[p11\ t11\ p15\ t12\ p11]$. The meanings of the added places are shown in Figure 12(e).

Consider two new PSPs: $\Pi_4 = [p8\ t9\ p14\ t10\ p8]$ and $\Pi_5 = [p11\ t11\ p15\ t12\ p11]$ and choose $t10$ from Π_4 and $t11$ from Π_5 to generate a new PSP, psp = $[t10\ p16\ t13\ p17\ t11]$. Note each of the transitions, $t10$ and $t11$, is in a cycle that was generated using Rule PP.1. Hence, now apply Rule TT.3 to pick up $t9$ from Π_4 and $t12$ from Π_5 and generate psp' = $[t12\ p18\ t14\ p19\ t9]$. It is easily verified that psp, psp,' Π_4, and Π_5 constitute a cycle. Furthermore, insert the tokens in $p19$ to represent the availability of raw material from Entry 1 to fulfill Rule TT.2. $H = \{p1, p8–p13, p19\}$.

Now apply Rule TT.2 to obtain the paths: $[t13\ p20\ t10]$, $[t11\ p21\ t13]$, and $[t14\ p22\ t12]$, which model the availability of Machines 4, 5, and AGV2. $H = \{p1, p8–p13, p19–p22\}$. This completes the modeling process and the final net is depicted in Figure 8.11 as a bounded, live, and reversible net if $\forall p \in H$, $M_0\ (p) > 0$ and $\forall p \in$ PN-H, $M_0\ (p) = 0$; i.e., the places in the net have no tokens, except those in H.

8.7   Enhancement of Synthesis Rules

The synthesized nets, however, are limited in classes because some generations are prohibited. Recently the author[34] has discovered new synthesis rules such that path generations previously forbidden are now allowed without incurring logical incorrectness. For instance, a TT-path generation (e.g., PSP $[t3\ p4\ t4]$ in Figure 8.13(a)) between two exclusive transitions was forbidden (Rule TT.0); the new rule is that it is acceptable with additional path generations (PSP $[t4\ p5\ t3]$) such that the two exclusive transitions ($t3$ and $t4$) become synchronized and hence *sequential* to each other. As a result, more classes of nets can be synthesized.

Another example is shown in Figure 8.14,[27] which models a machine/assembly shop with resource sharing and cannot be synthesized unless we allow TT-path generations between exclusive transitions and extend the rules to GPNs. In the sequel, we explain the model and develop the rules for both.

The Model

The net model consists of three parts: Components I and II and the interconnecting part between them. The system assembles two Type-1 parts and three Type-2 parts into a final product. Type-1 and Type-2 parts are processed in Workstations 1 and 2, respectively, from the corresponding raw stocks fed by Conveyors 1 and 2, respectively. Robot 1 (3) loads Type-1 (2) raw stock from Conveyor 1 (2) into Workstation 1 (2), performs some machining, and loads the finished part into the buffer. The above two subactivities are modeled by Components I and II, respectively. Component II also models the assembly process in

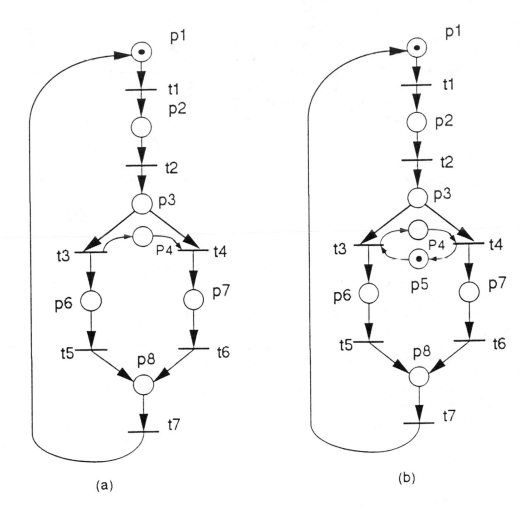

FIGURE 8.13   An example of rule TT.0. (a) [t3 p4 t4] is generated. The net deadlocks or is unbounded. (b) The addition of [t4 p5 t3] renders the net live and bounded.

Workstation 2. After the three Type-2 parts finish machining, they are assembled (along with the two Type-1 parts loaded by Robot 2 from the buffer in Component I) on the same Workstation 2 by Robot 3. After the assembly task is completed, the final product is placed on Conveyor 3 by Robot 3.

The interconnecting part between Components I and II models the loading process of two Type-1 parts by Robot 2 from the buffer to Workstation 2. To further understand the model, Table 8.4 lists the physical meaning of key transitions and places in Figure 8.14 where "WS" denotes a workstation, "R" a robot, and "C" a conveyor.

This net contains multiple-weighted arcs and hence cannot be synthesized using the current rules. We will extend the rules by including the synchronization rule for TT-path generations among exclusive PSPs and the arc-ratio rule for handling GPNs.

The Synchronization Rule

Here, we develop new Rule TT.4 for exclusive TT generations. We first define the following terms for the description of the synchronization rule:

Definition: A decision place p_d is the common generation place of a set of exclusive PSPs.

TABLE 8.4 The Physical Meaning of Key Transitions and Places in Figure 8.14

*t*1:	Start processing in WS1 with R1.
*t*2:	End processing in WS1 with R1.
*t*3:	Start loading Type-1 raw stock into WS1.
*t*4:	End loading WS1.
*t*5:	End loading buffer.
*t*6:	Start loading Type-1 part into buffer.
*t*7:	Start processing in WS1.
*t*8:	Input Type-1 raw stock to C1 (assuming unlimited raw stock available).
*t*9:	Start loading Type-1 into WS2.
*t*10:	End loading WS2.
*t*11:	End assembly task.
*t*12:	Start assembly task.
*t*13:	Start loading Type-1 parts into WS2.
*t*14:	End loading and start processing Type-2 parts in WS2.
*t*15:	Start processing Type-2 parts with R3.
*t*16:	End processing in WS2.
*t*17:	Start loading final product on C3.
*t*18:	Input Type-2 parts to C2 (assuming unlimited Type-2 parts available).
*t*18:	End loading C3.
*p*1:	Robot 1 available (shared resource).
*p*13:	Robot 2 available.
*p*15:	Robot 3 available (shared resource).

Note: WS: Workstation; R: robot; C: conveyor.

Definition: The control transitions t_c are the output transitions of a decision place.

For example, in Figure 8.13(a) $p3$ is a decision place and $t3$ and $t4$ are control transitions.

In the net as shown in Figure 8.13, before the generation of $[t_3 \ p_4 \ t_4]$ and $[t_4 \ p_5 \ t_3]$, $d_{12} = \infty$ because if we put a token in $p1$, there exists a firing sequence $\sigma = t1 \ t2 \ t3 \ t5 \ t7 \ t1 \ t2 \ t3 \ t7 \ldots$ in which σ ($t1$) $= \infty$ and σ ($t2$) $= 0$ and $p9$ will become unbounded. Therefore, a single TT-path generated between these two exclusive transitions with synchronic distance $d = \infty$ is insufficient. To synchronize $t3$ and $t4$, one can add a circle $[t3 \ p4 \ t4 \ p5 \ t3]$ (by Rule TT.4) and add a token in $p5$ (by Rule TT.2) and d now becomes 1. Hence, in any firing sequence σ, $t3$, and $t4$ must alternate; i.e., $\sigma = \ldots t3 \ldots t4 \ldots t3 \ldots t4 \ldots$. As a result, the temporal relationship "EX" no longer captures that between $t3$ and $t4$. "SX" is used instead, which stands for "sequential exclusive." Note that if we only generate a $\Pi^w = [t4 \ p5 \ t3]$, it violates Rule TT.0.

The Petri net after the synchronization generation is bounded, live, and reversible. Note that $t1$ and $t5$ fire twice and once, respectively, within one iteration, to restore the initial marking. Thus $\sigma_{51} = 2$. An arbitrary positive synchronic distance of v can be created by sequentializing (i.e., synchronizing) v exclusive transitions. In general, we should avoid a TT-generation between two transitions with synchronic distance greater than 1 in an iteration which may cause unboundedness. For instance, assuming that we generate a new TT-path $[t1 \ p10 \ t5]$, $p10$ becomes unbounded after an infinite number of iterations. One can avoid this unboundedness by applying Rule TT.3, considering that $t5$ is still exclusive to $t6$. That is, we also generate a VP from $p10$ to $t6$. As a result, the Rule TT.3 still applies. Note that we may put more than one token into Π^w without destroying the well behavedness. In this case, the SX structure relationship does not capture the true behavior among the transitions that are SX to each other. For this reason, we retain the EX for them, but enhance it with the synchronic distance identical to the total number of tokens added to Π^w. It takes linear time complexity to update the d_{ij} against the exponential one required for the analysis of the net.

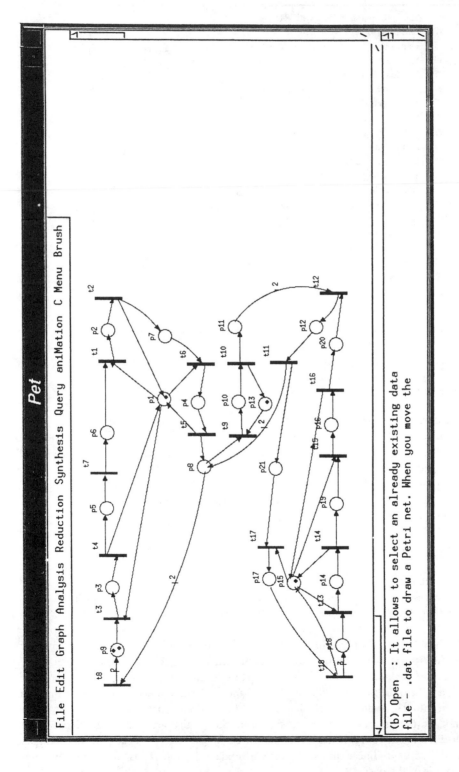

FIGURE 8.14   A GPN model of a machine/assembly shop with resource sharing. (From Reference 27. With permission.)

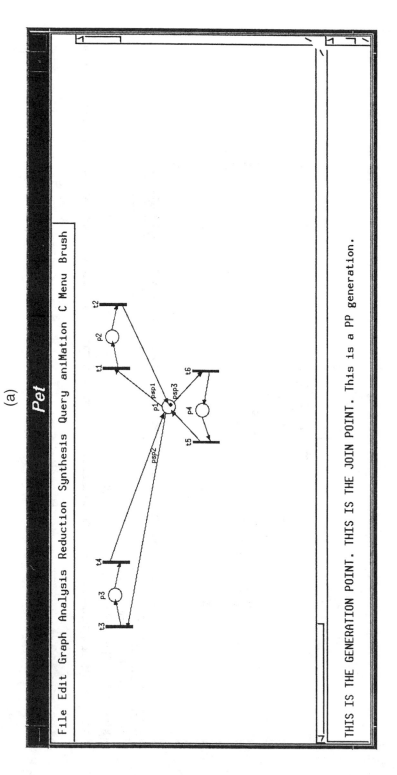

FIGURE 8.15(a)   Basic process: $[p1\ t1\ p2\ t2\ p1]$, two PP-generations: $\Pi_1^w = [p1\ t3\ p3\ t4\ p1]$ and $\Pi_2^w = [p1\ t6\ p4\ t5\ p1]$.

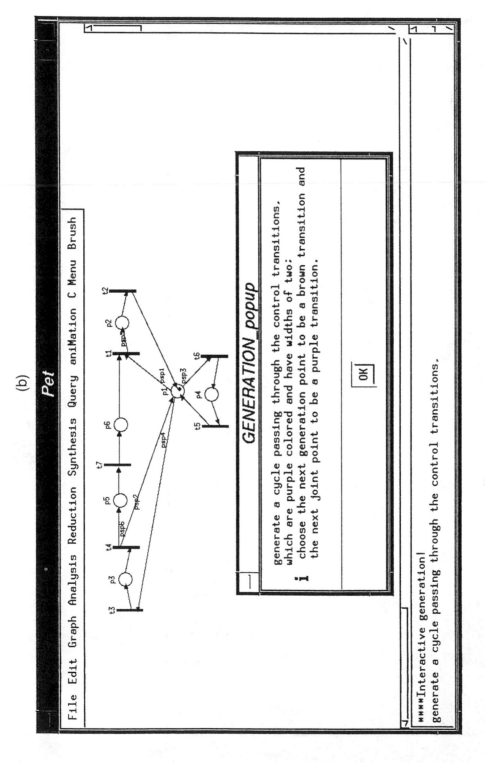

FIGURE 8.15(b) The first exclusive TT-generation. $\Pi^w = [t4\ p5\ t7\ p6\ t1]$.

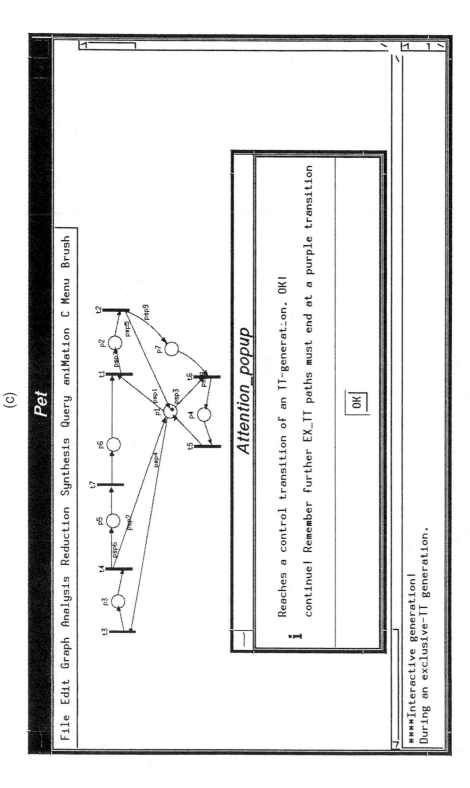

FIGURE 8.15(c) The second exclusive TT-generation. $\Pi^w = [t2\ p7\ t6]$.

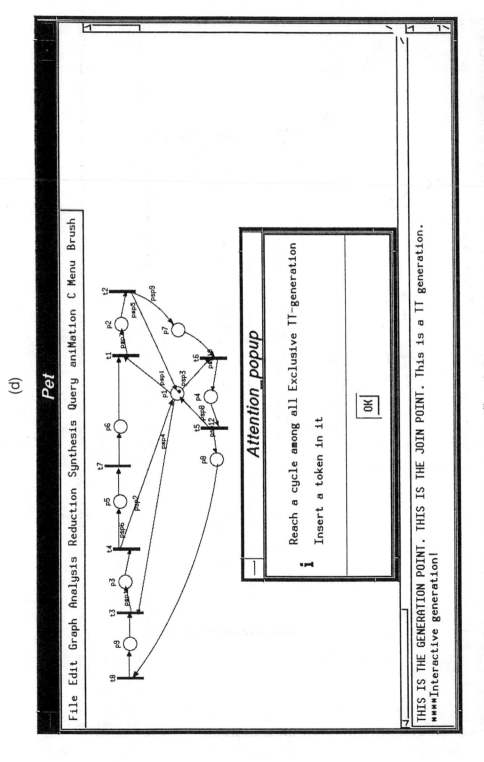

FIGURE 8.15(d)   The last exclusive TT-generation forming a cycle. $\Pi^w = [t5\ p8\ t8\ p9\ t3]$.

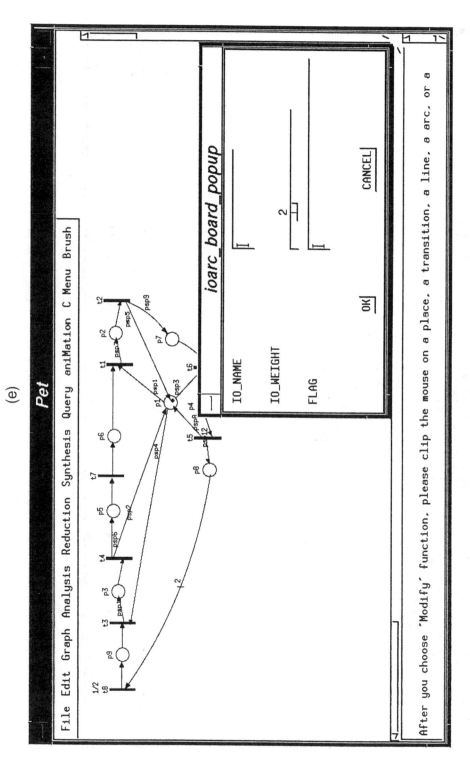

FIGURE 8.15(e)   Entering arc weight for the arc [*p8 t8*] by clicking the "Edit/Modify" button and a point on the arc.

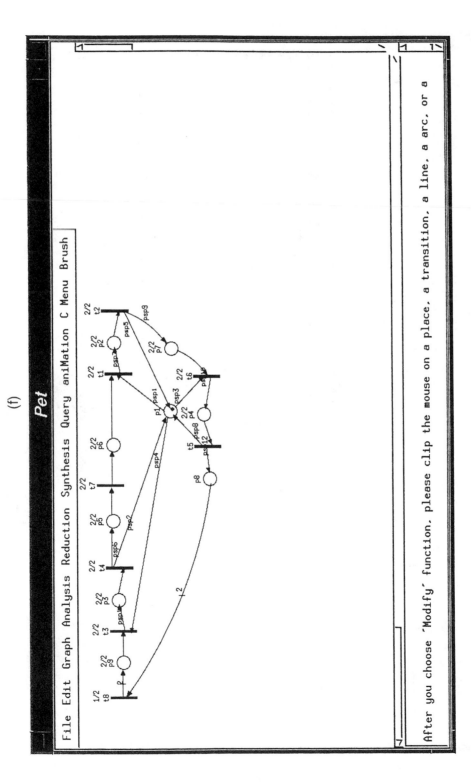

Operational Methods in Computer-Aided Design

FIGURE 8.15(f)   After entering the arc weight for the last arc [p9 t3] in the NP, all least ratios in reference to t4 in the cycle just formed by the three successive exclusive TT-generations are displayed.

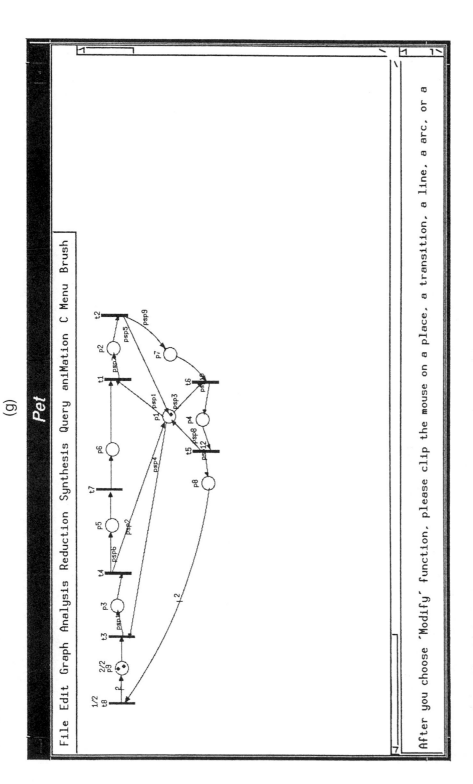

(g)

FIGURE 8.15(g) After entering two tokens in *p9* by clicking the "Edit/Modify" button and a point in *p9*, all least ratios (other than 1/1) in the cycle are displayed in reference to *p9*, the new home place.

(h)

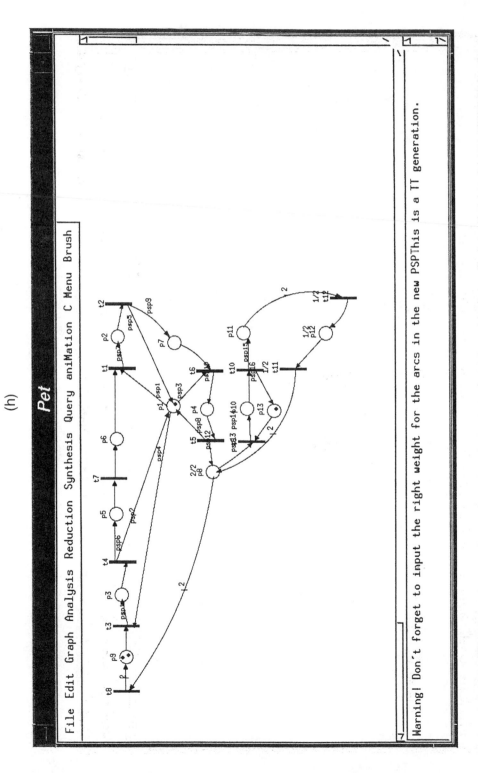

FIGURE 8.15(h)   Completion of Component I by a PP-path: [*t8 t9 p10 t10 p11 t12 p12 t11 p8*] and a backward TT-path: [*t10 p13 t8*].

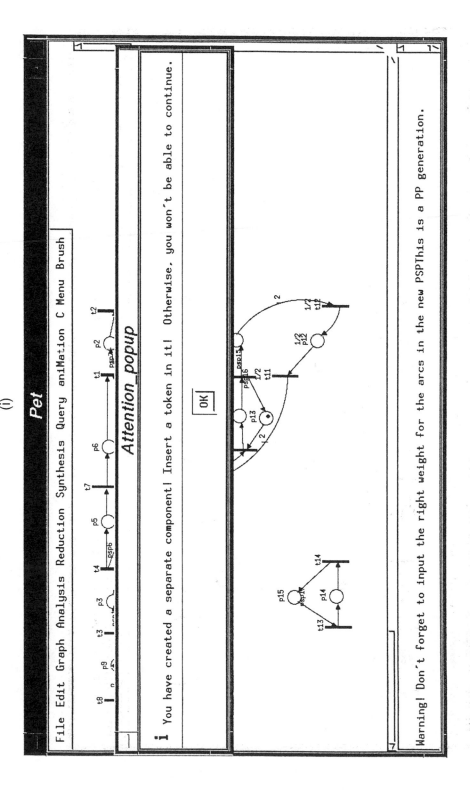

FIGURE 8.15(i)   The first step of the synthesis of Component II: a PP-cycle: psp 17 = [p15 t13 p14 t14] which is disconnected from Component I. The message box reminds us to insert a token in it.

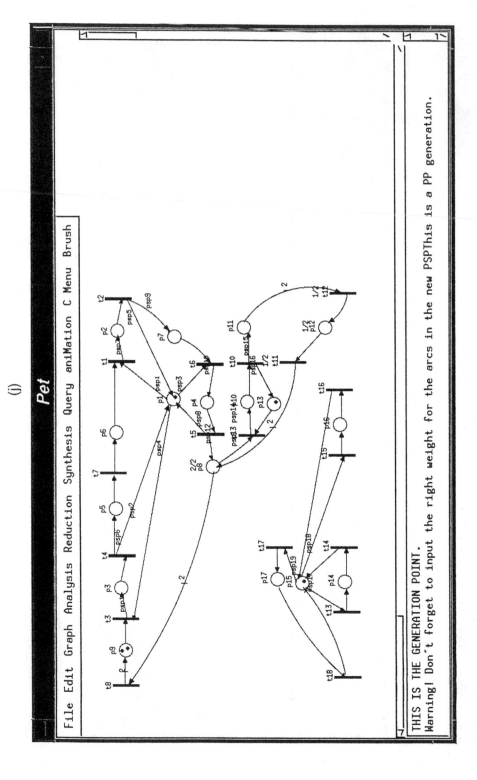

FIGURE 8.15(j)   Two PP-generations: psp18 = [*p*15 *t*15 *p*16 *t*16 *p*15] and psp19 = [*p*15 *t*17 *p*17 *t*18 *p*18].

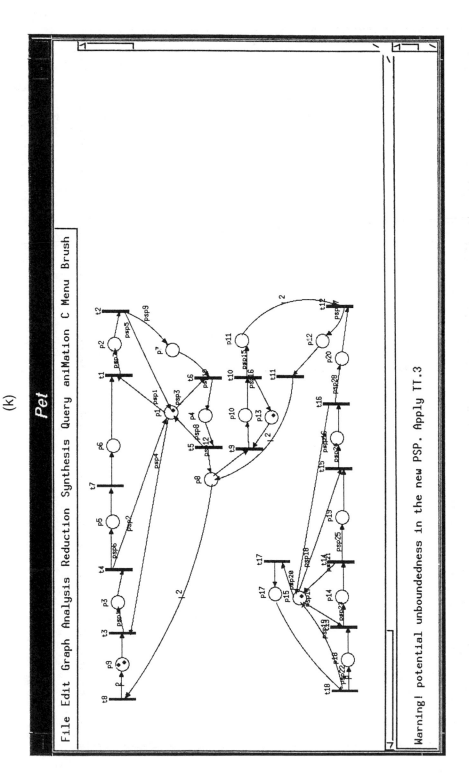

FIGURE 8.15(k) The first three exclusive TT-generations: psp22 = [*t*18 *p*18 *t*13], psp25 = [*t*14 *p*19 *t*15], and psp28 = [*t*16 *p*20 *t*12]. PSP28 connects Component II to I.

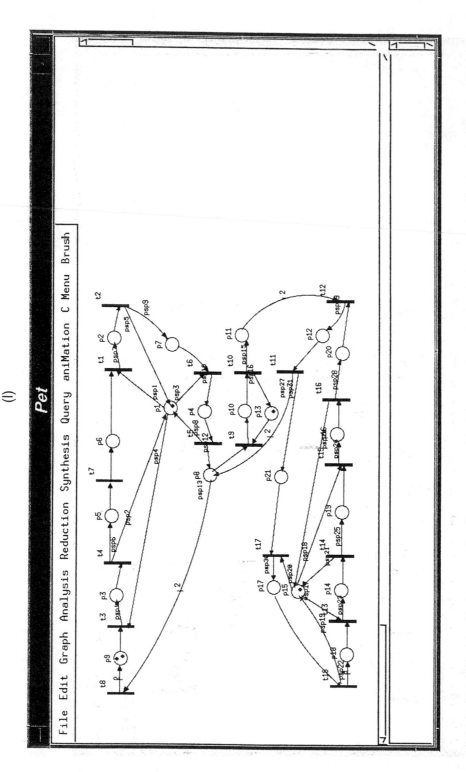

FIGURE 8.15(I)　The last exclusive TT-generation: psp31 = [t11 p21 t17] connects Component I to Component II and makes them synchronized. It forms a cycle.

Although the example in Figure 8.13 synchronizes only two mutually exclusive transitions, the same idea can apply to a set of more than two mutually exclusive transitions. As a result of synchronization, they are sequentialized; i.e., their mutual synchronic distance changes from ∞ to one. For the net shown in Figure 8.14, this set contains three mutually exclusive transitions.

Summarizing the above discussion, we have the following synchronization rule: If a TT-path (Π_1) is generated from t_g to t_j, $t_g|t_j$ and t_j is a control transition, then generate a series of TT-paths (Π_2, Π_3,...) from t'_g to t'_j such that all above Πs are in a cycle and each t'_j is also a control transition with the same decision place for t_g and t_j.

Note that t_g to t'_g need not be a control transition. On the other hand, if t'_j is not a control transition, the token in the decision place may flow to the input place of this control transition, not matching to the token in the cycle and hence resulting in a deadlock.

This rule must be checked concurrently along with other TT rules. For instance, after the above cycle (with a token) generation, the following TT-path generation may become a backward generation; then Rule TT.2 must be applied to insert a token in this TT-path.

The correctness of this rule can be proved by showing that it satisfies the two guidelines. We omit the proof here because it is not the main purpose of this chapter. Similar comments apply to the rule described in the next subsection.

Arc-Ratio Rules for General Petri Nets

The synthesis rules of the knitting technique for ordinary PN should be modified for GPN in order to meet the guidelines. This set of rules is referred to as **connection rules** since the absence of any Π^w may cause guidelines to be violated. The weight of arcs in the PN must satisfy some constraints. Consider the guideline of no-disturbance, i.e., if the NP is a TT-path from t_g to t_j, then the firing ratio between t_g to t_j of NP must equal that in N^1. Otherwise, the firing behavior of t_j in N^1 is disturbed. Similarly, if the NP is a PP-path from p_g to p_j, then the marking ratio between p_g and p_j of the NP must equal that in N^1. Otherwise, the marking behavior of t_j in N^1 is disturbed. Similar but slightly different conditions apply if the NP is a TP-path or a PT-path.

Now consider the second guideline of well-behaved NP. Examples are shown in Figure 8.16(a) and 8.16(b) where the second guideline is violated; both NPs are not well-behaved. These additional rules are called the arc-ratio Rules.

Arc-Ratio Rules

Again, one can develop the arc-ratio rules based on the two guidelines.

The following definitions of least ratios consider a pair of nodes, transitions, or places, and all the paths between them in isolation (i.e., all nodes and arcs not in these paths are deleted).

Definition: The *least firing ratio* between t_i and t_k, $R^f_{ik} = \frac{a}{b}$, where a is the least firing number of t_i for t_k to fire b times with no tokens left in paths between t_i and t_k.

Definition: The *least marking ratio* between p_i and p_k, $R^m_{ik} = \frac{a}{b}$, where a is a least number of tokens in p_i for p_k to get b tokens by firing all transitions between p_i and p_k with no tokens left in paths between p_i and p_k.

Definition: The *least marking-firing ratio* between p_i and t_k, $R^{mf}_{ik} = \frac{a}{b}$, where a is the least number of tokens in p_i such that t_k fires b times with no tokens left in paths between p_i and t_k.

Definition: The *least firing-marking ratio* between t_i and p_k $R^{fm}_{ik} = \frac{a}{b}$, where a is the least firing number of t_i such that p_k can get b tokens by firing transitions between t_i and p_k, with no tokens left in paths between t_i and p_k.

(a)

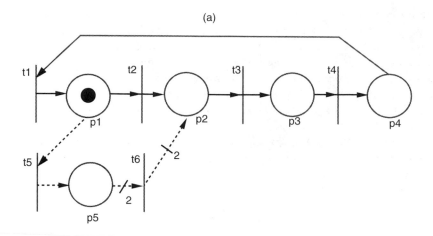

FIGURE 8.16(a)   An example of PP generation, which violates the arc-ratio rule ARR. 1.a. $R_{NP}^m = 2/2$, $R_{NP}^m >_{lit} R_{12}^m$.

(b)

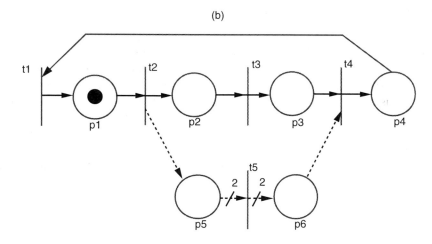

FIGURE 8.16(b)   An example of TT generation, which violates the arc-ratio rule ARR.1.a. $R_{12}^f = 1/1$, $R_{NP}^f = 2/2$, $R_{NP}^f >_{lit} R_{12}^f$.

These ratios are referred to as *least ratios q*. Let $q = \frac{a}{b}$ be one of these ratios, then $q^u = a$ and $q^l = b$. In these ratios, the numerator and the denominator may not be mutually prime. Hence, we also need to define the prime ratio of the least ratio q.

Definition: The prime ratio $[q]' = \frac{a'}{b'}$ where a' and b' are prime to each other, where $q \in \{R_{ik}^f, R_{ik}^m, R_{ik}^{fm}, \text{ or } R_{ik}^{mf}\}$.

Definition: If $[q_1]' = [q_2]'$ $q_1 = \frac{a_1}{b_1}$ and $q_2 = \frac{a_2}{b_2}$ and $a_1 > a_2$ and $b_1 > b_2$ then q_1 is *literally greater than* q_2, denoted as $q_1 >_{lit} q_2$.

Example: Figure 8.16(a) shows an NP ($p_1 t_5 p_5 t_6 p_2$) with $R_{12}^m = 2/2$, which is literally greater than ($>_{lit}$) that ($R_{12}^m = 1/1$) for the PP-path ($p_1 t_2 p_2$). Both $[R_{12}^m]'$ are 1/1. The net is nonlive. Figure 8.16(b) shows an NP ($t_2 p_5 t_5 p_6 t_4$) with $R_{24}^f = 2/2$, which is literally greater than ($>_{lit}$) than that (R_{24}^f) = 1/1 for the TT-path $t_2 p_2 t_3 p_3 t_4$]. Both $[R_{24}^f]'$ are 1/1. The net is nonlive.

Lemma 7: Let $n_1 \rightarrow n_2 \rightarrow n_3$, then $R_{13} = \dfrac{(R_{12})^u * gcd}{(R_{23})^l * gcd}$ where gcd is the greatest common denominator between $(R_{12})^l$ and $(R_{23})^u$.

The following observation regarding the relationship between two nodes x and y is useful to find the least ratios between nodes that are not sequential to each other.

Observation 2:

1. If $x \rightarrow y$, then $\exists \bar{P}_1$ from x to y, which does not pass through any home place.
2. If $x \parallel y$ or $x \mid y$, then the $\bar{P}_1$ from x to y contains the home place.
3. If $x \parallel y$ ($x \mid y$), then $\exists \bar{P}_1$ and $\bar{P}_2$ containing x and y, respectively, which do not share common PSPs, but share the same end transition (place).

Example: In Figure 8.12(a), $t1 \rightarrow t3$; $\bar{P}_1$ [$t1$ $p2$ $t2$ $p3$ $t3$] does not include the home place $p1$. $p3 \mid p7$; the $\bar{P}_1$ from $p3$ to $p7$ [$p3$ $t3$ $p4$ $t4$ $p5$ $t5$ $p6$ $t6$ $p1$ $t1$ $p2$ $t7$ $p7$] includes the home place $p1$. $\bar{P}_1$ [$p3$ $t3$ $p4$] and $\bar{P}_2$ [$p7$ $t8$ $p4$] join at $p4$. In Figure 8.16(b), $t3 \parallel t5$; the $\bar{P}_1$ from $t3$ to $t5$ [$t3$ $p3$ $t4$ $p4$ $t1$ $p1$ $t2$ $p5$ $t5$] includes the home place p_1.

Definition: A PP-path with a generation point, p_g (which has more than one output transition), is said to have *partial flow* if its t_g, the output transition of p_g on the path, must fire more than once to have no tokens blocked inside the path.

Partial flow may lose some tokens in a PP-path if these tokens are insufficient to enable some transition in the path and thus may cause a deadlock.

Definition: The input ratio of a PP-path is the ratio of $(R_{gj}^m)^u$ to the arc weight between p_g and t_g.

Example: Figure 8.17 illustrates an example of partial flow that causes a deadlock as shown in Figure 8.17(b). The input ratio of path $(p_2 t_2 p_3 t_3 p_4)$ is 2/1 and that of path $(p_2 t_5 p_5 t_6 p_4)$ is 4/1.

Because the transitions in $X_{j_1 g_1} = \text{LEX } (t_{j_1}, t_{g_1})$ are mutually exclusive, the number of transitions that are fired during one iteration may differ from that during another iteration. The *weighted firing* of transitions in $X_{g_1 j_1}$ is defined as

$$\sum_{t_{g_s} \in X_{g_1 j_1}} \sigma(t_{g_s}) R_{g_1 g_s}^f.$$

The following theorem implies that the ratio of weighted firings of $X_{g_1 j_1}$ to that of $X_{j_1 g_1}$ is $[R_{g_1 j_1}^f]'$, i.e., the prime ratio of the least firing between t_{g_1} and t_{j_1}, if only t_{g_1} in $X_{g_1 j_1}$ and t_{j_1} in $X_{j_1 g_1}$ fire during an iteration, respectively.

Theorem 7: If $t_{g_1} \parallel t_{j_1}$ in a synthesized strongly connected net, $\forall \sigma$ such that $M_0[\sigma > M_0$, then

$$\frac{\displaystyle\sum_{t_{g_s} \in X_{g_1 j_1}} \sigma(t_{g_s}) R_{g_1 g_s}^f}{\displaystyle\sum_{t_{g_s} \in X_{j_1 g_1}} \sigma(t_{j_s}) R_{j_1 j_s}^f} = [R_{g_1 j_1}^f]'.$$

Proof: Assuming that the paths containing t_{g_1} and t_{j_1}, respectively, intersect at t_s, when t_s fires w times, then each member t_{g_s} and t_{j_s}, respectively, of $X_{g_1 j_1}$ and $X_{j_1 g_1}$, may fire $(v_{g_s})(R_{g_s}^f)$ and $(v_{j_s})(R_{j_s}^f)$ times, respectively, where v_{α_s} $(\leq w)[\alpha = g, j]$ is the number of firings of t_s applied to the path containing α_s. We have

$$\sum_{\alpha_s} \sigma(t_{\alpha_s}) R_{\alpha_1 \alpha_s}^f = \sum_{\alpha_s} v_{\alpha_s} (R_{\alpha_s}^f)(R_{\alpha_1 \alpha_s}^f) = \left(\sum_{\alpha_s} v_{\alpha_s}\right)(R_{\alpha_1 s}^f), \alpha = g, j.$$

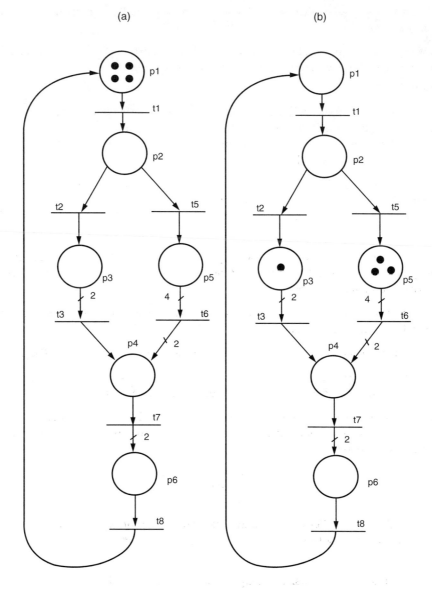

FIGURE 8.17   An example of partial flow (a), which causes a deadlock in (b). The input ratio of path [p2 t2 p3 t3 p4] is 2, and that of path [p2 t5 p5 t6 p4] is 4.

Since each PSP of $X_{g_1 j_1}$ and $X_{j_1 g_1}$ is exclusive to each other in its own set, the sum of v's must equal w; i.e.,

$$\sum_{t_{g_s} \in X_{g_1 j_1}} v_{g_s} = w \quad \text{and} \quad \sum_{t_{g_s} \in X_{g_1 j_1}} v_{j_s} = w.$$

We thus obtain

$$\frac{\displaystyle\sum_{t_{g_s} \in X_{g_1 j_1}} \sigma(t_{g_s}) R^f_{g_1 g_s}}{\displaystyle\sum_{t_{j_s} \in X_{j_1 g_1}} \sigma(t_{j_s}) R^f_{j_1 j_s}} = \frac{(w)(R^f_{g_1 s})}{(w)(R^f_{j_1 s})} = \frac{R^f_{g_1 s}}{R^f_{j_1 s}} = [R^f_{g_1 j_1}]'.$$

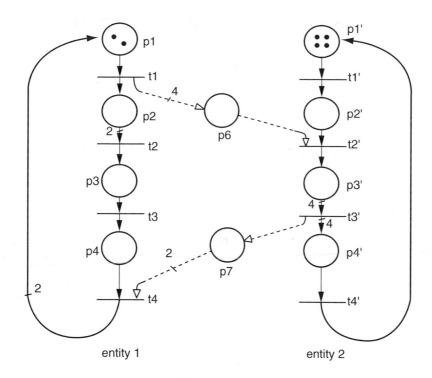

FIGURE 8.18 An example of the arc-ratio ARR.2. PN_1 consists of all solid lines and holes. There are no paths in PN_1 from $t1$ to $t2$. The NP from $t1$ to $t2'$ must be accompanied by the NP from $t3'$ to $t4$. There are two paths from $t1$ to $t4$ in PN_2; both $R_{12}^f = 2/1$.

The following theorems show that the ratio of maximum markings between two places equals their least marking ratio.

Theorem 8: If $p_i | p_k$ in a synthesized net, then $\exists M_1$ and M_2, where $M_1(p_i) \neq 0$, $M_1(p_k) = 0$, $M_2(p_k) \neq 0$, and $M_2(p_i) = 0$ such that

$$\frac{M_1(p_i)}{M_2(p_k)} = [R_{ik}^m]'.$$

Proof: Assume that $\bar{P}_1$ and $\bar{P}_2$ contain p_i and p_k, respectively, and start from the same place, p_s, which has w tokens. If all these w tokens flow to p_i and p_k, respectively, then

$$M_1(p_i) = (w)(R_{is}^m), M_1(p_k) = 0 \text{ and } M_2(p_k) = (w)(R_{ks}^m), M_2(p_i) = 0, \text{ respectively.}$$

We thus obtain

$$\frac{M_1(p_i)}{M_2(p_k)} = [R_{ik}^m]'.$$

The following theorems can be proved in a similar way.

Theorem 9: If $p_i \| p_k$ in a synthesized net, $\forall M$ such that $M(p_i) \neq 0, M(p_k) \neq 0$, then

$$\frac{M(p_i)}{M(p_k)} = [R_{ik}^m]'.$$

Theorem 10: If $p_i \rightarrow p_k$ in a synthesized net, then $\exists$ M_1 and M_2 such that $M_1(p_i) \neq 0$, $M_1(p_k) = 0$, $M_2(p_k) \neq 0$, $M_2(p_i) = 0$, and

$$\frac{M_1(p_i)}{M_2(p_k)} = [R_{ik}^m]'.$$

The arc-ratio rules are summarized as follows.

Arc-Ratio Rules:

Given an NP connecting from node g to node j,

(ARR.1)   If $\exists$ paths in N_1 between g and j, then

(ARR.1.a)   The weight in NP must be such that $q_{NP} \leq_{lit} q_{gj}$.

(ARR.1.b)   If g is a place, and some output transitions of g do not have a common output place, then all (except one with arbitrary input ratio) paths between g and j must have input ratio of one.

(ARR.1.c)   Same as Rule TT.2, except that the statement in Rule TT.2 should be replaced with "If, without firing t_j, there does not exist a σ to enable $t_g(R_{jg}^f)^l$ times, insert enough tokens in Π^w to enable $t_j(R_{jg}^f)^u$ times."

Possible sets of $\{g, j, q\}$ are {transition, transition, R^f}, {transition, place, R^{fm}}, {place, place, R^m}, and {place, transition, R^{mf}}.

(ARR.2)   Otherwise, there must be another accompanying NP (Rule TT.3) from g' to j' such that the NP from g, via j and g', to j' satisfies ARR.1.a.

Example: Figure 8.18 shows a PN with two originally disconnected subnets, *entity* 1 and *entity* 2. A TT-path connects from t_1 to t_2'. To synchronize *entity* 1 and *entity* 2, another TT-path (Rule ARR.2) is generated from t_3' to t_4. Note that $R_{14}^f = [R_{(14)'}^f] = \frac{2}{1}$ where subscript $(14)'$ indicates the path from t_1 to t_4 $(t_1 p_6 t_2' p_3' t_3' p_7 t_4)$ (Rule ARR.1.a).

Figure 8.19 shows that a Π^w $[t_2 p_6 t_3']$ is generated using the TT rule from t_2 of Π_1 to t_3' of $\Pi_2 . t_2 \parallel t_3'$ and their synchronic distance $d_{23'} = \infty$. We generate another Π^w $(t_4' p_7 t_4)$ such that d($\{\Pi_1, \Pi_3\}$, $\{\Pi_2, \Pi_4\}$) = 1. $R_{56}^{fm} = 2/8$ for the path from t_5 to p_6, $(t_5 p_5 t_6 p_4 t_4 p_1 t_1 p_2 t_2 p_6)$, indicating that p_6 gets 4 tokens for every firing of t_5. Thus we connect an arc, from t_5 to p_6, with weight $1/[R_{56}^{fm}]'$ (=4) by Rules TT.4.1 and ARR.1.a. $R_{66'}^{mf} = 4/4$ along the only path from p_6 to t_6': $(p_6 t_3' p_4' t_4' p_1'$ $t_1' p_2' t_5' p_5' t_6')$. Thus we connect an arc with weight $[R_{66'}^{mf}]$ (= 1), from p_6 to t_6', by Rule ARR.1.a.

In Figure 8.19, Π^w $(p_2 t_g p_6 t_j p_2')$ is generated using the PP rule from p_2 of Π_1 to p_2' of Π_2. $p_2 | p_2'$, $R_{22'}^m = 4/4$ along the only path from p_2 to p_2', $(p_2 t_2 p_3 t_3 p_4 p_1 t_1' p_2')$. $R_{2g}^m = 2/1$. We connect a VP with weight 2 from p_5 to t_g (Rules PP.2.2 and ARR.1.a). Based on Rule PP.2.1, we connect a TP-Path from t_j to p_5' with weight 2, since $R_{j5'}^{fm} = 1/2$.

For GPNs, the synchronic distance between any two transitions is no longer 1 or ∞ as the synthesized OPNs. Since synchronic distance depends on the marking, the determination of synchronic distance between all pairs of transitions may take exponential time complexity. Therefore, it is better that we compute them during the synthesis process.

Updating of Structure Synchronic Distance

The steps for updating the structure synchronic distance, d^s, is as follows:

1. For each basic process generation, the d^s between it and any other PSP is ∞.
2. For each new generation, the d^s between any two Π^w is one.

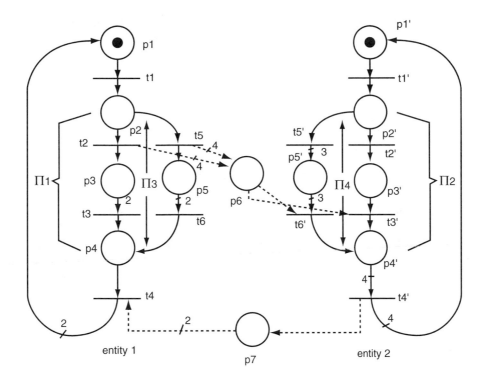

FIGURE 8.19 An example of the arc-ratio rule ARR.1.a. on top of the TT.4 rule.

3. For each backward PP-generation, $d^s = 1$ for any two Π^w in the same new circle. $d^s = \infty$ for any PSP in the new circle and any other PSP not in the same circle.
4. For TT-generation between two PSPs with $d^s = \infty$, their $d^s = 1$ after the application of TT.3 rule.
5. For all other generations, the d^s between any Π^w and any current PSP depends on their structure relationship. If they are *concurrent or sequential* (*exclusive*), then the value of d^s is one (∞).

The time complexity for the above steps is $O(n_1^2)$; hence, the total time complexity is $O(n_2^3)$.

Enhancement of the Synthesis Algorithm

(D.1) Generations Between Exclusive Transitions

Since now the generation where $t_g | t_j$ is permitted, Rule TT.0 must be modified to take out the corresponding phrase as follows:

(0) TT.0 If one of them is in a cycle which was solely generated using Rule PP.1, then signal "forbidden, delete the Π^{w}" and return.

In addition, we add a new rule:

(5) TT.5 If $t_g \mid t_j$, then follow the synchronization rule to serialize t_g and t_j.

Here we discuss the implementation. First, in practice, the designer may wish to serialize other transitions in addition to t_g and t_j. Thus, it is desirable that we flag all control transitions for the designer to select. Note that all control transitions must be mutually exclusive. As a result, they must have a common generation point, which is the decision place. Colors of all these control transitions must change to help the designer construct the cycle. Second, the designer needs to construct a cycle passing through the control transitions at his own choice. Whenever a control transition is included in the cycle, set its color to black. Put tokens into one place in the cycle and all nodes in the cycle will be automatically colored black (originally red).

Note that control transitions are output transitions of the prime start place, denoted p_{ps}^{gj}, of t_g and t_j, and t_j must be a control transition. Let C_T be the set of control transitions. Procedure "find_prime_start (t_g, t_j)" is to find p_{ps}^{gj}.

1. p_{ps}^{gj} = find_prime_start (t_g, t_j){
   Select one input place p_{in} of t_j such that
   $(p_{in} \rightarrow t_g)$,
   then p_{in} is the p_{ps}^{gj};
   }
2. find_C_T(){
   $C_T = \{t \mid t \in (p_{ps}^{gj}) \cdot \}$.
   }

The time complexities for "find_prime_start (t_g, t_j)" and "find_C_T()" are both $O(n)$.

During the process of constructing the cycle, each new TT-path must have its t_j, a control transition (i.e., $t_j \in C_T$). It is desired that we generate these new TT-paths in a continuous fashion; i.e., it is better not to construct the cycle in a piecewise fashion. Let $\Pi^{w'}$ be the PSP that has just been generated with t_g' and t_j'. The t_g and t_j for the next new PSP must satisfy the following conditions:

1. $t_j' \rightarrow t_g$.
2. t_j must be a control transition
3. $t_j | t_c$, $t_c \in C_T$ and $t_c \neq t_j$

All such t_g are colored "brown" for the designer to pick. Afterwards, all such brown transitions will be colored "black." Procedure "color_generation_transitions()" extracted from our source code is to color all above t_g.

```
void color_generation_transitions (num,num1)
int num,num1;
{ int i,j,k;
  for(i=1;i<α;i++)
    {
      /*find PSP that are sequential later than PSP num1 and
        concurrent to PSP num*/
      if(T_Matrix[i][num1]==1 || T_Matrix[i][num1]=='N'||
        T_Matrix[i][num1]=='L') {
        for(each Π_k∈C_T){
        if(T_Matrix [i][k]!='X')
            continue;/*failed try a new i*/
        }
        /*so far, Π_i | all PSP in C_T*/
        for(each transition t_j in PSP i)
          strcpy(color[j],"brown");/*color t_i brown*/
        }
    } }
```

where num1 are the PSP containing t_j', α_i the total number PSPs, and T_Matrix[a][b] the temporal relationship between Π_a and Π_b. The total time complexity of the above procedure is $O(\alpha^2)$.

(D.2) GPN

We now consider how to implement the arc-ratio rule. Upon the generation of an NP, its least ratio in the NP must be literally smaller than that in N^1. This causes extra computation of the least ratio in N^1. To do so, we have to search all paths between n_g and n_j, which is quite time consuming (called the *canonical* method for later reference). To avoid this, we select a reference node and choose it to be the home place of the basic process (called p_{bh}) from which the synthesis begins (called the *time-saving* method). We will briefly compare the two after the presentation of the latter method.

The algorithm then assigns a global variable (initialized to 1) to p_{bh}. This variable H_c indicates the minimum tokens required in p_{bh} for the synthesized net to be live. It also assigns two variables to each node n_i in the net: upper[i] and lower [i] representing $(R_{ih})^u$ and $(R_{ih})^l$, respectively. Their ratio is called the *least home ratio*. Lower[i] indicates the minimum tokens, among all reachable markings, required in p_{bh} for n_i, if a place, to gain at least one token (actually upper [i]), or, if a transition, to fire at least one (actually upper[i]) time by firing transitions along any path from p_{bh} to n_i.

Whenever we generate a new arc and a new node at its end, we calculate the upper and lower of the new node using the following rule.

Rule New-Ratio

Let $[n_1 n_2]$ be an arc with only two nodes with weight w and the least home ratio of n_1 is known, then

(1) if n_2 is a place

$$\frac{upper[2]}{lower[2]} = \frac{upper[1]w}{lower[1]},$$

i.e., upper[2] = upper[1]*w,

lower[2] = lower[1], and H_c unchanged.

(2) if n_2 is a transition

$$\frac{upper[2]}{lower[2]} = \frac{1}{lower[1]}\left(\frac{upper[1]}{w}\right),$$

i.e; lower[2] = lower[1]*w/gcd, and

upper[2] = upper[1]/gcd, and H_c = H_c*w/gcd, where gcd is the greatest

common denominator of upper[1] and w.

(1) is obvious; (2) is to avoid tokens left in the place between n_1 and n_2 when all H_c tokens in p_{bh} are consumed and proceed to fire n_2.

When we stop at an n_j, which is in N^1, we require that its *least home ratio* be primarily equal, but not necessarily literally smaller (unlike Rule Arr.1.a) than that in N^1. This is to reduce the time to find the firing ratio between n_g and n_j, which cannot be deferred from the mere least home ratios of n_g and n_j. As a result, extra time complexity would be incurred to trace all paths between n_g and n_j to find the firing ratio between them.

By "primarily," we mean the resulting ratio after the cancellation of the GCD (greatest common denominator) between upper and lower. By "literally," we mean we do not perform the above cancellation of the GCD between upper and lower.

The two ratios must be primarily equal and there are no restrictions on the magnitudes of upper and lower in N^1. They may be smaller or greater than those in the NP. In the former (smaller) case, there is no need to update all uppers and lowers. In the latter (greater), there are two cases: forward and backward generations.

For a backward generation, there is no need to update for a TT-generation, since it creates a new home place as will be discussed later. In the former case of a forward TT-generation plus backward PP-generations, the updates at t_j are needed for any node n_1, $n_j \rightarrow n_1$ must have its upper and lower updated if its lower is smaller than that of t_j. And H_c must be updated also if any updated lower is greater than the current H_c. Otherwise, it will not be able to get a token or fire at least once if the amount of tokens in the home place equals the old lower (not updated) at the node. If the lower of the above n_1 is smaller than the lower at t_j, it indicates that the amount of tokens at the home place required for n_1 is smaller than that for t_j. This is impossible, because if t_j cannot fire once, n_1 will not either if it is a transition.

For a forward or backward PP-generation, there is no need for the update at p_j; however, H_c must be updated. This is because if the tokens do not flow through the NP, the least amount of tokens at p_{bh} for p_j to get at least one token remains unchanged. H_c must be updated, otherwise, some transitions in the NP may be not live.

In the case of multiple home places, there are multiple H_c's and the above update of H_c must be extended to other home places being are sequential to t_j. Home places resulting from single exclusive TT-generations, as well as exclusive to the above home places, need not have their H_c updated. As a matter of fact, there should be more than one upper and lower for t_g and t_j involved in a series of exclusive TT-generations.

Note that by removing the restriction of "literally smaller" from Rule ARR.1.a, we violate Guideline 1 as N^1 is disturbed by the above updates even though N^2 remains well behaved. This is because Rule ARR.1.a restricts us to synthesizing a net in a certain order, i.e., the path with the largest literal ratios between two nodes must be constructed prior to all other paths. This restriction is certainly inflexible and certain nets may be unable to be synthesized because paths with literally greater ratios than existing paths can never be constructed.

Upon a backward TT-generation, Rule ARR.1.b entails that we insert enough tokens into the new PSP Π^w per ARR.1.b. The number of tokens to be inserted into a place p_n in Π^w is stated in the following rule.

Updated Rule ARR.1.b

Upon a backward TT-generation, insert $upper_n$ tokens into p_n in Π^w.

The rationale comes from the fact that $upper_n$ represents the maximum number of tokens that it will obtain by firing t_g. This same amount of tokens will be completely consumed by firing t_j. Consequently, there will be neither continuous token accumulations nor token depletion at p_n. However, we do not need to perform updates even though the new ratios at t_j may be literally greater than the existing ratio prior to the generation. This is because the backward TT-generation creates a new home place, and the tokens in the new PSP support the firing of t_j unlike the forward TT-generation case where the firing of t_j must be supported by the tokens in the home place.

Although the computation and updating of upper and lower incurs extra computations compared to that for OPNs, the associated time complexity is linear to the total number of arcs in the net and does not increase the total time complexity for the synthesis.

It is easy to see that the nets synthesized by the *canonical* method are a subset of those synthesized by the *time-saving* method. However, the NP by the former alters the behavior of N^1.

Synthesis Steps for the Net in Figure 8.14

Now we return to the synthesis of the net example in Figure 8.14. At various stages, message boxes will be popped up to provide warning or error messages or guidelines for the next generations. Figure 8.15(a) is constructed using the PP.1 rule. Note that least-home-ratios are 1/1; Rule arc-ratio need not apply here and we do not display the ratios. Figure 8.15(b)–(d) shows the three consecutive exclusive TT-generations making up a cycle (all nodes in the cycle turn their color from red into black) along with the least home ratios; the new home place is $p9$. Using the arc-ratio rule, $n \geq 2$ for $p9$. Note the three control transitions t_1, t_3, and t_5 are in the same cycle.

Note that we enter arc weights only after the NP is completed and we must enter them in order from n_g. If we enter the arc weight for an arc with its start node's least home ratio not yet computed, a warning message box will appear.

Figure 8.15(h) is constructed using Rules PP.1 and TT.2 (a token inserted in $p13$). This forms Component I. Steps to synthesize Component II are shown in Figure 8.15(i)–(j) for the subactivity of robot 2 on Conveyor 2. Note that the two Components I and II are not connected. The final net is obtained using the rule for exclusive-TT generation, which requires constructing a cycle containing all control transitions. During the construction process, part of the cycle comes from Component I. Hence, Rule

TT. 3 must be used upon the generation [$t16$ $p20$ $t12$] from Component II to I, to generate another TT-path [$t11$ $p21$ $t17$] from Component I to II.

In Figure 8.15(b), the message box tells us that it is an exclusive TT-generation and we should pick a brown node and a purple node for the generation point and joint, respectively. In Figure 8.15(c), the message box tells us that we are in the midst of the exclusive TT generation and a cycle has not been formed. In Figure 8.15(d), the message box tells us that we reach a cycle and we should enter tokens in it. When we generate the first exclusive TT-path, the firing ratio between t_g ($t4$) and t_j ($t1$) is 1/1, and so too is the next TT-generation between $t2$ and $t6$. Because of the exclusive TT-generations, the lesat home ratios at this step must be expressed relative to $t4$ rather than to the home place $p1$. Afterwards, we enter 2 tokens to $p9$ since its upper variable is 2 via the "Modify" button.

In the last exclusive TT-generation, the arc weight between $p8$ and $t8$ is 2, the least home ratio at $t8$ is 1/2, which implies that in order to fire $t8$ once, $t4$ must fire twice. Now the next arc weight between $t8$ and $p9$ is 2; hence, the least home ratio at $p1$ is 2/2, implying that after $t4$ fires twice, $p9$ will get two tokens. Finally, the last arc weight between $p9$ and $t3$ is 1; hence, the least home ratio at $t3$ is 2/2, as well as at $t4$. Since we insert tokens at the new home place, $p9$, whose least home ratio is 2/2, we need to insert at least two tokens; i.e., $n \geq 2$. Figure 8.15(f) shows the least home ratios for all nodes in the cycle in reference to $t4$.

After entering two tokens at $p9$, the least home ratio will be recomputed in reference to $p9$ rather than to $t4$ (Figure 8.15(g)).

Similar statements apply to the synthesis of Component II. We insert 3 tokens at $p18$ (Figure 8.15(k)) whose "upper" variable equals 3; hence, $m \geq 3$. Note that the requirement in Rule ARR.1.a (i.e., equal prime ratio along any paths between two nodes) does not hold for exclusive TT-generations. There are two different paths from $p15$ to $t13$ [$p15$ $t13$] and [$p15$ $t17$ $p17$ $t18$ $p18$ $t13$] with different prime ratios of 1/1 and 3/1, respectively. As a matter of fact, there may be two different least-home-ratios for all t_g and t_j that are involved in an exclusive-TT generation. In this case, the tool will display the one that is relevant to the current generation.

In Figure 8.15(h), we generate a PP-cycle [$p8$ $t9$ $p10$ $t10$ $p11$ $t12$ $p12$ $t11$ $p8$] followed by a backward TT-generation [$t10$ $p13$ $t8$]. The least ratios in reference to $p8$ are displayed for nodes in the PP-cycle. Note that the new ratio for $p8$ in reference to itself is 2/2 indicating that $p8$ must have at least two tokens in order to complete the firings of all transitions in the PP-cycle. This, in turn, implies that $t3$ must fire at least twice. However, we do not need to update the ratio at $t8$ since its lower is no smaller than 2, which implies that the least amount of firings of $t3$ (the reference node) for $t8$ to fire at least once remains at 2.

In Figure 8.15(k), we perform two successive exclusive TT-generations and generate another TT-path between two separated components; a message is shown in the bottom window to warn us about the potential problem of unboundedness and remind us that we can create another path from $t12$ to $t16$. In the next step (Figure 8.15(1)), we generate another TT-path to synchronize the two components (see the message in the bottom window) and complete the cycle for the exclusive TT-generations.

8.8   Other Features of the Tool

The tool can apply to many fields as Petri nets are a powerful modeling and analysis tool for

- Distributed operating systems,
- Distributed data base systems,
- Concurrent and parallel processes,
- Flexible manufacturing/industrial control processes,
- Discrete event systems,
- Multiprocessor systems,
- Data flow computing systems,
- Fault-tolerant systems,

- Programming logic and VLSI arrays,
- Communication networks and protocols,
- Neural networks,
- Digital filters, and
- Decision models.

However, in most of these applications, a different graphical model is used instead, such as data flow for digital signal processing and state diagrams for communication protocols. As a result, it is desired that the tool can also draw and analyze these graphical models. The tool has been enhanced in this respect: for rapid development, the tool converts these graphs into Petri nets internally for analysis. Therefore, there is no need to develop new algorithms and codes for each type of graph. Features of the tool include:

A. **Data Communication:** We have developed a very powerful CAD tool for designing protocols and Petri nets.[11] Currently, the tool is able to draw Petri nets, state diagrams, data flow graphs, finite state machines, and general graphical objects. Once the graph is drawn, the tool can analyze, simulate, reduce, and synthesize it. Few existing tools are capable of such an integration. The tool has been enhanced to synthesize error-recovery protocols with great time-complexity reduction, simulate extended finite state machines, hierarchically model and simulate Petri nets, automatically generate "C" codes, and generate unique I/O (UIO) conformance test patterns. We are extending the tool to

- Simulate queuing networks,
- Software engineering for protocol design implementation and documentation,
- Advanced synthesis and reduction of communication protocols using PN,
- Implementation of synthesis of local entities,
- Complexity reduction in synthesizing multiparty protocols, and
- Animation of protocols with real objects.

B. **Network Simulation:** We[66] have enhanced our protocol design tool for object-oriented simulations. After drawing network objects and interconnecting them, we can perform simulation. The "Step" mode allows easy debugging and detailed trace of network behavior, which is useful for computer-aided education. To illustrate, we have applied the tool to the bitonic sorting part of the Batcher-bayan network. Automatic code generation helps produce efficient "C" code for non-X-Window environments after verification via simulation.

C. **Parallel Processing:** The above tool has been extended[65] to analyze and simulate performance of DFGs. It incorporates a unique, efficient algorithm that we developed to do rate-optimal steady state scheduling without initial transients. We have extended the tool to draw multiple rate DFGs and simulate. It can find critical and subcritical loops and perform rate-optimal scheduling. The user can randomly pick a node to check its input and output values against its functional type after the simulation. The critical loops, scheduling ranges, and processor assignments can be displayed in a graphical fashion. It can also display critical paths and Gann charts for DFGs without any loops. Few tools are capable of such integration and ease of use.

The final matrix theory developed is the first of its kind and is used throughout the design. It calculates the iteration bound, finds critical loops, derives formulas for scheduling ranges, and performs a fast processor assignment with rate-optimal scheduling. This eliminates the need for inequality charts for finding the scheduling ranges by Heemstra et al.[67] Most approaches repetitively update scheduling ranges (which is very time consuming), and our theory can eliminate many of these updates. Benchmark testing indicates remarkable (thousandfolds) speedup of scheduling, especially for large DFGs that take less than 1 second compared with other approaches that take several hundred seconds. In addition, the resulting scheduling needs fewer numbers of processors than other approaches.

The theory has been updated to include the case of resource constraints and iteration periods greater than iteration bounds.

D. **Inference Engine for Expert Systems:** This allows both forward and backward chaining.[11] Coding and the associated compilation are not required, thus avoiding grammatical and typographical errors. In addition, the user can easily modify the designed inference network in a graphical fashion. An example of the inference network for car diagnosis has been demonstrated. Rules can be represented by transitions and preconditions, or symptoms can be represented by tokens in places. Clicking the "Auto" button of the "Simulate" submenu of the "Analysis" menu fires the production rules (transitions) and, when the firing terminates, the places holding tokens indicate the cause of the problem.

8.9 Conclusions

The tool, based on the synthesis rules and the algorithm, helps designers to construct large PNs interactively and to synthesize an automated manufacturing system in a user-friendly fashion.

None of the existing tools integrate drawing, file manipulation, analysis, simulation, animation, reduction, synthesis, and property query in one software package. Furthermore, because PNs model discrete-event systems, the tool finds applications in communication protocols, flexible manufacturing systems, (extended) finite state machines, expert systems, interactive parallel debuggers,[11] digital signal processing,[5,7,11] etc.

We have enhanced the tool to include models not only of PNs but also of state diagrams and data flow graphs (DFGs) with few code changes. Thus a designer can choose the model with which he is familiar. For instance, DSP professionals do not know PNs well. They can, however, draw DFGs and obtain iteration bounds, critical loops, rate-optimal scheduling, etc. by just clicking a button.[4,5,65]

We have also implemented a reduction algorithm based on the rules; the code is very simple, containing less than 100 lines. The distinct point of this approach is that, besides the possibility of continuous enhancement, while reducing, it can discover wrong designs and suggest how to fix the problem based on the knitting rules.

This work overcomes some drawbacks of most existing synthesis approaches; i.e., they do not

- Deal with the algorithm and CAD tool using graphical user interface for synthesis explicitly
- Show how to continuously update their synthesis techniques
- Indicate how to extend the synthesis for analysis
- Show temporal relationships among processes after synthesis
- Find the maximum concurrency of the synthesized net.

References

1. Agerwala, T. and Y. Choed-Amphai, A synthesis rule for concurrent systems, *Proc. of Design Automation Conference,* 1978, pp. 305–311.
2. Berthelot, G., Checking properties of nets using transformations, in *Advances in Petri Nets,* G. Rozenberg (ed.), 1985, Springer-Verlag, pp. 19–40.
2a. Chao, D.Y., Application of a synthesis algorithm to flexible manufacturing systems, *Journal of Information Science and Engineering,* Vol. 14, No. 2, June 1998, pp. 409–477.
3. Chao, D. Y. and D. T. Wang, Synchronized choice ordinary Petri net, (Invited) *Proc. 1995 IEEE Int'l Conf. SMC,* Vancouver, Canada, October 22–25, pp. 1442–1447.
4. Chao, D. Y. and D. T. Wang, XPN-FMS: A modeling and simulation software for FMS using Petri nets and X windows," *International Journal of Flexible Manufacturing Systems,* Vol. 7, No. 4, October 1955, pp. 339–360.
5. Chao, D. Y. and D. T. Wang, Iteration bounds of single-rate data flow graphs for concurrent processing, *IEEE Trans. Circuits Syst.,* CAS-40, No. 9, September 1993, pp. 629–634.
6. Chao, D.Y. and D. T. Wang, Two theoretical and practical aspects of knitting techniques—invariants and a new class of Petri net, *IEEE Trans. SMC_27,* No. 6, December 1997, pp. 962–977.

7. Chao, D. Y. and D. T. Wang, Application of final matrix to data flow graph scheduling using multiprocessors, *MIS Review,* Vol. 5, December 1995, pp. 65–80.

8. Chao, D. Y. and D. T. Wang, X-Window implementation of an algorithm to synthesize ordinary Petri nets, *Journal of National Cheng Chi University,* Vol. 73, October 1996, pp. 451–496.

9. Chao, D. Y., M. C. Zhou, and D. T. Wang, Extending knitting technique to Petri net synthesis of automated manufacturing systems, *The Computer Journal,* Oxford University Press, Vol. 37, No. 1–2, 1994, pp. 1–10.

10. Chao, D. Y. and D. T. Wang, A synthesis technique of general Petri nets, *J. Systems Integration,* Vol. 4, No. 1, 1994, pp. 67–102.

11. Chao, D. Y. and D. T. Wang, An interactive tool for design, simulation, verification, and synthesis for protocols, *Software-Practice and Experience, an International Journal,* Vol. 24, No. 8, August 1994, pp. 747–783.

12. Chao, D. Y. and D. T. Wang, Petri net synthesis and synchronization using knitting technique, *Proc. 1994 IEEE Int'l Conf. SMC,* San Antonio, TX, October 2–5, pp. 652–657.

13. Chao, D. Y. and D. T. Wang, Knitting technique and structural matrix for deadlock analysis and synthesis of Petri nets with sequential exclusion, *Proc. 1994 IEEE Int'l Conf. SMC,* San Antonio, TX, October 2–5, pp. 1334–1339.

14. Chao, D. Y. and D. T. Wang, Knitting technique with TP-PT generations for Petri net synthesis, (Invited) *Proc. 1995 IEEE Int'l Conf. SMC,* Vancouver, Canada, October 22–25, pp. 1454–1459.

15. Chao, D. Y. and D. T. Wang, Linear algebra based verification of well-behaved properties and P-invariants of Petri nets synthesized using knitting technique, *MIS Review,* Vol. 5, December 1995, pp. 27–48.

16. Chen, Y., W. T. Tsai, and D. Y. Chao, Dependency analysis—a compositional technique for building large Petri net, *IEEE Trans. on Parallel and Distributed Systems,* PDS-4, No. 4, 1993, pp. 414–426.

17. Chu, W. W. and K. K. Leung, Task response time model and its application for real-time distributed processing systems, *Proc. 1984 IEEE Real Time Syst. Symp.,* pp. 225–236.

18. Datta, A. and S. Ghosh, Synthesis of a class of deadlock-free Petri nets, *Journal of ACM,* Vol. 31, No. 3, 1984, pp. 486–506.

19. Datta, A., Modular synthesis of deadlock-free control structures, in *Foundations of Software Technology and Theoretical Computer Science,* G. Goos and J. Hartmanis, (eds.), Springer-Verlag, Vol. 241, 1986, pp. 288–318.

20. Dong, T., The Modeling, Analysis, and Synthesis of Communication Protocols, Ph.D. Dissertation, Computer Science Division, EECS.

21. Esparza, J. and M. Silva, Circuits, handles, bridges, and nets, in *LNCS, Advances in Petri Nets 1991,* Springer-Verlag, pp. 210–242.

22. Esparza, J. and M. Silva, On the analysis and synthesis of free choice systems, in *LNCS, Advances in Petri Nets 1991,* Springer-Verlag, pp. 243–286.

23. Huang, J. P., Modeling of software partition for distributed real-time application, *IEEE Trans. on Software Eng.,* SE-11, October 1985, pp. 1113–1126.

24. Hyung, L-K. and J. Favrel, Analysis of Petri nets by hierarchical reduction and partition, in *IASTED Modelling and Simulation,* Acta Press, Zurich, Switzerland, 1982, pp. 363–366.

25. Hyung, L-K. and J. Favrel, Hierarchical reduction method for analysis of decomposition of Petri nets, *IEEE Trans. Syst., Man, Cybern.,* SMC-15, No.2, March 1985, pp. 272–280.

26. Hyung, L-K., Generalized Petri net reduction method, *IEEE Trans. Syst., Man, Cybern.,* SMC-17, No. 2, 1987, pp. 297–303.

27. Koh, I. and F. DiCesare, Modular transformation methods for generalized Petri nets and their application to automated manufacturing systems, *IEEE Trans. Syst., Man, Cybern.,* SMC-26, No. 6, 1991, pp. 1512–1522.

28. Jeng, M. D. and F. DiCesare, A review of synthesis techniques for Petri nets with applications to automated manufacturing systems, *IEEE Trans. Syst., Man, Cybern.,* SMC-23, No. 1, 1993, pp. 301–312.

29. Jeng, M. D. and F. DiCesare, A modular synthesis technique for Petri nets, *Proc. 1992 Japan-USA Symp. on Flexible Automation,* pp. 1163–1170.

30. Johnsonbaugh, R. and T. Murata, Additional method for reduction and expansion of marked-graphs, *IEEE Trans. Circuits Syst.,* CAS-28, No. 10, October 1981, pp. 1009–1014.

31. Koh, I. and F. DiCesare, Transformation methods for generalized Petri nets and their applications in flexible manufacturing systems, *IEEE Trans Syst., Man, Cybern.,* SMC-21, No. 6, 1991, pp. 963–973.

32. Krogh, B. H. and C. I. Beck, Synthesis of place/transition nets for simulation and control of manufacturing systems, *Proc. 4th IFAC/IFORS Symp.* Large Scale System, Zurich, 1986.

33. Kwong, Y. S., On reduction of asynchronous systems. *Theorit. Comput. Sci.,* Vol. 5, 1977, pp. 25–50.

34. Chao, D. Y., Petri net synthesis and sychronization using knitting technique, *Journal of Information Science and Engineering,* Vol. 15, No. 4, 1999, pp. 543–568.

35. Molloy, M. K., On the integration of delay and throughput measures in distributed processing models, Ph.D Thesis, Computer Science Dept., UCLA, Los Angeles, 1981.

36. Murata, T. and J. Y. Koh, Reduction and expansion of live and safe marked-graphs, *IEEE Trans. Circuits Syst.,* CAS-27, January 1980, pp. 68–70.

37. Murata, T. et al., A Petri net based controller for flexible and maintainable sequence control and its applications in factory automation, *IEEE Trans. on Industrial Electronics,* IE-33, 1986, pp. 1–8.

38. Murata, T., Petri nets: properties, analysis and application, *IEEE Proceedings,* Vol. 77, No. 4, April 1989, pp. 541–580.

39. Murata, T., Modeling and analysis of concurrent systems, in *Handbook of Software Engineering,* C. Vick, and C. V. Ramamoorthy, (eds.), Van Nostrand Reinhold, 1984, pp. 39–63.

40. Murata, T., Circuit theoretic analysis and synthesis of marked graphs, *IEEE Trans. Circuits Syst.,* CAS-24, No. 7, 1977, pp. 400–405.

41. Murata, T., Synthesis of decision-free concurrent systems for prescribed resources and performance, *IEEE Trans. Software Engineering,* SE-6, No. 6, 1977, pp. 400–405.

42. Narahari, Y. and N. Viswanadham, A Petri net approach to the modeling and analysis of flexible manufacturing systems, *Annals of Operations Research,* 3, 1985, pp. 449–472.

43. Peterson, J. L., *Petri Net Theory and the Modeling of Systems,* Prentice-Hall, Englewood Cliffs, NJ, 1981.

44. Ramamoorthy, C. V., Y. Yaw, and W. T., Tsai, A Petri net reduction algorithm for protocol analysis, *Computer Communication Review (USA),* Vol. 16, No. 3, August 1986, pp. 157–166.

45. Ramamoorthy, C. V. and H. So, Software requirements and specifications: status and perspectives, *IEEE Tutorial: Software Methodology,* 1978.

46. Ramamoorthy, C. V., S. T. Dong, and Y. Usuda, The implementation of an automated protocol synthesizer (APS) and its application to the X.21 protocol, *IEEE Trans. Software Engineering,* XE-11, No. 9, September 1985, pp. 886–908.

47. Ramamoorthy, C. V., Y. Yaw, W. T. Tsai, R. Aggarwal, and J. Song, Synthesis of two-party error-recoverable protocols, *Computer Communication Review (USA),* Vol. 16, No. 3, August 1986, pp. 227–235.

48. Ramamoorthy, C. V., Y. Yaw, W. T. Tsai, R. Aggarwal, and J. Song, Synthesis and performance evaluation of two-party error-recoverable protocols, *Proc. COMASC Symp.,* October 1986, pp. 214–220.

49. Silva, M., Toward a synchrony theory for P/T nets, in *Concurrency and Nets,* K. Voss, H. J. Genrich, and G. Rozenberg (eds.), Springer-Verlag, pp. 435–460.

50. Suzuki, I. and T. Murata, A method for stepwise refinements and abstraction of Petri nets, *J. Comp. Syst. Sci.,* 27, 1983, pp. 51–76.

51. Valette, R., Analysis of Petri nets by stepwise refinement, *J. Comp. Syst. Sci.* 18, 1979, pp. 35–46.

52. Valvanis, K. S., On the hierarchical analysis and simulation of flexible manufacturing systems with extended Petri nets, *IEEE Trans. Syst., Man, Cybern.,* SMC-20, No. 1, pp. 94–100.

53. Villaroel, J. L., J. Martinez, and M. Silva, GRAMAN: a graphic system for manufacturing system design, *Proc. IMACS International Symp. on Syst. Model and Simul.*, Cetraro, Italy, 1988.
54. Wang, D. T., and D. Y. Chao, Enhanced knitting technique to Petri net synthesis, *Proc. 1994 IEEE Int'l. Conf. SMC*, San Antonio, TX, October 2–5, pp. 658–663.
55. Wang, D. T., and D. Y. Chao, New knitting technique for large Petri net synthesis with automatic preservation of liveness, boundedness and reversibility, (Invited) *Proc. 1995 IEEE Int'l. Conf. SMC*, Vancouver, Canada, October 22–25, pp. 1460–1465.
56. Wilson, R. G. and B. H. Krogh, Petri net tools for the specification and analysis of discrete event controllers, *IEEE Trans. Software Engineering.*, SE-16, No. 1, 1990, pp. 39–50.
57. Yau, S. S., and Caglayan, Distributed software system design representation using modified Petri nets, *IEEE Trans. Software Engineering*, SE-9, No. 6, November 1983, pp. 733–745.
58. Yaw, Y., Analysis and Synthesis of Distributed Systems and Protocols, Ph.D. Dissertation, Dept. of EECS, U.C. Berkeley, 1987.
59. Yaw, Y., C. V. Ramamoorthy, and W. T. Tsai, A synthesis technique for designing concurrent systems, *Second Parallel Processing Symposium*, April 1988, pp. 143–166.
60. Yaw, Y., C. V. Ramamoorthy, and W. T. Tsai, Synthesis rules for cyclic interactions among processes in concurrent systems, *Proc. COMSAC Symp.*, October 1988, pp. 496–504.
61. Yaw, Y. and F. L. Foun, The algorithm of a synthesis technique for concurrent systems, *Proc. 1989 IEEE Int'l. Workshop on Petri Nets and Performance Models*, Tokyo, pp. 266–276.
62. Zhou, M. C. and F. DiCesare, Parallel and sequential mutual exclusions for Petri net modeling for manufacturing systems with shared resources, *IEEE Trans. on Robotics and Automation*, RA-7 No. 4, 1991, pp. 515–527.
63. Zhou, M. C., F. DiCesare, and A. A. Desrochers, A hybrid methodology for Petri net synthesis of manufacturing systems, *IEEE Trans. on Robotics and Automation*, RA-8, No. 3, 1992, pp. 350–361.
64. Zhou, M. C., K. McDermott, and A. Patel, Petri net synthesis and analysis of a flexible manufacturing systems cell, *IEEE Trans. SMC*, Vol. 23, No. 1, March 1993, pp. 524–531.
65. Chao, D. Y., Final-matrix based and fast implementation of recurrent DSP scheduling, *Proc. ICSPAT '96/DSP World Expo*, Boston, October 1996, pp. 171–175.
66. Chao, D. Y., A CAD tool for network simulation based on a protocol design CAD tool, *Proc. ICSPAT '96/DSP World Expo*, Boston, October 1996, pp. 836–840.
67. Heemstra, S. M. et al., Range-chart-guided iterative data-flow graph scheduling, *IEEE Trans. Circ. Syst.*, CAS-39, No. 5, 351–364, 1992.

Appendix

A_{ab}: entries of stucture matrix
$a_{jj_1}^-$: the weight on the arc from p_j to t_{j_1}
AC: asymmetric-choice nets
AGV: automatic guided vehicles
α: the total number of final PSPs in the final system
β: the total numbr of nodes in the final system
B: a bridge
CL (o): *cyclic*
C: local concurrent set
C_{ik}: LCN (Π_i, Π_k)
C_T: the set of control transitions
CN (‖): *concurrent*
d: synchronic distance
d^s : structure synchronic distance
DEP: directed elementary path
DFG: data flow graph

DSP: digital signal processing
EFC: extended free-choice nets
EX (|): *exclusive*
FC: free-choice nets
G (*J*): the set of Π_g (Π_j) involved in a single application of the TT or PP rule
GPN: general Petri nets
H: the set of home places
H_c: the least amount of tokens required in home place p_{bh} for the synthesized net to be live
I: an input function defining the set of directed arcs from P to T
IG: interactive generation
L: language of Petri net
LCN: local concurrent set of *PSP*s
LEX: local exclusive set of *PSP*s
M: marking
$M^2(N^1)$ or $M^{1/2}$: the marking of N^1 in N^2
N: a PN
N^a: a synthesized PN with every home place holding only one token
N^b: a PN after adding tokens to N^a
N^{1a}: an N^1 and N^a
N^{2a}: an N^2 and N^a
$N^{1/2}$: the subnet N^1 in N^2
n: the number of *PSP*s in a Petri net
n_i: the number of *PSP*s before the *i*th generation
$n_g(n_g)$: the (next) generation point
$n_j(n_j)$: the (next) joint
$n_{ps}(p_{ps}, t_{ps})$: the prime start node (place, transition)
NP: new path
O: an output function defining the set of directed arcs from T to P
OPN: ordinary Petri nets
p: place
p_{bh} : a home place in the basic process
p_d: a decision place
$\bullet p(p\bullet)$: the set of input (output) transitions of place *p*
$p_s(p_{ps})$: the (prime) start place for a pair of exclusive *PSP*s
p_{ps}^{gj} : a prime start place relative to n_g and n_j
P: the set of places in net N
$\bar{P}$: directed path
PG: pure generation
π: process
Π: *PSP*
Π_g : the *PSP* containing n_g
Π_j : the *PSP* containing n_j
Π^w : new *PSP*
PN: Petri net
PP generation: place-place path generation
PP-path: place-place path
PT-Path: place-transition path
PSP: pseudo-process
q: least ratio
$[q]'$: the fraction $\frac{a'}{b'}$, where *q* is any ratio, $\frac{a}{b}$ and *a*' and *b*' are relatively prime
q^1 : the denominator of the least ratio

q^u: the numerator of the least ratio

R or R (N, M_0): reachable set of markings of N with initial marking M_0

R^f: least firing ratio

R^{fm}: least firing-marking ratio

R^m: least marking ratio

R^{mf}: least marking-firing ratio

SC: synchronized choice nets

SPN: Stochastic PNs

t: transition

t_c: a control transition

$\cdot t$ ($t\cdot$): the set of input (output) places of transition p

$t_s(t_{ps})$: the (prime) start transition for a pair of concurrent *PSPs*

T: the set of transitions in net N

T-Matrix: temporal matrix

TP-path: transition-place path

TT generation: transition-transition path generation

SE or E ($\rightarrow$): *sequentially earlier*

SL or L ($\leftarrow$): *sequentially later*

SN or N ($\leftarrow$): *sequentially next*

SP or P ($\leftarrow$): *sequentially previous*

SQ: *sequential*

subscript g : related to generation point

subscript j : related to joint

superscript 1: for the net prior to the generation, i.e., $N^1(T^1, P^1, M^1, R^1)$ refers to the net (T, P, M, and R) prior to the generation

superscript 1a: for the net prior to the generation and every home place has only one token, i.e., N^{1a} $(T^{1a}, P^{1a}, M^{1a} R^{1a})$

superscript 2: for the net after the generation, i.e., $N^2(T^2, P^2, M^2, R^2)$ refers to the net (T, P, M, and R) after the generation

superscript w: for the NP, i.e., (T^w, P^w, M^w) refers to the (T, P, M) of the *NP*. σ^w denotes a complete firing sequence through the *NP*, which includes both n_g and n_j.

σ: transition firing sequence

σ (t): the number of firings of t in σ

VP: *virtual PSP* which has only the generation point and the joint

X: local exclusive set

X_{ik}: *LEX* (Π_i, Π_k)

$>_{lit}$: literally greater than

Γ: a circuit

γ : a handle

9

Computer-Aided Design Techniques for Development of Automated Product Assembly Systems in Manufacturing Systems

Bart O. Nnaji
University of Pittsburgh

Hsu Chang Liu
Solid Works

9.1   Introduction

Most CAD modules are developed independently on their applications and run on a specific platform. The complexity of integrating all design modules into an all directions CAD system requires a big investment in both software and hardware and has long been the burden of small-scale companies trying to move from computer-aided drafting to computer-aided design.

Recent growth in computer technology development means that computers with higher performance and lower costs have been designed and have become more affordable. Also, by applying network technologies, more and more centralized systems have been broken down into cheaper distributed units. This shared schema has improved the computation speed, increased the memory storage size, and increased communication among end users. The CAD development has also extended its vision from geometric modeling, engineering analysis, and manufacturing to the design of the life cycle of the product. The life cycle includes, for example, the stages of functional specifications development, conceptual design (Kang, 1996), component design, assembly modeling, and, finally, manufacturing.

In most cases, assembly is the most complex process in manufacturing, occupying approximately 50% of total manufacturing cost (Boothroyd, 1995). In order to reduce assembly cost, a designer has to come up with a design that has a minimun number of assembly processes; each assembly process has to be simple so that it can be easily achieved by manual or automatic assembly systems. This requires a great deal of communication between design engineers and manufacturing engineers. The requirements for design engineers have to agree with manufacturing methods. Evaluations have to be iterated between these two systems until an optimal design solution has been reached.

In general, design criteria mainly are product specifications and manufacturing methods that can be carried by geometric constraints (Anantha, 1996; Liu, 1991). This information has to be uniformly interpreted and represented throughout the product life cycle. In this chapter, we will introduce spatial relationships (Liu, 1991), a structural representation scheme embedded with geometric information, and nongeometric constraints that carry design criteria. Spatial relationships compose the common language that communicates between design engineers and manufacturing in design considerations.

Also in this chapter, we will begin with the introduction of Product Modeler (ProMod) (Nnaji, 1993), a modeling system that applies design with spatial relationships and is able to capture and represent the design evaluation criteria through assembly modeling processes (Nnaji, 1994). This is followed by discussion of subsequent developments in automatic assembly that are listed as follows:

- Stability analysis of assembly (Vishnu, 1992)
- Feasible approach directions and precedence constraints (Yeh, 1992)
- Kinematic modeling (Aguwa, 1997; Prinz, 1994).

9.2 ProMod: A Concurrent Conceptual and Product Design System for Mechanical Assemblies

Design process is a refinement of abstract representations where the product functionality, manufacturability, and all other life cycle issues are optimized. Existing computer-aided design systems focus on the design and analysis of components in isolation. Since these design systems are unable to support early conceptual design, a designer's intentions (i.e., functional and geometric constraints) cannot be captured, represented, and propagated to down-stream activities. The consequence of this inability is that existing design systems are not capable of optimizing the product against life-cycle constraints, thus rendering the effort to integrate design and manufacturing unlikely to bear fruit. In this section, we introduce a new type of mechanical product modeling system, ProMod, (Nnaji, 1993; Rembold, 1991-1; Rembold, 1991-2) which is not only capable of capturing and integrating the designer's intent at conceptual-design level, but also is capable of propagating these intentions as constraints to guide the development and detailing of product design. Thus early conceptual design and product design are merged in the same computational environment (Kang and Nnaji, 1993).

In ProMod, the design process begins with modeling of the abstract representation of the components. While connecting these abstract components, the designer uses spatial relationships as the peg on which to hang the representation of the designer's intentions. In the initial realization of form, the constraints may be violated. In this case, a refinement process is applied to change the topology and geometry of the components so as to satisfy these constraints. Three subsystems, *module design*, *connectivity design*, and *detailed design*, as shown in Figure 9.1, are the kernel of the ProMod system. The designer uses *module design* to create an initial design, which may be unique or standardized, and is saved in part library. The *connectivity design* defines the functional and geometric constraints upon the refinement and configuration of components to meet product specifications. The *detailed design* contains modeling

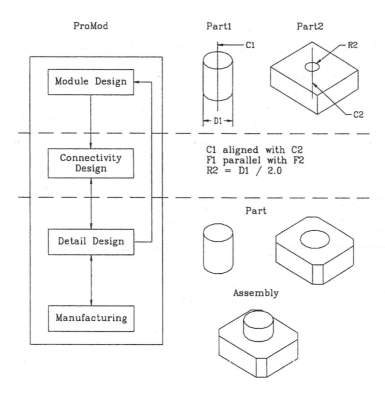

FIGURE 9.1 ProMod.

functions that modify model geometry in order to meet constraints issued from connectivity design; it also allows designers to create new geometry or modify the existing geometry manually.

Module and detail design are the application of conventional CAD modeling techniques. Module design uses CAD to develop an initial design. Detail design uses CAD to modify the initial design in order to satisfy connectivity constraints. We will have a general discussion of CAD modeling techniques, followed by the introduction of methods to modify geometry.

9.3 Computer-Aided Design and Revision

Computer-aided design has taken the place of drafting boards, design analysis tools, and prototyping. It has been designed to replace each single module in the design processes. The integration of CAD modules faces the same problem as bridging the communication gap between design and process engineers: a common language is required to represent CAD data and to constrain the data to satisfy all the engineering requirements. In this section, we will introduce modeling techniques that are used to build, as well as to modify, in order to satisfy the design constraints.

The least design and revision requirement is "undo process." Primitive operations are adopted by modern CAD modeling systems to construct a design; they use Eular operator to build topology, which is a reversible process. Primitive operation can be applied to modeling methods ranging from boundary construction, Boolean, sweeping, rounding, etc.

A design can be modified locally without changing topology, such as lifting of a vertex, edge, face, or feature. Any change made by Boolean operation will cause change in both geometry and topology, e.g., using a cutting plan to cut away geometry, adding/removing material from Boolean of two bodies. These alteration methods may achieve design requirements, but cannot be uniquely represented and thus cause difficulty in automation.

In dimension driven or parametric design systems (Lin, 1983; Serrano, 1984), dimensions are registered along with modeling operations. The design revision process is to override existing dimension value, and then remodel by repeating modeling operations with the new dimension value. This revision method requires recorded modeling history. Dimensions and parameters in parametric design are determined by designers during the component construction stage. These dimensions or parameters may not meet design constraints, so the modeling construction dimensions will be different from the dimensions shown on final drawings. This is also a burden for automatic revision, which requires manual interpretation of the relation between design constraints and model construction dimensions and parameters.

In the next section, we will explain how design with spatial relationships can be used for connectivity design and followed by 3-D variational geometry, an assembly-based, constraints-modeling system. In this system, the parametric dimensions used for geometry revision are derived from design specifications so that the revision processes are independent of modeling history. This will allow design or manufacturing engineers to add constraints with specific considerations to the product, to modify design without violating existing constraints, and, thus, to communicate the designer's intent through constraints.

9.4   Design with Spatial Relationships

Today, researchers focus on bridging the gap between product design and manufacturing. The major obstacle to this is the lack of communication among designers, process planners, and inspectors. It is impractical to require that designers possess knowledge of both design and manufacturing. Design engineers tend to ignore the manufacturing and other product life-cycle requirements. Manufacturing engineers are sometimes forced into modifying a design or the manufacturing environment because their requirements were not captured as constraints in the design. In order to link design and manufacturing processes automatically, a common product's specifications must be carried from the beginning of design to the final stage of inspection. For example, Figure 9.2(a) shows a product with two parts: one is a plate **P**, with two holes, and the other one is a bent shaft, **S**. Figure 9.2(b) shows the features of the two parts in Figure 9.2(a). The features of part **P** are extracted and classified as a **Cvex_block_base**, **C_hole1**, **C_hole2**, **Fillet1**, and **Fillet2**. We have **C_shaft1**, **C_shaft2**, and **Elbow** features for part **S**. In Figure 9.2(c) and (d), the product is made by snapping **C_shaft1** of **S** into the **C_hole1** of **P**. Traditionally, these mating criteria are annotated in engineering drawings, but this requires trained humans to create and interpret the notes. Also, certain questions arise here, such as

1. How are the specifications of components precisely determined from the product specifications?
2. Can these specifications be represented efficiently?
3. Can existing design procedures be simplified and automated?

In Figure 9.2(c), the plate and bent shaft are assembled with the geometric constraint that the shaft is inserted, not completely, but with **Da** $\pm$ **ta** offset from the tip of the shaft to the bottom of the plate and with the other side of the shaft having **Db** $\pm$ **tb** offset with the plate. There are three general stumbling blocks to assembly evident in this assembly process that involve snapping two components together:

1. It is difficult to specify how deep the **C_shaft1** should go along the length of **C_hole1** during individual part modeling.
2. The tolerance specification of $\pm$**tb** has to be manually *distributed* to tolerance of $\pm$**tc** of part **P** and $\pm$**td** of part **S** as shown in Figure 9.2(d).
3. The resultant tolerance $\pm$**tc** and $\pm$**td** should be equal to $\pm$**tb**.

These issues are the drawbacks of conventional modeling systems that are made for design and machining of individual parts and do not consider the general specifications of a product.

In this chapter, we apply the fact that an assembly model can represent more information than an individual part model and product specifications can be captured during the stage of assembly. Once

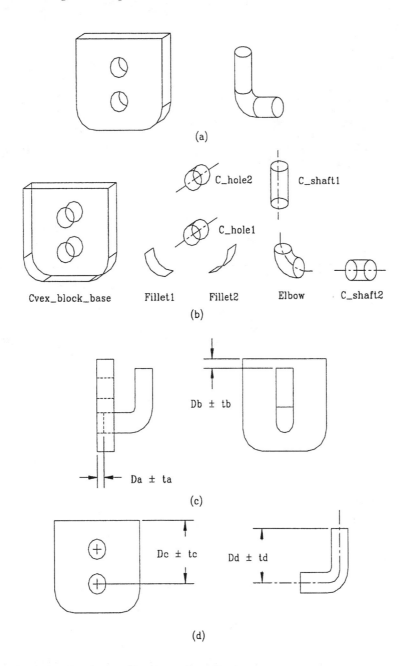

FIGURE 9.2 The relationships between features and specifications.

product specifications are captured, this information will be analyzed and propagated into the detailed design of individual components. Spatial relationships will be introduced to capture designers' intents through assembly processes and to propagate them as constraints to determine the form and functions of the design in general, as well as in automatic assembly (Liu, 1990; Liu, 1991; Nnaji, 1988).

Types of Spatial Relationships

The types of spatial relationships adopted make the kinematic aspect primary, relegating the contact aspect to being something that can be verified (and captured as a designer's intention) once a definite

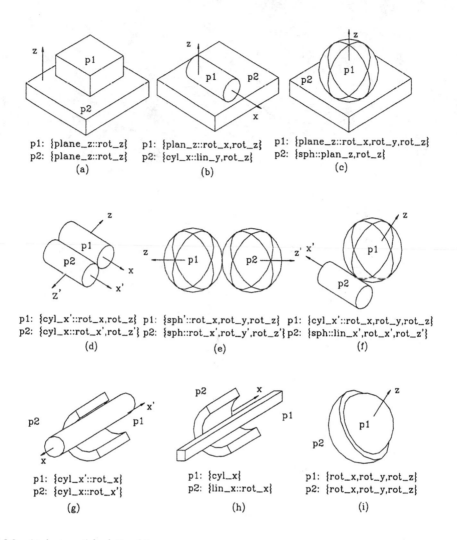

FIGURE 9.3   Against spatial relationships.

configuration of bodies has been determined. Five basic types of spatial relationships are listed as follows:

- *Against*: As illustrated in Figure 9.3, this relation means that the faces touch at some point. Any two features can bear this relation to each other. For user convenience, we may also allow speaking of a plane as being *against* a plane.

- *Parallel-offset and parax-offset*: Figure 9.4 illustrates the *parallel_offset* relation holding between planar faces and cylindrical and spherical features. In the case of two parallel planar faces, the outward normals point in the same direction, and F_2 is offset a distance from F_1 in the direction of the outward facing normal. *Parax-offset* (Figure 9.5(a)) differs from parallel offset in that the direction of the normal to the second body feature is reversed.

- *Aligned*: Two features are *aligned* if their center lines are collinear or, when one feature is spherical, its center lies on the center line of the other. This is illustrated in Figure 9.5(b).

- *Incline-offset*: This relation allows us to specify that the parts are related with any position and orientation. It is illustrated in the case of plane-faces in Figure 9.6.

- *Include-angle*: As illustrated in Figure 9.7, *include angle* specifies the angle between the outward normals between two planar faces F_1 and F_2 pointed by z and z'.

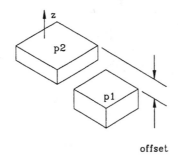

p1: {plane_z::rot_z}
p2: {plane_z::rot_z}
(a)

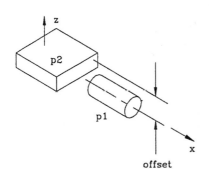

p1: {plane_z::rot_x,rot_z}
p2: {cyl_x::lin_y,rot_z}
(b)

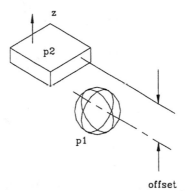

p1: {plane_z::rot_x,rot_y,rot_z}
p2: {sph::plan_z,rot_z}
(c)

FIGURE 9.4   Parallel-offset spatial relationships.

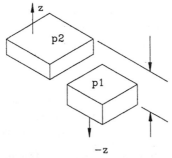

p1: {plane_z::rot_z}
p2: {plane_z::rot_z}
(a)

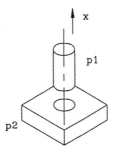

p1: {lin_x::rot_x}
p2: {lin_x::rot_x}
(b)

FIGURE 9.5   (a) Parax-offset, (b) aligned.

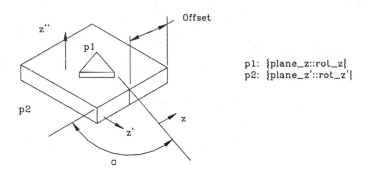

p1: {plane_z::rot_z}
p2: {plane_z'::rot_z'}

FIGURE 9.6   Incline-offset.

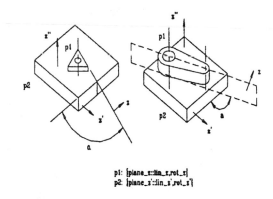

p1: {plane_x::lin_x,rot_x}
p2: {plane_z'::lin_x',rot_x'}

FIGURE 9.7   Include-angle.

Geometric Interpretation of Spatial Relationships

A spatial relationship can be interpreted as a constraint imposed on the degrees of freedom between mating or interacting features. In every mating feature (Nnaji, 1990-1; Nnaji, 1990-2), a mating-coordinate frame is attached. A mating-coordinate frame is derived to represent the coordinate frame needed for expressing degrees of freedom. The origin and axes of the mating coordinate frame are determined by mating condition or degrees of freedom of mating.

In plane feature, a mating-coordinate frame is on the plane with **Z**-axis point out of the plane (Figure 9.3(a)). In cylindrical feature, the **X**-axis of the mating-coordinate frame is coincident with the center line (Figure 9.3(b)) and, in spherical feature, the origin of the mating frame is coincident with the center of the sphere (Figure 9.3(c)). The degrees of freedom between mating components can be geometrically represented and interpreted in accordance with mating geometry. The element of degrees of freedom to describe spatial relationships can be described as follows:

1. **lin_n**: linear translation along **n**-axis, where **n** contains a fixed point and a vector;
2. **rot_n**: rotation about **n**-axis, where **n** contains a fixed point and a vector;
3. **plane_n, cyl_n, sph**: translating along a planar, cylindrical, or spherical surface, etc.

The degrees of freedom of a part can now be expressed as {**degrees of freedom of moving along the feature of relative mating part :: degrees of freedom of moving the part with respect to itself**}, where the relative mating part is fixed. For example, in Figure 9.3(b), part **p1** is *against* part **p2** over a line. The degree of freedom for **p1** with respect to **p2** describes the relative motion that maintains a line contact. **p1** may move along the mating planar face of **p2** while maintaining fixed orientation. This can be described as the equation for an infinite plane, **Ax+By+Cz=D**. For **p1** itself, **p1** can rotate about the **z-axis**, which gives a new orientation about the contact line, and **p1** can rotate about **x-axis** to generate a new contact line with respect to itself. So, the degree of freedom for **p1** with respect to **p2** is {**plane_z::rot_x,rot_z**}. For **p2** with respect to **p1**, **p2** can translate along the cylindrical surface of **p1**, and **p2** can move itself along the **x-axis** and rotate about the **z-axis**. The degree of freedom for **p2** is {**cyl_x::lin_y,rot_z**}.

A rule-based system is adopted for assisting in the process of selecting an appropriate set of spatial relationships and for inferring the final state of degrees of freedom. From the above definitions, surfaces characterize degrees of freedom. For a **plane_n_a** degree of freedom, a body may move on a planar surface along two **lin_n** directions. When a new **plane_z_b** is introduced, the remaining degree of freedom is derived by intersecting these two planes. For example, if **plane_z_a** is parallel to **plane_z_b**, then the degree of freedom may be either **plane_z_a** or **plane_z_b**; if they are not parallel, then **lin_n** is the degree of freedom obtained from the intersection of **plane_z_a** and **plane_z_b**. The intersection degrees of freedom are obtained by intersection of two geometry entities, (e.g., **circle_n** is the result of intersecting **plane_n** with **cyl_n** (or **sph_n**) together). In the intersection of two rotational degrees of freedom, e.g., **rot_z_a** and **rot_z_b**, if they share the same rotational axis, then **rot_z_a** or **rot_z_b** remains; if not, they cancel each other. In the above examples, we demonstrate some simple rules for the reduction of degrees of freedom, but in some cases, such as combining the spatial relationships with numbers of rotational degrees of freedom, the solution becomes complicated. We show some general reduction rules based on our earlier work as follows:

1. {{**plane_z_a::rot_z_a**}∩{**plane_z_b::rot_z_b**}||vec_of(z_a) · vec_of(z_b)| < 1} = {**lin_n_c**| lin_n_c = **intersect(plane_z_a, plane_z_b)**}

2. {{**plane_z_a::rot_z_a**}∩{**lin_v_b**}| | vec_of(z_a) · vec_of(z_a)| < 0} = {**fix_c**}

3. {{**plane_z_a::rot_x_a,rot_z_a**}∩{**plane_z_b::rot_x_b,rot_z_b**}| x_a = x_b, |vec_of (z_a) · vec_of (z_b)|≠1} = {**lin_x_c::rot_x_c**| lin_x_c = cent_of(x_a) = cent_of(x_b), rot_x_c = rot_x_a = rot_x_b}

4. {{**plane_z_a::rot_z_a**} ∩ {**lin_x_b::rot_x_b**}||vec_of(z_a) · vec_of (z_b)| = 1} = {**fix_c::rot_x_c** | rot_x_c = rot_x_b}

5. {{**plane_z_a::rot_x_a,rot_y_a,rot_z_a**}∩{**plane_z_b::rot_x_b,rot_y_b,rot_z_b**}| cent_of(x_a,y_a, z_a) ≠ cent_of (x_b,y_b,z_b), vec_of (z_a) = vec_of(z_b), |vec_of (rot_z_a) × vec_of(rot_z_b)| = 0} = {**plane_z_c::rot_z_c, rot_x_c**| plane_z_c = plane_z_a or plane_z_b, rot_z_c = rot_z_a or rot_z_b, x_c = axis (cent_of(rot_x_a), cent_of(rot_x_b))}

6. {{**plane_z_a::rot_x_a,rot_z_a**}∩{**plane_z_b::rot_x_b,rot_y_b, rot_z_b**}|vec_of(z_a) = vec_of (z_b), vec_of(rot_z_a) × vec_of(rot_z_b) = 0} = {**plane_z_c::rot_z_c**|plane_z_c = plane_z_a: or: plane_z_b, rot_z_c = rot_z_a: or :rot_z_b}

7. {{**plane_z_a::rot_z_a**} ∩ {**plane_z_b::rot_x_b, rot_y_b,rot_z_b**}| 0<|vec_of (z_a)cdot vec_of (z_b)|<1} = {**lin_n1_c::rot_n2_c**| lin_n1_c = intersect (plane_z_a, plane_z_b), point_of(n2_c) = cent_of(x_b,y_b,z_b), vec_of(n2_c) = vec_of(z_a)}

8. {{**lin_n1_a::rot_n2_a**} ∩ {**plane_z_b::rot_x_b,rot_y_b,rot_z_b**}| vec_of(n2_a) cdot vec_of(z_b) = 0} = {**lin_n1_c** | lin_n1_c = lin_n1_a}

9. {{**lin_n1_a**} ∩ {**plane_z_b::rot_x_b,rot_y_b,rot_z_b**}| vec_of(n1_a) cdot vec_of(z_b) = 0} = {**fix_c**}

where

Axis is a translation or rotation axis that contains a start and a vector.
point_of(Axis) returns start point of **Axis**.
vec_of(Axis) returns vector of **Axis**.
cent_of(Axis_x, Axis_y, Axis_z) returns origin of the **Axis_x, Axis_y, Axis_z** coordinate frame. (Also, this origin is a center point of a sphere.)
axis(Point_a, Point_b) returns an **Axis** with **Point_a** as start point and **Point_b-Point_a** as direction.
intersect(Dof_a, Dof_b) returns intersection of two degrees of freedom.
fix is the origin of degrees of freedom representation.

Figure 9.8 shows an example of assembly with spatial relationships and the intersection of degrees of freedom. Degrees of freedom are the allowable relative motion between mating components derived from

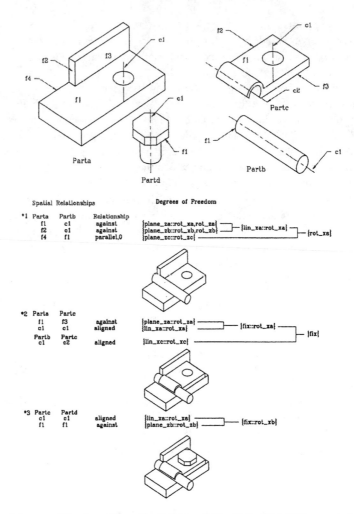

FIGURE 9.8   Intersection of degrees of freedom. (From Nnaji, B.O. and Liu, H.C., 1994. With permission from the Society of Manufacturing Engineers.)

assembly constraints. Mating frames are the coordinate frames attached to mating components to describe the degrees of freedom between them (Nnaji, 1994).

A mating frame is a coordinate frame of a part relative to the base coordinate frame of that part. Each component, as well as each subassembly, has its own mating frame (see Figure 9.9). We would like to find the mating frames by studying two approaches where one is to locate the origin of a mating frame and the other one is to find its orientation.

Origin of mating frames can be derived from the intersection of the features that have spatial relationships assigned to them, which is the same as the intersection of degrees of freedom. The orientation of a mating frame can be derived from the normal vector of features. For three planar face-**against** relationships, if the mating face normals of one of the part's faces are **fa_1**, **fa_2**, and **fa_3**, and none of them is parallel, then the **Xm_a**, **Ym_a**, and **Zm_a** axes of a mating frame can be decided upon as follows:

$$Zm\_a = fa\_1 \times fa\_$$
$$V\_a = fa\_1 \times a\_3$$
$$Xm\_a = Zm\_a \times V\_a$$
$$Ym\_a = Zm\_a \times Xm\_a$$

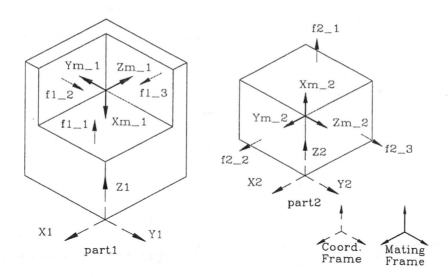

FIGURE 9.9 Derivation of mating frames.

With the same sequence of calculations, one can find the direction of mating frame axes of the mating part. Relationships such as **parallel-offset, incline-offset,** and **include-angle** may use this method of reversing one of the normal directions of a face of a part. For a cylindrical (spherical) feature, the normal directions of the feature are infinite; the orientation can be finite through the intersection of other features' spatial relationships.

Product Specification Attributes

Product specifications describe the constraints of form and function of a product. Tolerances are form specifications that specify the available range for position, orientation, and form of a product or component. Function specifications describe the functional performance of a product or its individual components. Form specifications influence the functional performance; function is the product of interaction of forms. For example, in order to fasten two blocks with a screw, the bolt of the screw must have the same type of thread as the threaded hole of the block below. Based on design principles, the hole for the block above has to be a smooth cylindrical surface. The outside diameters should be the same for the threaded holes and the depth of thread of the bolt should move far enough to make contact with the threaded hole of the bottom block. It can be seen that the form specifications needed are: thread type, thread length, and the diameter tolerance. In assembling the forms, the **fasten** function can thus be achieved. In this example, if designers assign the screw-fasten function between threaded hole and bolt with a specified load, then they provide enough information for searching the form specifications for both bolt and threaded hole. We propose that the most appropriate time to assign the specifications to a product is in the assembly mode, which saves human effort and clarifies the intention of the designer.

Product specification attributes such as functional descriptions and positional specifications can be specified by the designers. Others can be automatically (or interactively, when the default information is insufficient) generated through the specifications assigned. These specification attributes contain the information about the methods of assembly (such as glue, weld, snap fit, etc.) and the tolerances associated with positioning the parts. Through assignment of the spatial relationships, these attributes are inferred. For example, in Figure 9.2, the positional tolerances $Da \pm ta$ and $Db \pm tb$ are assigned with association of **parallel-offset**, and then the tolerance $Dc \pm tc$ and $Dd \pm td$ of components are decided upon. With complete specifications represented and assigned, this information can be propagated to a postprocessor such as tolerance propagation and process planning. In the next section, we will introduce a way to propagate product specification into dimensional constraints of the part model. These dimensions are shown on the final drawings, and they are also the constraints used by 3-D variational geometry.

Tolerance Propagation and Control

In this section, we will briefly outline Bjorke's (1989) theories on tolerance analysis. This will lead to the basis of tolerance propagation and control that help a designer attach dimensions to components. Tolerance control is the process of fitting a sum dimension tolerance to that specified by the designer; tolerance propagation allocates the specified tolerance to individual dimension tolerances. Tolerance propagation and control both require the identification of one or more sum dimensions. A sum dimension, or functional dimension, is any dimension that affects the functionality of the assembly more than the other dimension (Bjorke, 1989).

The sum dimension tolerance (T_s) will be influenced by the tolerances on any individual dimension in the assembly, which affect the sum dimension. The relationship between these individual tolerances and the sum dimension tolerance is known as the fundamental equation. For a particular assembly, there may be several sum dimensions, each of which will have a fundamental equation of the form:

$$T\_i \ = \ f\_i(T\_1, T\_2, \ldots T\_j, \ldots T\_n)$$

where

   T_i is the ith sum dimension tolerance,
   T_j is the tolerance of the jth dimension influencing the sum dimension.

The fundamental equations can be used to generate tolerance chains that can be shown visually using graphs in which arcs represent chain links and vertices represent the connections between the chain links. There are two types of tolerance chains of interest, namely the simple chain, in which no element is encountered twice, and the interrelated chain, which covers all other cases in which at least one endpoint is encountered twice.

An example of an assembly generating an interrelated tolerance chain is shown in Figure 9.10. The dimensions **S1** and **S2** in the diagram denote sum dimensions, while the dimensions **D1** to **D3** show part dimensions. The data **A** and **B**, the tolerance on the sum dimensions, and the geometric tolerance of parallelism between faces **A-F_1** and **B-F_2** are originally specified by the designer when the assembly model is generated. Note that, due to the against- and clearance-fit spatial relationships and the riveting of parts **B** and **C** into position, any eccentricity between the holes in part **A** and the shafts of parts **B** and **C** will not affect the sum dimensions. This tolerance chain is interrelated. We see that the tolerances **T_S1** and **T_S2** are propagated into **T_D1**, **T_D2** and **T_D3**, while the tolerance **TP_S1** is propagated into **TP_D1** and **TP_D2** such that **TP_S1 = f(TP_D1, TP_D2)**. Proper datum selection by the designer is important since this is the basis of the determination of the individual dimensions affecting the sum dimension and hence the fundamental equations.

The steps required in order to carry out tolerance propagation are as follows:

1. Break down sum dimension to component dimension by analyzing dimension data and spatial relationships within the assembly that affect sum dimension.
2. Establish a cost function by identifying the contribution of each individual tolerance to the manufacturing cost of the assembly.
3. Prepare fundamental equations of sum tolerances.
4. Solve fundamental equations that minimize cost function.

It is envisaged that the ability to represent traditional and geometric tolerances will enable the analysis of the effect of a given geometric tolerance on a particular sum dimension, thus providing this tolerance analysis module.

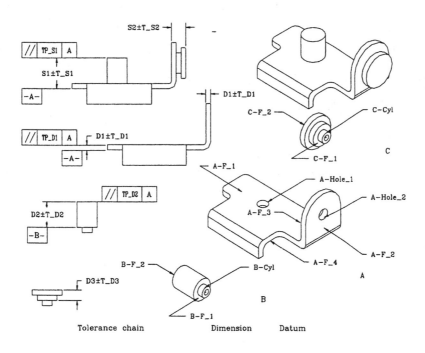

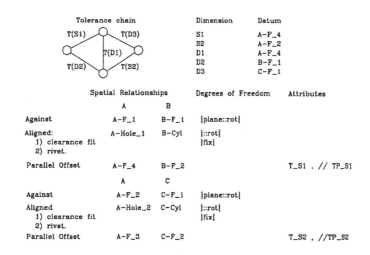

FIGURE 9.10 Example of an assembly generating an interrelated tolerance chain.

9.5 3-D Variational Geometry

An object model can be represented as a set of geometric entities topologically connected. Each entity has degrees of freedom on its defined space while maintaining the topological relationships. Dimensions are the constraints that restrain the degrees of freedom of geometric entity into a fixed place. Variational geometry is a mathematical model that calculates new positions of entities when the values of dimensions are changed.

Before solving the variational geometry model, constraints have to be set and not over-defined, i.e., no one entity will be constrained with different criteria, such as two entities dimensioned twice with different values. In the next section, a method will be introduced to show how to detect under- or over-constrained system.

Completeness of Dimensions

Completeness of dimensioning is determined by entities' degrees of freedom, such as face, edge, vertex, centerline of a cylindrical feature, and the center point of a spherical feature, where the degrees of freedom are represented by a set of vectors that define the basis of a space.

A face has a single degree of freedom, with its directions aligned with face normal direction, which is a basis of an $\mathbf{E}^1$ space. An edge can move along two joined faces, thus, it has two degrees of freedom on an $\mathbf{E}^2$ space. (The $\mathbf{E}^2$ space is defined as a plane perpendicular to the edge; the vectors may not lie on the joined faces.) The degrees of freedom of a vertex depend on the number of connected edges. The largest number of vectors that form the basis of an $\mathbf{E}^3$ space is three, so the degrees of freedom of a vertex is equal to three. A centerline has two degrees of freedom, which form an $\mathbf{E}^2$ basis describing the position of this centerline. A centerpoint has three degrees of freedom to locate its position in an $\mathbf{E}^3$ space. (The basis vectors may not be along any one of the edges.) Thus, we have the degrees of freedom of geometric entities listed as follows:

1. face = 1;
2. edge = 2;
3. vertex = 3;
4. centerline = 2; and
5. centerpoint = 3.

Each geometric entity is assigned the initial degrees of freedom as described above. In the case of the dimension assigned between entities, the degrees of freedom are deducted. The deduction of degrees of freedom according to the type of dimensioning scheme can be categorized as follows:

1. Size dimension between two faces: Faces are constrained in a relative position. The degree of freedom of each face becomes zero. Each edge and vertex has one degree of freedom deducted from its total degrees of freedom in the normal direction of the face.
2. Size dimension between a face and an edge (centerline): The degrees of freedom of all the edges, vertices, and face are reduced by one as in 1 above.
3. Size dimension between two edges (centerlines): Each edge and vertex has one fewer degree of freedom in the direction of a vector that is perpendicular to the edges and that is on a plane by joining these two edges.
4. Size dimension between a face and a vertex (centerpoint): The degrees of freedom of all edges, vertices, and face are reduced by one as in 1 above.
5. Size dimension between an edge (centerline) and a vertex (centerpoint): Each edge and vertex has one fewer degree of freedom in the direction of a vector that is perpendicular to the edge and that is on a plane obtained by joining the edge and vertex.
6. Angular dimension between two planes: The same as 1 above.

Over-dimensioning is detected when dimensioning on two geometric entities both have zero degree of freedom. After all the dimensions have been put on a part, if there is any geometric entity with non-zero degrees of freedom, then this is the under-dimensioning that acquires more dimensions on this entity to reduce the degrees of freedom to zero. Figure 9.11 gives an example to deduct the degrees of freedom according to the dimensions applied.

Once a feasible dimension scheme is obtained, the next step is to interpret these dimensions into geometric constraints, which is described in the following section.

3-D Variational Geometry Constraints

Figure 9.12 illustrates types of constraints in dimensions of size and angle. The size dimensions are point-to-plane, line-to-plane, plane-to-plane, and line-to-line; the angular dimensions are line-to-line, line-to-plane, and plane-to-plane, where these constrained entities have to be valid and fully represented in a geometric model. In practice, dimension and tolerance are drawn based on implicit or explicit data. (A datum is not explicitly specified on drawing.) Size, angle, form, and correlative tolerances are

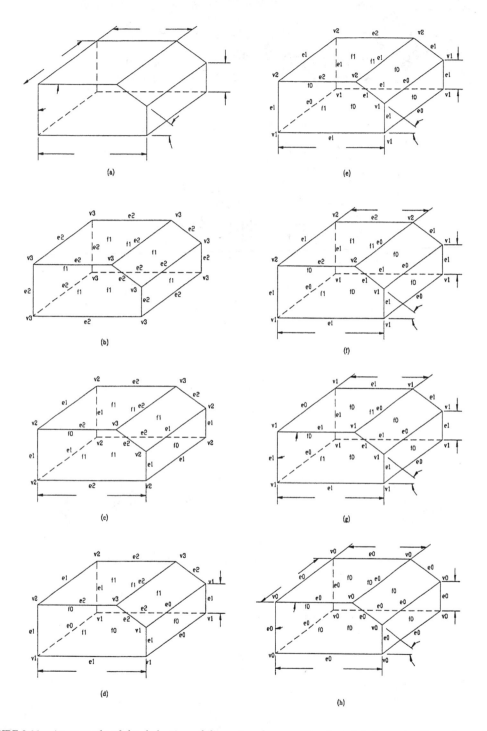

FIGURE 9.11 An example of the deduction of dimension degrees of freedom; (a) shows the dimensions will be attached, (b) shows the initial degrees of freedom of geometric entities, and (c), (d), (e), (f), (g), and (h) display the remaining degrees of freedom after a new dimension is assigned. Face = 1, edge = 2, vertex = 3, centerline = 2, and centerpoint = 3.

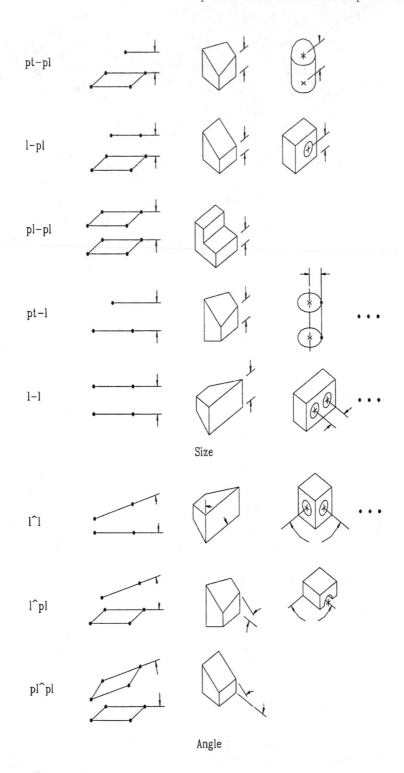

FIGURE 9.12　3-D constraints.

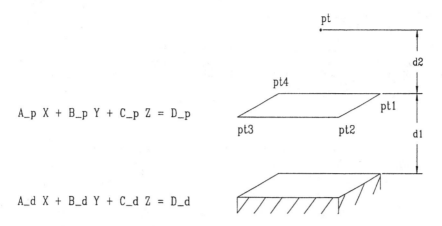

FIGURE 9.13 The point-to-plane constraints.

measured from implicit data. Position, orientation, and runout tolerances are measured from explicit data. Based on this point of view, the variational constraints should be derived from implicit or explicit data, accordingly.

Figure 9.13 shows an example of how to derive the point-to-plane constraints with an explicit datum. Given

1. a datum which is $A\_d\,X + B\_d\,Y + C\_d\,Z = D\_d$,
2. a plane with four points, pt_1, pt_2, pt_3, and pt_4,
3. point pt,
4. the distance between datum and plane is d_1, and
5. the distance between plane to point is d_2, six geometric constraints can be derived.

Plane function $A\_p\,X + B\_p\,Y + C\_p\,Z + D\_p$, can be generated from points of $pt\_1 = (x\_1, y\_1, z\_1)$, $pt\_2 = (x\_2, y\_2, z\_2)$, and $pt\_3 = (x\_3, y\_3, z\_3)$. We have,

$$
\begin{aligned}
A\_p &= (-y\_2 \cdot z\_1 + y\_3 \cdot z\_1 + y\_1 \cdot z\_2 - y\_3 \cdot z\_2 \\
&\quad -y\_1 \cdot z3\_ + y\_2 \cdot z\_3)/Norm\_p \\
B\_p &= (x\_2 \cdot z\_1 - x\_3 \cdot z\_1 - x\_1 \cdot z\_2 + x\_3 \cdot z\_2 + x\_1 \cdot z\_3 \\
&\quad -x\_2 \cdot z\_3)/Norm\_p \\
C\_p &= (-x\_2 \cdot y\_1 + x\_3 \cdot y\_1 + x\_1 \cdot y\_2 - x\_3 \cdot y\_2 - x\_1 \cdot y\_3 \\
&\quad + x\_2 \cdot y\_3)/Norm\_p \\
D\_p &= (-x\_3 \cdot y\_2 \cdot z\_1 + x\_2 \cdot y\_3 \cdot z\_1 + x\_3 \cdot y\_1 \cdot z\_2 - x\_1 \\
&\quad \cdot y\_3 \cdot z\_2 - x\_2 \cdot y\_1 \cdot z\_3 + x\_1 \cdot y\_2 \cdot z\_3)/Norm\_p \\
Norm\_p &= \sqrt{A\_p^2 + B\_p^2 + C\_p^2}
\end{aligned}
$$

The plane constraints thus are

$$
\begin{aligned}
A\_p &= A\_d \\
B\_p &= B\_d \\
C\_p &= C\_d \\
D\_p &= D\_d + d\_1
\end{aligned}
$$

Point $pt\_4 = (x\_4, y\_4, z\_4)$ is on plane $A\_p\,X + B\_p\,Y + C\_p\,Z = D\_p$ which is

$$
A\_px\_4 + B\_py\_4 + C\_pz\_4 = D\_p
$$

The distance constraint from $pt = (x, y, z)$ to plane is

$$A\_px + B\_py + C\_pz = D - d\_2$$

Given a distance d, the positive or negative of d will determine the direction of variation along with (A_d, B_d, C_d) or $(-A\_d, -B\_d, -C\_d)$.

Systems of Equations

Once constraints are translated into a set of equations, the next step is to solve for the variables part of the equations. For a simplified example, to solve the system equation of point-to-plane (as described in the previous section), if datum plane d is coincident with plane p, pt_1, pt_2, pt_3, pt_4, and d_2 are given, then the value of point pt (x, y, z) can be solved directly. When multiple dimensions are applied and with fewer fixed variables, the solution scheme becomes complicated. Two methods are available for solving a system of equations. One is a direct solution using substitution algorithm and the other is numerical iteration. The direct solution has the advantage that, in general, it is faster and more precise. The direct solution has the main disadvantage that it may not be used to solve all systems of equations because it may not be possible to define a variable explicitly in an expression for its subsequent substitution in other expressions (Serrano, 1984). The Newton-Raphson method that can solve simultaneous nonlinear equations does not have substitution problems. The disadvantage of the Newton-Raphson method is that it requires the user to input the initial guess close to the actual solution in order to converge to a solution, and it may not converge to the desired solution if multiple solutions are possible (Lin, 1981).

Direct Substitution Method

In order to determine if a direct solution is possible, it is necessary to determine if all the variables may be defined explicitly by an expression. Given a set of system equations,

$$y = y(z)$$
$$b = b(x,z)$$
$$x = x(y,z)$$
$$a = a(x,y,z)$$

to solve this set of equations we need the same number of equations and unknowns, and explicit definition for each unknown in terms of other variables; therefore, no slings should appear in the resulting digraph (directed graph). A sling would be formed if an unknown is defined in terms of itself. The system must be acyclic because the process of determining the order of execution is analogous to that of scheduling a network of activities. An activity originating at a node cannot start before all activities terminating at that node are completed. Using network path scheduling, it is possible to solve systems with more than one root, i.e., systems with subsets of expressions that are not related among themselves. An adjacency matrix as shown in Figure 9.14 is constructed to describe the digraph of the above system equations.

	y	b	x	a	z
y	0	0	0	0	1
b	0	0	1	0	1
x	1	0	0	0	1
a	1	0	1	0	1
z	0	0	0	0	0

FIGURE 9.14   Adjacent matrix.

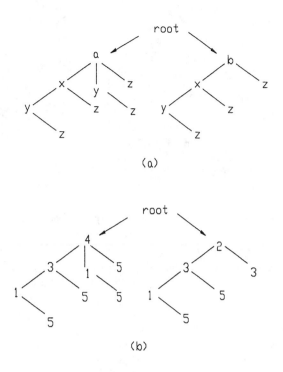

FIGURE 9.15 (a) Nodes represented by expressions, (b) nodes represented by pointers.

In this example, variable z is fixed in order to have as many equations as unknowns.

$$z = \text{Constant}$$

Two tree structures are built to have fixed value at the end of branches as shown in Figure 9.15. The equivalent tree structure is exhibited in Figure 9.15(a) and includes all parameters and variables. Notice that constants occupy the terminal nodes. In Figure 9.15(b), all nodes are numbered in pointerindices. In determining the execution sequence, the problem is breaking down into smaller subproblems. Subproblems are solved in a sequence such that any subproblems of the same type will not be solved more than once. An "and/or" graph algorithm is applied here that makes a list of all pointers to unknowns for each node in a given level. These lists are then sorted according to tree levels and merged roots a and b, because the adjacency matrix has all the information regarding both roots; it is possible to evaluate an individual root as well as all roots present in the system. To determine the execution sequence for this example, we have for nodes

a:

Level	unknown	pointers
4	none	...
3	y	1
2	x, y	3, 1
1	a	4

b:

level	unknown	pointers
4	none	...
3	y	1
2	x	3
1	b	2

therefore:

$$\text{root } a: (1,3,1,4)$$
$$\text{root } b: (1,3,2)$$

Now merge the lists and eliminate the repeated pointers thus:

$$\text{list}: (1,3,1,4,1,3,2)$$
$$\text{final}: (1,3,4,2).$$

   The pointers are eliminated from left to right because they are listed with the pointers in the deepest level occupying the leftmost position in the list. Therefore, to solve this simple example requires evaluating the expressions in the order 1-3-4-2.

   Because not all systems may be solved using this technique, it is necessary to check that the system satisfies the necessary conditions. If one of the following two conditions is not satisfied then the iterative solution will be invoked. The first condition that is checked is the presence of slings, which are detected when an expression defining a variable contains the same variable as part of its definition. The second condition requires an acyclic system. In the next section, Newton-Raphson will be introduced when either of these two conditions is not satisfied.

Iterative Solution

There are various methods of finding roots or solutions for nonlinear equations, of which Newton's method is one of the most commonly used because it has better convergence properties than direct iteration methods. The Newton-Raphson technique is a generalization to Newton's method, and it is used to determine the zeros of a set of simultaneous nonlinear equations.

   The general formulation for nonlinear systems can be stated in the following form. Given n-functions f_i in terms on n-unknowns x_i:

$$f_1(x_1, x_2,..., x_n) = 0$$
$$f_2(x_1, x_2,..., x_n) = 0$$
$$.$$
$$.$$
$$.$$
$$f_n(x_1, x_2,..., x_n) = 0$$

   The basis for the Newton-Raphson method is a Taylor's expansion of the n equations about some point (x_0, y_0):

$$f_1(x_1 + x,...x_n + x) = f_1(x_1,..., x_n) + x_1 f_x + \cdots + x_n f_n + \text{Higher order terms.}$$
$$.$$
$$.$$
$$f_n(x_1 + x,...x_n + x) = f_n(x_1,..., x_n) + x_1 f_x + \cdots + x_n f_n + \text{Higher order terms.}$$

If the higher order terms are dropped, the problem becomes one of finding the roots of the linear system:

$$\begin{bmatrix} f_{11} & f_{12} & . & . & f_{1n} \\ f_{21} & f_{22} & . & . & f_{1n} \\ . & & & & \\ . & & & & \\ f_{n1} & f_{n2} & & & f_{nn} \end{bmatrix} \begin{bmatrix} X_1 \\ X_2 \\ . \\ . \\ X_n \end{bmatrix} = \begin{bmatrix} -f_1 \\ -f_2 \\ . \\ . \\ -f_n \end{bmatrix}$$

The system may be written in matrix notation as:

$$J \cdot \Delta x = r$$

where

$x = x^{n+1} - x^n$ is the vector of displacements of iteration n

$f_{ij} = \partial f_i / \partial x_j$

J is the Jacobian matrix

r is the vector of residuals.

When formulating the equations system, all equations are satisfied, but if a change in the value of one or more variables is made, the residuals are no longer zero. The original values of the variables are used as initial guesses, and a new set of values of the variables is generated and used to solve the above equation. The process is iterative until the residuals tend to zero within a tolerance value.

9.6   Other Researches and Developments

In previous sections, we focused on the introduction of the connectivity design module of ProMod, in short, it is an assembly modeling technique that uses spatial relationships to constrain and maintain assembly relationships between components. In the following sections, we will show other related developments derived from ProMod, i.e., stability analysis of assembly, feasible approach directions and precedence constraints, and, finally, kinematic and world modeling.

Stability Analysis of Assembly

Stability analysis is a mathematical model that simulates the stability of assembly placement during each assembly stage. Research has been carried out in production planning, robot path planning, grasp planning, and fixture planning, but the problem of subassembly stability planing has been continually ignored. It has been implicitly assumed that parts and subassemblies will retain their configuration all through the manufacturing process. For instance, when a one-hand robot picks up a part and places it in some locations, it is assumed that the configuration of the part will not change until the robot moves it. In cases where the parts "looked unstable" during design, either 1. fixturing was taken as the default without even considering changes in the part or assembly design that would make the system stable and eliminate unnecessary fixturing, or 2. changes were made to the design to make the subassembly stable, without giving a thought to what minimal changes in the design would bring in stability.

An assembly process is basically a sequence of steps in which the parts are to be placed on a subassembly to achieve the final goal. However, it is necessary to satisfy certain constraints such as geometric and manufacturing resource constraints to realize the assembly. In addition, it is also important to consider stability constraints. At every instant of the manufacturing process, the components of the subassembly must be stable. Analyzing for stability is also helpful in determining the limiting values of the assembly forces and torques. If excessive forces or torques are applied while mating a certain component with a partial assembly, then that can result in instability of the subassembly.

There are quantitative and qualitative considerations in assembly stability analysis methods. Numerical simulation technique and equivalent parameter identification are considered in quantitative analysis. In assembly stability simulation, the system is excited with the assembly torques and forces. The response of the system varies with its structural properties, such as inertia, mass, damping, and stiffness of the joints. In addition, the response is also dependent on geometric characteristics, such as the location of the mass centers of the bodies. The equivalent parameter identification is to develop a system to evaluate the behavior of the mechanism at the point of interest with equivalent in mass and inertia or equivalent in stiffness and damping. With this in view, it is convenient to reduce the complex dynamics of the multiDOF assembly to that for a single body, namely, a scalar quantity for mass. The inertial matrix with body is known as the virtual mass (Vishnu, 1992).

Quantity analysis methods have the drawback of not giving the designer an overall picture of the stability of the assembly and its components. In order to create stable designs, the designer needs to have a global picture. He has to know the directions in which the subassembly is quite stable, and the directions in which it is marginally stable or totally unstable. The quality method, which can be incorporated in the design phase, is proposed to analyze the stability.

While manipulating objects, humans instinctively adopt arm configurations that very efficiently utilize the motion and strength capabilities of the arm. In the same spirit, while performing a task, one can exploit the motion and load capabilities of a kinematic mechanism by choosing cofigurations that maximize or minimize the kinematic and dynamic transmission characteristics, depending on the task requirements. The optimal direction for effecting a force is the direction in which the transmission ratio of the mechanism to the force is at a maximum. This direction also corresponds to the direction of application of the assembly force while maintaining a stable system. During the design process, a stability index can be derived based on variations in the velocities, forces, inertia, stiffness, and damping in the system. The stability index enables comparison of stabilities of various subassemblies and then choosing the best assembly design.

Feasible Approach Directions and Precedence Constraints

Feasible approach directions is a set of vectors that describe the approaching direction of removing a part from its assembled pace without interference. In general, the disassembly directions are the reverse of assembly directions since the problem of finding how to assemble a set of parts can be converted into an equivalent problem of finding how the same parts can be disassembled (Woo, 1987).

In this section, we will introduce how to derive the feasible approach directions from a polyhedral-represented model (Nnaji, 1992; Yeh, 1992). The inputs are mainly the mating faces. Wherever there is a planar contact between two parts, the parts can be approached in any direction in the half space created by the mating face. Thus if $F = \{f_1, f_2, ..., f_j\}$ represents a face set of all planar mating faces for a part, then the set of approach directions due to the ith mating face is given by:

$$R_i = \{r | r \cdot n_i \geq 0\}$$

Here n_i is the unit normal vector to the ith mating face. The set of resultant approach directions due to all the j mating faces is

$$\left[R = \bigcap_{i=1}^{j} R_i \right]$$

Once the final assembly is configured, the assembly precedence constraints can be obtained by analyzing feasible-approach directions. A part cannot be assembled or disassembled if there is an object crossing the assembly directions. In disassembling objects, once an object covers another object, the assembly direction of the object below will be blocked until the object above is removed. In the assembly structure, if the parts have the spatial relationships with the same ancestors and do not block the assembly of its offspring, then these parts have the same precedence. For example, in surface mounting, chips are mounted on a PCB, so the chips will always have the spatial relationships with PCB only. Consequently, the assembly precedence is the same for these chips. With the feasible-approach directions and precedence constraints, the exploded view of the assembly model can be automatically generated and the precedence of assembly/disassembly sequences can be automatically verified.

Kinematic Modeling with Spatial Relationships

Spatial relationships configure the assembly by constraining the degrees of freedom between mating components. The assembly configuration and remaining degrees of freedom are the input to kinematics analysis.

In this application, the objectives of computational kinematics comprise the following aspects:

- Automatic formulation of the governing kinematic equations
- Solving nonlinear equations for kinematic response
- Analysis of the kinematic characteristics of mechanical systems.

Kinematic modeling of a mechanism involves the definition of the geometry of the bodies that make up the mechanism, the selection of kinematic constraints that act between pairs of bodies, and the specification of time-depended drivers to reduce the number of degrees of freedom of the mechanism to zero. Intuitively, a CAD system appears to be an appropriate tool to model not only static mechanical systems, but also assemblies that contain moving parts. However, certain requirements must be met by the CAD system in order to be useful in the process of kinematic modeling and analysis. It is essential that the constraints on the relative motion of bodies in the mechanism imposed by the physical joints can be captured in the CAD knowledge base. The formulation of the governing kinematic equations of the system can then be achieved automatically after the desired input motion is specified. Thus, the user is freed from the tedious and error-prone process of translating the physical characteristics of a mechanism into a valid mathematical model. It is important to note that, rather than creating an animation of the mechanism's motion on a computer screen after the input of relevant data by the user, this approach is based on placing the CAD data into the center of the kinematic analysis process. The advantages of this philosophy derive from the flexibility of utilizing neutral data to perform a variety of tasks without any redundancy of information stored in a data base.

Design with spatial relationships is ideally suited to support computer-aided kinematic modeling and analysis of mechanisms. By assigning spatial relationships between individual components, the designer not only defines the relative position of a component in the final assembly, but he also specifies the degrees of freedom implied by a joint. Since spatial relationships constitute a pair-wise operation under assumption that both components, initially, are unconstrained free bodies, it is not able to close up an open-chain assembly where the degrees of freedom of components are retrained. In a four-bar linkage example, the first three joints can be assembled with spatial relationships in order to assemble the fourth joint to generate a closed chain. This has to be achieved by an inverse kinematics operation (Prinz, 1994).

The following example illustrates the process of assembling a closed kinematic chain using spatial relationships. The four-bar mechanism that is to be assembled and its kinematic diagram are shown in Figure 9.16. The assembly of the open kinematic chain consisting of base a, crank b, coupler c, and follower d is considered first. Each of the revolute kinematic pairs $\{a,b\}$, $\{b,c\}$, and $\{c,d\}$ can be modeled by using the spatial relationships **aligned** and **against**. Then the position of base e is specified by **parallel offset** spatial relationships. In order to close the kinematic chain, base e and follower d must be connected

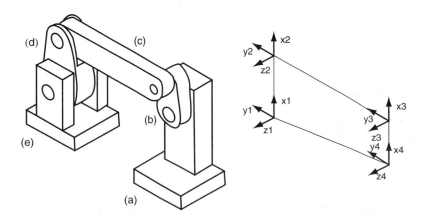

FIGURE 9.16 Assembled four-bar mechanism and its kinematic diagram with joint coordinate frames.

by a revolute joint. By selecting **aligned** and **against** as post-spatial relationships, the user defines the desired kinematic pair {d,e} and initiates the solution process for this inverse kinematics problem. Further input required is the selection of those parts that are fixed to the ground, i.e., the bases a and e.

Once the user has specified the spatial relationships between the components of the mechanism, local coordinate frames are automatically attached to each joint in the mechanism (see Figure 9.16). The 4 × 4 transformation matrix $A_{k(k+1)}$ relates the position and orientation of two adjacent joints k and $k+1$ in the kinematic loop that describes the mechanism. In order to close a loop, the coordinate system at the end of the loop must be congruent with the one at the beginning. That is, for a kinematic loop consisting of n joints, the product of the transformation matrices must be the identity matrix:

$$A_{00} = A_{01}A_{12}...A_{(n-1)n}A_{n0} = I$$

This equation is called the loop-closure equation. It is important to note that this matrix equation is automatically derived from the set of spatial relationships that define the kinematic constraints between the parts of the mechanism. The joint variables in this equation may be considered as belonging to one of two groups. We have simplified the expression here in order to show that the formulated system equations can be solved by the same techniques introduced in sections above, the direct substitution method and the iterative solution. The detailed solutions can be found in Prinz (1994) and Aguwa (1997).

9.7 Conclusion

Existing computer-aided design systems focus on the design and analysis of components in isolation. Since these design systems are unable to support early conceptual design, designer's intent cannot be captured, represented, and propagated to down-stream activities. The consequence of this inability is that existing design systems are not capable of optimizing the product against life-cycle constraints. Assembly is the final manufacturing stage and is also the most complicated and cost-effective process. The early evaluation of the assembly process that is in the design stage may reduce the time and cost of production.

In this chapter, we introduce a new type of mechanical product modeling system that is not only capable of capturing and integrating the designer's intent at conceptual design level, but also of propagating these intentions as constraints to guide the development and detailing of product design and manufacturing processes. We introduced spatial relationships, an assembly representation scheme embedded with geometric information and nongeometric constraints, to carry design criteria and to serve as the common language between design engineers and manufacturing engineers in design considerations.

We also introduced the 3-D variational geometry modeling technique, a constraint-based modeling revision technique that simplifies the design alternation processes. Constraints are derived from design and manufacturing specifications such that the violation will be detected automatically and communicated to both design and manufacturing engineers.

Finally, we showed three applications of ProMod in computer-aided assembly: stability analysis of assembly, feasible-approach directions and precedence constraints, and kinematic modeling.

Further Information

The development of ProMod originated at the Automation and Robotics Laboratory at the University of Massachusetts at Amherst, and continued at the University of Pittsburgh where B. Nnaji works at the time of this writing.

References

Aguwa, C., Spatial Kinematics Modeling and Analysis of Spherical and Cylindrical Joints and Applications in Automated Assembly Systems, Master's Thesis, University of Massachusetts, Amherst, February 1997.

Ambler, A. P. and Popplestone, R. J., Inferring the positions of bodies from specifed spatial relationships, *Artificial Intelligence* 6(2), pp. 157–174, 1975.

Anantha, R., Kramer, G. A., and Crawford, R. H., Assembly modeling by geometric constraint satisfaction, *Computer-Aided Design*, Vol. 28, No. 9, pp 707–772, 1996.

Bjorke, O., *Computer-Aided Tolerancing*, 2nd Ed., ASME Press, New York, 1989.

Boothroyd, G., *Assembly Automation and Product Design*, Marcel Dekker, Inc., NY, 1995.

Kang T. S. and Nnaji, B. O., Feature representation and classification for automatic process planning, *Journal of Manufacturing Systems*, Vol. 12, No. 2., 1993.

Lin, V. C., Gossard, D. C., and Light, R. A., Variational geometry in computer-aided design, *Computer Graphics (Proc. SIGGRAPH)*, pp. 17–179, August 1981.

Liu, H. C. and Nnaji, B. O., Feature reasoning for automatic robotic assembly and machining in polyhedral representation, *International Journal of Production Research*, Vol. 28, No.3, pp. 517–540, 1990.

Liu, H. C. and Nnaji, B. O., Design with spatial relationships, *Journal of Manufacturing Systems*, Vol. 10, No. 6, pp. 449–463, 1991.

Kang, J-H., Realization of Conceptual Design for Mechanical Products, Ph.D. Dissertation, University of Massachusetts, Amherst, September 1996.

Mantyla, M., *Intoduction to Solid Modeling*, Computer Science Press, Inc., 1988.

Nnaji, B. O., A framework for CAD-based geometric reasoning for robot assembly planning, *The International Journal of Production Research*, Vol. 26, No. 5, pp. 735–768, 1988.

Nnaji, B. O. and Liu, H., Feature reasoning for automatic robotic assembly and machining in polyhedral representation, *The International Journal of Production Research*, Vol. 28, No. 3, pp. 517–540, 1990.

Nnaji, B. O. and Kang, T., Interpretation of CAD models through neutral geometric knowledge, *AIEDAM*, Vol. 4, No. 1, pp. 15–45, 1990.

Nnaji, B. O., Kang, T., Yeh, S., and Chen, J. P., Feature reasoning for sheet metal components, *International Journal of Production Research*, Vol. 29, No. 9, pp. 1867–1896, 1991.

Nnaji, B. O., Sadrach, J. B., and Yeh, S., Spanning vector for assembly directions and other applications, *Journal of Design and Manufacturing*, Vol. 2, No. 4, pp. 211–224, 1992.

Nnaji, B. O., Liu, H. C., and Rembold, U., A product modeler for discrete components, *The International Journal of Production Research*, Vol. 31, No. 9, pp. 2017–2044, 1993.

Nnaji, B. O. and Liu, H. C., PAM: A product modeler for assembly, *Proceedings of the 1994 NSF Design and Manufacturing Grantees Conference*, MIT, pp. 67–69, 1994.

Popplestone, R. J., Ambler, A. P., and Bellos, I., An interpreter for a language for describing assemblies, *Artificial Intelligence*, Vol. 14, No. 1, pp. 79–107, 1978.

Prinz, M., CAD-Based Kinematic Modeling and Analysis with Spatial Relationships, Master's Thesis, University of Massachusetts, Amherst, May 1994.

Rembold, U. and Nnaji, B. O., CAD-based feature geometry and function reasoning for robotic assembly, North Atlantic Treaty Organization (NATO), Research project funded from July 1989 to January 1991.

Rembold, U. and Nnaji, B. O., The role of manufacturing models for the information technology of the factory of the nineties, *International Journal of Design and Manufacturing*, Vol. 1, No. 2, pp. 67–87, 1991.

Serrano, D., MATHPAK: An Interactive Preliminary Design Package, Master's Thesis, MIT, January 1984.

Vishnu, A. and Nnaji, B. O., Analysis of subassembly stability for automated assembly, *Robotics & Computer-Integrated Manufacturing*, Vol. 9, No. 6, pp. 447–470, 1992.

Woo, T. C., Automatic disassembly and total ordering in three dimensions, *Proceedings of the Winter Annual Meeting of the American Society of Mechanical Engineers*, December 13–18, 1987, Boston, MA, pp. 291–303.

Yeh, S. C., Sadrach, J. B., Jagtap, P. B., and Nnaji, B. O., Automatic Precedence and Spanning Vector Generation, *Journal of Design and Manufacturing*. Vol. 2, No. 4, pp. 211–224, 1992.

Index